Simulation Foundations, Methods and Applications

Series editor

Louis G. Birta, University of Ottawa, Canada

Advisory Board

Roy E. Crosbie, California State University, Chico, USA
Tony Jakeman, Australian National University, Australia
Axel Lehmann, Universität der Bundeswehr München, Germany
Stewart Robinson, Loughborough University, UK
Andreas Tolk, Old Dominion University, USA
Bernard P. Zeigler, University of Arizona, USA

More information about this series at http://www.springer.com/series/10128

David J. Barnes · Dominique Chu

Guide to Simulation and Modeling for Biosciences

Second Edition

 Springer

David J. Barnes
University of Kent
Canterbury, Kent
UK

Dominique Chu
University of Kent
Canterbury, Kent
UK

ISSN 2195-2817   ISSN 2195-2825   (electronic)
Simulation Foundations, Methods and Applications
ISBN 978-1-4471-6898-0   ISBN 978-1-4471-6762-4   (eBook)
DOI 10.1007/978-1-4471-6762-4

David dedicates this book to Ruby

Preface

In this book we seek to provide a detailed introduction to a range of simulation and modeling techniques that are appropriate for use in the biosciences. The book is primarily intended for bioscientists, but will be equally useful for anybody wishing to start simulation and modeling in related fields. The topics we discuss include agent-based models, stochastic modeling techniques, differential equations, and Gillespie's stochastic simulation algorithm. Throughout, we pay particular attention to the needs of the novice modeler. We recognize that simulation and modeling in science in general (and in biology, in particular) require both *skills* (i.e., programming, developing algorithms, and solving equations) and *techniques* (i.e., the ability to recognize what is important and needs to be represented in the model, and what can and should be left out). In our experience with novice modelers we have noticed that: (i) both skill and technique are equally important; and (ii) both are normally lacking to some degree.

The philosophy of this book, therefore, is to discuss both aspects—the technical side and the side that concerns being able to identify the right degree of abstraction. As far as the latter area is concerned, we do not believe that there is a set of rules that, if followed, will necessarily lead to a successful modeling result. Therefore, we have not provided a list of such rules. Instead, we adopt a practical approach which involves walking the reader through realistic and concrete modeling projects. In doing so, we highlight and comment on the process of abstracting the real system into a model. The motivation for this approach is that it is akin to apprenticeship, allowing the reader both to observe practical expertise and to generate personal understanding and intuition, which will ultimately help them to formulate their own models.

Included in the book are practical introductions to a number of useful tools, such as the Maxima computer algebra system, the PRISM model checker, the Repast Simphony agent modeling environment, and the Smoldyn cellular modeling environment. Some of the chapters also include exercises to help the reader sharpen their understanding of the topics. The book is supported by a website, http://www.cs.kent.ac.uk/imb/, which includes source code of many of the example models we discuss.

Canterbury, UK   David J. Barnes
July 2015   Dominique Chu

Contents

Foundations of Modeling

1

Until not so long ago, there was a small community of, so-called, "theoretical biologists" who did brilliant work, no doubt, but were largely ignored by the wider community of "real biologists" who collected data in the lab. As with most things in life, times changed in science, and mathematical modeling in biology has now become a perfectly respectable activity. It is no longer uncommon for the experimental scientist to seek help from the theoretician to solve a problem. Theoreticians are, of course, nice people, and always happy to help their experimental colleagues with advice on modeling (and other questions in life). Most of the theoreticians are specialists in a particular field of modeling where they have years of experience formulating and solving models.

Unfortunately, in biological modeling there is no one modeling technique that is suitable for all problems. Instead, different problems call for different approaches. Often it is even helpful to analyze one and the same system using a variety of approaches, to be able to exploit the advantages and drawbacks of each. In practice, it is often unclear which modeling approaches will be most suitable for a particular biological question. The theoretical "expert" will not always be able to give unbiased advice in these matters. What this tells us is that, in addition to experts specializing in particular modeling techniques, there is also a need for generalists; i.e., researchers who know a reasonable amount about many techniques, rather than very much about only a single one.

This book is intended for the researcher in biology who wishes to become such a generalist. In what follows we will describe the most important techniques used to model biological systems. By its very nature, an overview like this must necessarily leave out much. The reader will, however, gain important insights into a number of techniques that have been proven to be very useful in providing understanding of biological systems at various levels. And the level of detail we present will be sufficient to solve many of the modeling problems one encounters in biology.

In addition to presenting some of the core techniques of formal modeling in biology, the book has two additional objectives. Firstly, by the end the reader should have developed an understanding for the constraints and difficulties that different

© Springer-Verlag London 2015
D.J. Barnes and D. Chu, *Guide to Simulation and Modeling for Biosciences*,
Simulation Foundations, Methods and Applications,
DOI 10.1007/978-1-4471-6762-4_1

modeling techniques present in practice—in other words she should have acquired a certain degree of literacy in modeling. Even if the reader does not herself embark on a modeling adventure, this will facilitate her interaction and communication with the specialist modeler with whom she collaborates, and also help her decide who to approach in the first place. Secondly, the book also serves as an introduction to jargon, allowing the reader to understand better much of the primary literature in theoretical biology. Assumed familiarity with basic concepts is perhaps one of the highest entry barriers to any type of research literature. This book will lower the barrier.

The primary goal of this book, however, is to equip the reader with a basic array of techniques that will allow her to formulate models of biological systems and to solve them. Modeling in biosciences is no longer performed exclusively using pen-and-paper, but increasingly involves simulation modeling, or at least computer-assisted mathematical modeling. It is now a commonplace that computers act as efficient tools to help us achieve insights that would have been impossible even just 30 years ago, say. Thanks to the Internet and advances in information technology, there is no shortage of software tools encapsulating specialist knowledge to help the modeler formulate and solve her scientific problems. In fact, some cynics say that there are more tools than problems out there! As with all things involving choice, variety poses its own problems. Separating the wheat from the chaff—the useful software from the useless—is extremely time consuming. Often the weaknesses and strengths of a software package only appear after intense use, when much time and energy has been expended sifting through pages of documentation and learning the quirks of a particular tool. Or, even worse, there is this great piece of software, but it remains inaccessible through a lack of useful documentation—those who love to write software often have no interest in writing the accompanying documentation.

In this book we will introduce the reader to high quality software tools and modeling environments that have been tried and tested in practical modeling enterprises. The main aim of these tool descriptions will be to provide an introduction to how to use the software and to convey some of their strengths and shortcomings. We hope this will provide enough information to allow the reader to decide whether or not the particular package will likely be of use or not. None of the software packages is described exhaustively; by necessity, we only present a small selection of available options. At the time of writing, most if not all of the software tools we describe are available to download for free and can be installed and run on most common operating systems.

This introductory chapter has two main goals. Firstly, we will give a brief overview of the basic concepts of modeling and various types of models. Secondly, the chapter deals with the fundamental question of how to make a model. The specifics of this process will depend on the particular application at hand, of course. However, there are a number of rules that, if followed, make the process of modeling significantly more efficient. We have often found that novice modelers struggle precisely because they do not adhere to these guidelines. The most important of these rules is to *search for simplicity*.

While this chapter's contents are "softer" than those of later chapters—in the sense that it does not feature equations or algorithms as the later chapters do—the message it contains is perhaps the most important one in the entire book. The reader is therefore strongly encouraged to read this chapter right at the outset, but also to consider coming back to it at a later stage as well, in order to remind herself of the important messages it contains.

1.1 Simulation Versus Analytic Results

The ideal result of any mathematical modeling activity is a single, closed form formula that states in a compact way the relationship between the relevant variables of a system. Such analytic solutions provide global insight into the behavior of the system. Unfortunately, only in very rare cases can such formulas be found for realistically-sized modeling problems. The science of biology deals with real-world complex systems, typically with many interactions and non-linear interdependencies between their components. Systems with such characteristics are nearly always hard to treat exactly. Suitable approximations can significantly increase the range of systems for which analytic solutions can be obtained, yet even these cases will remain a tiny fraction of all cases.

In the vast majority of cases, mathematical models in biology need to be solved numerically. Instead of finding one single general solution valid for all parameters, one needs to find a specific solution for a particular set of parameter values. This normally involves using some form of computational aid to solve for the independent variables. Numerical procedures can solve systems that are far too complicated for even approximate analytic methods, and are powerful in this sense. The downside is, of course, that much of the beauty and generality of analytic results becomes lost when numerical results are used. In particular, this means that the relationship between variables, and how this depends on parameters, can no longer be seen directly from an individual solution, but must be inferred from extensive sweeps through the parameter space.

For relatively small models it is often possible to explore the space of parameters exhaustively—at least the space of *reasonable* parameters. This is a tedious exercise but it can lead to quite robust insights. Exhaustive exploration of the parameter space quickly becomes much harder as the number of parameters increases beyond a handful. In those cases, one could try to switch strategy; instead of exploring the entire parameter space, one could concentrate on experimentally measured values for parameters. This sounds attractive, but is often not a solution. It may be that nobody has ever measured these parameters and, even if they have been measured, they are typically afflicted by large errors, which reduces their usefulness. Moreover, mathematical models are often highly simplified with respect to real systems, and for this reason some of their parameters may not relate in an obvious way to entities in the real world.

In these situations one typically has to base models on guesses about parameters. In many cases one will find that there are only a few parameters that actually make a qualitative difference to the behavior of the system, whereas most parameters do not have a great influence on the model. The sensitivity of the model to changes of parameters must be explored by the modeler.

A common strategy for dealing with unknown parameters is to fit the model to measured data. This can be successful when there are only a few unknown parameters. However, there are also significant dangers. Firstly, if a model is complicated enough, it could well be the case that it could be fitted to nearly anything. A good fit with experimental data is not a sufficient (or indeed necessary) condition for the quality of a model. The fitted parameters may, therefore, be misleading or even meaningless. This does not mean to say that fitting is always a bad thing, but that any results obtained from fitted models have to be treated with appropriate caution.

There are situations in modeling when even numerical solutions cannot be obtained. This can be either because the system of equations to be solved is too complicated to even be formulated (let alone be solved), or because the modeling problem is not amenable to a mathematical (i.e., equation based) description. This could be the case for evolutionary systems which are more easily expressed using rules rather than equations (e.g., "when born, then mutate"). Similarly, it is also difficult to capture stochastic behavior using mathematical formulas. True, there are methods to estimate the statistical properties of stochastic systems using mathematical tools, and some of these methods will be discussed in this book. However, what mathematical methods struggle to represent are concrete examples of stochastic behavior—the noise itself rather than just its properties. We will have more to say on this in subsequent chapters.

Many of these problems can be addressed by computer simulations. Simulation can be powerful tool, able to capture accurate models of nearly limitless complexity, at least in principle. In practice there is, of course, the problem of finding the correct parameters, as in the case of the numerical mathematical models. In addition, there are two more serious limitations. The first is that simulation models must be specified in a suitable form that a computer can understand—often a programming language. There are a number of tools to assist the modeler for specific types of simulations. One of these tools (Repast Simphony) will be described in some detail in this book in Chap. 3. Yet, no matter how good the tool, specifying models takes time. If the model contains a lot of detail—many interactions that are so different from one another that each needs to be described separately—then the time required to specify the model can be very long. For complex models it also becomes harder to ensure that the model is correctly specified, which further limits simulation models.

The second, and perhaps more important limitation in practical applications is that arising from run-time requirements. The run time of even relatively simple models may scale unfavorably with some parameters of the system. A case in point is the simulation of chemical systems. For small numbers of molecules such simulations can be very rapid. However, through increases in the number of molecular interactions, once moderate to high numbers are reached, simulations on even very powerful computers will be limited to smallest periods of simulated time. Simplifications of a model can be valuable in those circumstances. A common approach is to remove spatial arrangements from consideration. This is often called the case of *perfect mixing*, where it is assumed that every entity interacts with every other entity with equal probability. All objects of the simulated world are, so to speak, in a soup without any metric. Every entity is equally likely to bump into every other. Simulation models often make this assumption for good reasons. Perfect mixing reduces the complexity of simulations dramatically, and consequently leads to shorter run times and longer simulated times. The difference can be orders of magnitude.

When spatial organization is of the essence then discrete spaces—spaces that are divided into perfectly mixed chunks—are computationally the cheapest of all spatial worlds. In many cases they make efficient approximations to continuous space whose explicit simulation requires significantly greater resources. Another determining factor is the dimension of the space. Models that assume a 3-dimensional world are usually the slowest to handle. The difference in run time compared to 2d can be quite dramatic. Therefore, whenever possible, spatial representations should be avoided. Unfortunately, often the third dimension is an essential feature of reality and cannot be neglected.

1.2 Stochastic Versus Deterministic Models

The insight that various phenomena in nature need to be described stochastically is deeply rooted in many branches of physics; quantum mechanics or statistical physics are inherently about the random in nature. Stochastic thinking is even more important in biology. Perhaps the most important context in which randomness appears in biology is evolution. Random alterations of the genetic code of organisms drive the eternal struggle of species for survival. Clearly, any attempt to model evolution must ultimately take into account randomness in some way.

Stochastic effects also play an important role at the very lowest level of life. The number of proteins, particularly in bacteria, can be very low even when they are expressed at the maximum rate. If the experimenter measures steady-state levels of a given protein, then this steady state is the dynamic balance between synthesis and decay; both are stochastic processes and hence a source of noise. Macroscopically, this randomness will manifest itself through fluctuations of the steady-state levels around some mean value. If the absolute number of particles is very small then the relative size of these fluctuations can be significant. For large systems they may be barely noticeable. Sometimes, these fluctuations are a design feature of the system, in the sense that the cell actively exploits internal noise to generate randomness. An example of this is the *fim* system in *E.coli* which essentially implements a molecular random bit generator. This system will be described in this book in Chap. 2. More often than not, however, noise is a limitation for the cell. Understanding how the cell copes with randomness is currently receiving huge attention from the scientific community. Mathematical and computational modeling are central to this quest for understanding.

Nearly all systems in nature exhibit some sort of noise. Whether or not noise needs to be taken into account in the model depends on the particular question motivating the model. One and the same phenomenon must be modeled as a stochastic process in one context, but can be treated as a noise-free system in a different one. The latter option is usually the easier. One common approach to the modeling of noise-free systems is to use systems of *differential equations*. In some rare cases these equations can be solved exactly; in many cases, however, one has to resort to numerical methods. There is a well developed body of theory available that allows us to infer properties of such systems by looking at the structure of the model equations. Differential equations are not the only method to formulate models of noise-free systems, although a very important one. Yet, it is nearly always easier to formulate and analyze a system under the assumption that it behaves in a deterministic way, rather than if it is affected by noise. Deterministic models are, therefore, often a good strategy at the start of a modeling project. Once the deterministic behavior of a system is understood, the modeler can then probe into the stochastic properties. Chapter 4 will provide an introduction to deterministic modeling in biology.

There are methods to model stochastic properties of systems using equation-based approaches. Chapter 6 introduces some stochastic techniques. Stochastic methods can provide analytic insight into the noise properties of systems across the parameter space. Unfortunately, the cases where analytic results can be obtained are rare. For even moderately complicated systems, approximations need to be made, most of which are beyond the scope of this book. In the majority of cases, the stochastic behavior of systems must be inferred from simulations. There are a number of powerful high-quality tools available to conduct such simulations.

1.3   Fundamentals of Modeling

There are two vital ingredients required for modeling: skill and technique. By technique we mean the ability to formulate a model mathematically, or to program computational simulation models. Technique is a *sine qua non* of any modeling enterprise in biology, or indeed any other field. However, technique is only one ingredient in the masala that will eventually convince reviewers that a piece of research merits publication. The other ingredient is modeling skill.

Skill is the ability to ask the correct, biological question; to find the right level of abstraction that captures the essence of the question while leaving out irrelevant detail; to turn a general biological research problem into a useful formal model. While there is no model without technique, unfortunately, the role of skill and its importance are often overlooked. Sometimes models end up as masterpieces of technical virtuosity, but with no scientific use. In such pieces of work the modeler demonstrates her acquaintance with the latest approximation techniques or simulation tools, while completely forgetting to address clearly a particular problem. In many cases, ground breaking modeling-based research can be achieved with very simple techniques; the beauty arises from the modeler's ability to ask the right question, not from the size of her technical armory.

The authors of this book think that modeling skill is nearly always acquired, not something that is genetically determined. There may be some who have more inclination towards developing this skill than others, but in the end everybody needs to go through the painful process of learning how to use techniques to answer the right questions. Depending on the educational background of the novice modeler, developing the right modeling skill may require her to go against the ingrained instincts that she has been developing through years of grueling feedback from examiners, tutors and peer reviewers. These instincts sometimes predispose us to apply the wrong standards of rigour to our models, with the result that the modeling enterprise is not as successful as it could have been, or the model does not provide the insight we had hoped for.

One of the common misconceptions about modeling is the principle that, "More-detailed models are necessarily better models". Let us assume that we have a natural system S that consists of N components and interactions (and we assume for the sake of argument that this is actually a meaningful thing to say). Assume, then, that we have two models of S: M_1 and M_2. As is usual in models, both will represent only a subset of the N interactions and components that make S. Let us now assume that M_2 contains everything that M_1 contains, and some more. The question is now, whether or not that necessarily makes M_2 the better model?

Let us now suppose that we can always say that M_2 is better than M_1, simply because it contains more. If this is so, then we can also stipulate that there is an even better model, M_3, that contains everything M_2 contains and some more. Continuing this process of model refinement, we would eventually reach a model that has exactly N components and interactions and represents everything that makes our natural system S. This would then be the best model. This best model would be equivalent to S itself and, in this sense, S is its own best model. Since we have S available, there would be no point in making any model as we can directly inspect S. Hence, if we assumed that bigger models are always better models, then we have to conclude that we do not need any models at all.

There will be situations (although not typically in biological modeling) where it is indeed desirable, at least in principle, to obtain models that replicate reality in every detail. In those cases, there will then really be this hierarchy of models, where one model is better than the other if it contains more detail. This will typically be the case in models that are used for mission planning in practice, for example in epidemiological modeling. Yet, in most cases of scientific modeling, the modeler struggles to understand the real system because of its many interactions and its high degree of complexity. In this case the purpose of the model is precisely to represent the system of interest S in a *simplified* manner, leaving out much of the irrelevant, but distracting detail. This makes it possible for the modeler to reason about the system, its basic properties and fundamental characteristics. The system S is always its own best model if accuracy and completeness are the criteria. Yet, they are not. A model is nearly always a rational simplification of reality that allows the modeler to ask specific questions about the system and to extract answers.

There are at least two reasons why simplification is a virtue in modeling. Modelers are (to borrow a term from economics), agents with "bounded rationality." We use this description in a wide sense here, but essentially it means that the modeler's ability to program/formulate detailed models is limited. The design process of models normally is done by hand, in the sense that a modeler has to think about how to represent features of the real system and how to translate this representation into a workable model. Typically this involves some form of programming or the formulation of equations. The more components there are the longer this process will take. What is more, the larger the model the longer it will take to determine model parameters and the longer it will take to analyze the model. Particularly in simulation models, run-time considerations are important. The size of a model can quickly lead to computational costs that prevent any analysis within a reasonable time frame. Also, quality control becomes an increasingly challenging task as the size of models increases. Even with very small models it can be difficult to ensure that the models actually do what the modeler intends. For larger models, quality control may become impossible. It is not desirable (and nearly never useful) to have a model whose correctness cannot be ensured within reasonable bounds of error. Hence, there are practical limits to model size, which is why more detailed models are not always better models.

In a sense, the question of model size, as presented above, is an academic one anyway. A modeler always has a specific purpose in mind when embarking on a modeling project. Models are not unbiased approximations of reality, but instead they are biased towards a specific purpose. In practice, there is always a lot of detail that would complicate the model without contributing to fulfill the purpose. If one is interested in the biochemistry of the cell, for instance, then it is often useful to assume that the cell is a container that selectively retains some chemicals while being porous to others. Apart from its size and maybe shape, other aspects of the "container" are irrelevant. It is not necessary to model the details of pores in the cell membrane or to represent its chemical and physical structure. When it comes to the biochemistry of the cell, in most cases there is no need to represent the shape of proteins. It is sufficient to know their kinetic parameters and the schema of their reactions. If, on the other hand, one wishes to model how, on a much finer scale, two proteins interact with one another, then one would need to ignore other aspects and focus on the structure of these proteins.

Bigger models are not necessarily better models. A model should be fit for its specific purpose and does not need to represent everything we know about reality. Looked upon in this abstract setting, this seems like an evident truth, yet it illustrates what goes against the practice many natural scientists have learned during their careers. In particular, biologists who put bread on their tables by uncovering the minutiae of molecular mechanisms and functions in living systems are liable to over-complicate their models. After all, it is not surprising that a scientist who has spent the last 20 years understanding how all the details of molecular machinery fit together wants to see the beauty of their discoveries represented in models. Yet, reader be warned: Do not give in to such pressure (even if it is your own); rather, make simplicity a virtue! Any modeling project should be tempered by the morality of laziness.

The finished product of a modeling enterprise, with all its choices and features, can sometimes feel like the self-evident only solution. In reality, to get to this point much modeling and re-modeling, formulating and re-formulating, along with a lot of sweat and tears, will have gone into the project. The final product is the result of a long struggle to find the right level of abstraction and the right question to ask, and of course the right answer to the right question. There are no hard and fast rules on how to manoeuvre through this process.

A generally successful principle is to start with a bare-bones model that contains only the most basic interaction in the system and is just about not trivial. If the predictions of this bare-bones model are in any way realistic or even relevant for reality, then this is a good indicator that the model is too complicated and needs to be stripped down further. Adhering to the morality of laziness, the bare-bones model should be of minimal complexity and must be easy to analyze. Only once the behavior and the properties of the bare-bones model are fully understood should the modeler consider extending it and adding more realism. Any new step of complexity should only be made when the consequences of the previous step are well understood.

Such an incremental approach may sound wasteful or frustrating at first. Surely it seems pointless to consider a system that has barely any relevance for the system under investigation? In fact, the bare-bones model, (i) often contains the basic dynamical features that come to dominate the full system. Yet, only in the bare-bones model does one have a chance to see this, whereas in the full model one would drown in the complexity hiding the basic principles. Then also, (ii) this approach naturally forces the modeler to include into the model only what needs to be included, leaving out everything extraneous. Moreover, this incremental approach, (iii) provides the modeler with an intuition about how model components contribute to the overall behavior.

By now it should be clear that simplicity is a virtue in modeling. Yet the overriding principle should always be fitness for purpose. A modeling project should always be linked to a clear scientific question. Any useful model should directly address a scientific problem. A common issue, particularly with technical virtuosos, is that the modeling enterprise lacks a clear scientific motivation, or research question. Computer power is readily available to most and the desire to use what is available is strong. Particularly for modelers with a leaning towards computer science, it is often very tempting to start to code a model and to test the limits of the machine. This unrestrained lust for coding often seduces the programmer into forgetting that models are meant to beget scientific knowledge and not merely to satisfy the desire to use one's skills. Hence, alongside the morality of laziness, a second tenet that should guide the modeler is: Be guided by a clear scientific problem. The modeling process itself should be the ruthless pursuit to answer this problem, and nothing else.

A principle that is often used to assess the quality of a model is its ability to make predictions. Indeed, very often models are built with the express aim of making a prediction. Among modelers, prediction has acquired something of the status of a holy cow, and is revered and considered the pinnacle of good modeling. Despite this, the reader should be aware that prediction (depending on how one understands it) can actually be quite a weak property of a model. Indeed, it may well be the case that more predictive models are less fit for purpose than others that do not predict as well.

One aspect of "prediction" is the ability of models to reproduce experimental data, which is one aspect of predictability. Rather naively in our view, some seem to regard this as a gold standard of models. Certainly, in some cases, it is but in others it might not be. Particularly in the realm of biology, many (even most) parameters will be unknown. In order to be able to reproduce experimental data it is therefore often necessary to fit the unknown parameters to the data. This can either succeed or fail. Either way, it does not tell us much about the quality of the model, or rather its fitness for its particular purpose. For one, the modeler is very often interested in specific qualitative aspects of the system under investigation. Following the morality of laziness, she has left out essential parts of the real system to focus on the core of the problem. These essential parts may just prevent the system from being able to be fitted to experimental data. This does not necessarily make the model less useful or less reliable. It just means that prediction of experimental data is, in this case, not a relevant test for the suitability and reliability of the model. Often these models can, however, make qualitative predictions; for instance, "If this and that gene are mutated, then this and that will happen." These qualitative predictions can lend as much (or even more) credibility to the model as a detailed reproduction of experimental data.

Secondly, given the complexity of some of the models and the number of unknown parameters, one can wonder whether some dynamical models cannot be fitted to nearly any type of empirical data. As such, model fitting has the potential to lend the model a false credence. This is not to say that fitting is always wrong, it is only to say that one should be wary of the suggestibility of perfectly reproduced experimental data. Successful reproduction of experimental data does not make a model right, nor does it make a model wrong or useless if it cannot reproduce data.

Once a modeler has a finished model, it is paramount that she is able to give a detailed justification as to why the model is relevant. As discussed above, all models must be simplified versions of reality. While many of the simplifying assumptions will be trivial in that they concern areas that are quite obviously irrelevant for the specific purpose at hand, models will normally also contain key simplifications whose impact on the final result is unclear. A common example in the context of biochemical systems is the assumption of perfect mixing, as mentioned above. This assumption greatly simplifies mathematical and computational models of chemical systems. In reality it is, of course, wrong. The behavior of a system that is not mixed can deviate quite substantially from the perfectly-mixed dynamics. In many practical cases it may still be desirable to make the assumption of perfect mixing, despite its being wrong; indeed, the vast majority of models of biochemical systems do ignore spatial organization. In all those cases, as a modeler one must be prepared to defend this and other choices. In practice, simplifying assumptions can sometimes become the sticking point for reviewers who will insist on better justification.

One possible way to justify particular modeling choices is to show that they do not materially change the result. This can be done by comparing the model's behavior as key assumptions are varied. In the early phases of a modeling project, such variations can also provide valuable insights into the properties of the model. If the modeler can actually demonstrate that a particular simplification barely makes any difference to the results but yields a massively simplified model, then this provides a strong basis from which one can pre-empt or answer referees' objections.

Apart from merely varying the basic assumptions of the model it is also good modeling practice to vary the modeling technique itself. Usually one and the same problem can be approached using more than one method. Biochemical systems, for example, can be modeled in agent-based systems, differential equations, using stochastic differential equations, or simulated using the Gillespie [1] and related algorithms. Each of these approaches has its own advantages. If the modeler uses more than just a single approach, then this will quite naturally lead to varying assumptions across the models. For example, differential equation models usually rest on the assumption that stochastic fluctuations are not important, whereas stochastic simulations using Gillespie's algorithm are designed to show fluctuations. Specifically during early stages of a modeling project, it is often enlightening to play with more than one technique. Maintaining various models of the same system is, of course, also a good way to cross-validate the models and overall can lead to a higher confidence in the results generated by them. Of course, building several models requires considerably more effort than building just one!

1.4 Validity and Purpose of Models

A saying attributed to George E.P. Box [2] is: "Essentially, all models are wrong, but some are useful." It has been discussed already that a model that is correct—in the sense that it represents every part of the real systems—would be mostly useless. Being "wrong" is not a flaw of a model but an essential attribute. Then again, there are many models that are, indeed, both wrong and useless. The question then is, how can one choose the useful ones, or better, the most useful one from among all the possible wrong ones?

A comprehensive theory of modeling would go beyond the scope of this introductory chapter, and perhaps also over stretch the reader's patience. However, it is worth briefly considering some types of models classified according to their usefulness, though the following list is certainly not complete. The reader who is interested in this topic is also encouraged to see the article by Groß and Strand on the topic of modeling [3].

A simple, but helpful way to classify models is to distinguish between, (i) predictive, (ii) explanatory and (iii) toy models. The latter class is mostly a subset of explanatory models, but a very important class in itself and, hence, worth the extra attention. As the name suggests, predictive models are primarily used for the purpose of predicting the future behavior of a system. Intuitively, one would expect that predictive models must adhere to the most exacting standards of rigour because they have to pass the acid test of correctness. There is no arguing with data. Therefore, the predictive model, one might think, must be the most valid one; in some sense the most correct one. In reality, of course, the fact that the model does make correct predictions is useful, but it is only *one* criterion, not always the most important one, and never sufficient to make a model useful.

A well known class of models that are predictive, but otherwise quite uninformative, are the so-called empirical formulas. These are quite common in physics and are models that have been found empirically to correctly predict the results of experiments. Sometimes, these empirical models even make assumptions that are known to be completely incorrect, but they are still used in some circumstances, simply for predicting certain values. An example of such a model is the "liquid drop model" in nuclear physics which crudely treats the nucleus as an incompressible drop of fluid. Despite its crude nature, it does still have some useful predictive properties. There are many other such models. Purely predictive models do not tell us anything about the nature of reality, even though they can be quite good at generating numerical values that correspond well to the real world. This is unsatisfactory in most circumstances. Models should do more than that—they should also explain reality in some way.

Explanatory models *do* explain reality in some way, and are very important in science. Unlike predictive models, their primary purpose is to show how certain aspects of nature that have been observed experimentally can be made sense of. Often explanatory models are also predictive models; in these cases, the predictive accuracy can be a criterion for the usefulness of a model. However, one could easily imagine that there are explanatory models that do not predict reality correctly. Particularly in biology, this case is quite common because of the chronic difficulty of measuring the correct parameters of systems. A quantitative prediction is then often impossible. In those cases one could decide to be satisfied with a weaker kind of prediction, namely qualitative prediction. This means that one merely demonstrates that the model can reproduce the same kinds of behavior without insisting that the model reproduces some measured data accurately. Again, such qualitative prediction is not necessarily inferior to a quantitative prediction; this is particularly true when the quantitative prediction relies on fitting the model to empirical data, which risks being the equivalent of an empirical formula.

The main evaluation criterion for explanatory models is how well they illuminate a particular phenomenon in nature. Prediction is one way to assess this, but another is the structural congruence between model and reality. Explanatory models are often used to ask whether a certain subset of ingredients is sufficient to explain a specific observed behavior. A good example in this respect is models in game theory. A typical objective of research in game theory is to understand under which conditions co-operation can evolve from essentially selfish agents. Game-theoretical models practically never intend to predict how evolution will continue, but simply attempt to understand whether or not a specific kind of behavior has the potential to evolve given a set of specific conditions. Nearly all models in this field are "wrong" but they are productive, or, as Box said, "useful." They tell us something about the basic properties of evolution and, as such, add to human understanding and the progress of science, without, however, allowing us to predict the future, or (in many cases) retrodict the past. Within their domain, these and many other explanatory models are of equal usefulness as predictive models. Explanatory models are not the poor relation of their more glorious predictive cousins, and their value can be independently justified.

A sub-class of explanatory models are "toy models." These are models that are primarily used to demonstrate some general principle without making specific reference to any particular natural system. The main advantage of toy models is that they are very general in scope, while at the same time abstracting away from all the complications of reality that normally make the life of a modeler difficult. Toy models take parsimony to the extreme. Due to their simplicity, such toy models can often be formulated mathematically and, as such, provide general insights that cannot be obtained from more specific models. Nonetheless, insights extracted from such models can often be transferable to real cases.

A particularly famous example of such a toy model is Per Bak's sand pile model described in his very well written book, "How nature works" [4]. The basic idea of this model is to look at a pile of sand on which further grains of sand are dropped from time to time. At first, the pile grows, but after some time it stops growing and newly dropped grains may fall down the sides of the pile, potentially causing "avalanches". There are events, therefore, where adding a single grain causes a large number of existing grains to become unstable and glide down the slope of the pile. Bak's original model was a computer model and does not actually describe the behavior of real sand. However, an experimental observation with rice [5] has been found to behave in a similar way. So, Bak's computer model is wrong—dramatically so—yet, it is useful.

The beauty of Bak's models is that it provided an explanatory framework that allowed the unification of a very large number of phenomena observed in nature, ranging from extinction in evolution, to earthquakes and stock market crashes. None of these events were accurately modeled by Bak's sand pile, but ideas from the model could be applied to them. The avalanche events had their counterparts in real systems, be it waves of extinction events in evolution or earthquakes that suddenly relax tensions that have become built up in the Earth's tectonic plates. As such, Bak's toy model opened up a new way to look at phenomena, and allowed novel scenarios to be considered.

Not all toy models are of the same generality as Bak's. Even in more restricted scope, it is good idea to start modeling enterprises using toy models which can then be refined. The criterion of the usefulness of such toy models is clearly the extent to which they help generate new understanding. They also harbor a danger, however; if taken too far then toy models can actually obscure rather than illuminate.

In summary: There is no single criterion to assess the quality or usefulness of a model. In every modeling enterprise, the specific choices made to produce the model must be justified each time. Predictive power in a model is a good thing, but it is not the only criterion, neither should it be used as the acid test of the quality of a model—at least not in all cases. Sometimes, the most powerful models are highly simplified and do not predict anything, at least not quantitatively.

Having now established what we believe to be the most fundamental guiding principles for developing good modeling skills, it is now time to get going with the practice. The rest of the book primarily concerns itself with technique. What we mean is that it will introduce the reader both to the underlying ideas and the scope of modeling techniques, but it will also present details of practical tools and environments that can be used in modeling. In addition to the focus on technique, walk-through examples will be provided to show the reader how the various guidelines of good modeling established in this chapter translate into actual modeling practice.

References

1. Gillespie, D.: Exact stochastic simulation of coupled chemical reactions. J. Phys. Chem. **81**(25), 2340–2361 (1977)
2. DeGroot, M.H.: A conversation with George Box. Stat. Sci. **2**(3), 239–258 (1987). doi:10.1214/ss/1177013223
3. Gross, D., Strand, R.: Can agent-based models assist decisions on large-scale practical problems? A philosophical analysis. Complexity **5**(5), 26–33 (2000)
4. Bak, P.: How Nature Works. Oxford University Press, Oxford (1997)
5. Frette, V., Christensen, K., Malthe-Sørenssen, A., Feder, J., Jøssang, T., Meakin, P.: Avalanche dynamics in a pile of rice. Nature **379**, 49–52 (1996)

Agent-Based Modeling

2

Traditionally, modeling in science has been *mathematical modeling*. Even today, the gold standard for respectability in science is still the use of formulas, symbols and integrals to communicate concepts, ideas and arguments. There is a good reason for this. The language of mathematics is concise and precise and allows the initiated to say with a few Greek letters what would otherwise require many pages of text. Mathematical analysis is important in science, but it would be wrong to make its use an absolute criterion for good science.

Many of the phenomena modern science studies are complicated, indeed so complicated that the brightest mathematicians have no hope of successfully applying their craft to describe these phenomena. Most of reality is so complex that even formulating the mathematical model is close to impossible. Nevertheless, many of these phenomena are worth our attention and meaningful knowledge can be obtained from studying them.

This is where computer models become useful tools in a scientist's hands. Simulation models can be used to describe and study phenomena even when traditional mathematical approaches fail. In this chapter we will introduce the reader to a specific class of computer models that is very useful for exploring a multitude of phenomena in a wide range of sciences: the class of *agent-based models*.

Computers are still a relatively recent addition to the methodological armory of science. Initially their main use was to extend the range of tractability of mathematical models. In the pre-computer era, any mathematical model had to be solved by laborious manual manipulations of equations. This is, of course, time intensive and error prone. Computers made it possible to out source the tedious hand-manipulations and, more importantly, to generate numerical solutions to mathematical models. What takes hours by hand can be done within an instant by a computer. In that way, computers have pushed the boundaries of what could be calculated.

© Springer-Verlag London 2015
D.J. Barnes and D. Chu, *Guide to Simulation and Modeling for Biosciences*,
Simulation Foundations, Methods and Applications,
DOI 10.1007/978-1-4471-6762-4_2

2.1   Mathematical and Computational Modeling

In what follows, we are not going to be interested in computer-aided mathematical modeling, i.e., methods to generate numerical solutions. Instead, we will be looking at a class of computer models that is *formal* by virtue of being specified in a programming language with precise and unambiguous semantics. Yet the models we are interested in are also non-mathematical, in the sense that they represent and model phenomena that go well beyond what can even be formulated mathematically, let alone be solved.

Computer models are formal models even when they are not mathematical models. They are written in a precise language that is designed not to leave any room for ambiguity in its implementation. Every detail of the model must be expressed in this language and no aspect can be left out. All the computer does, once the model is formulated, is to mercilessly draw conclusions from the model's specification. Formal analysis of this kind is a useful tool for generating a deep understanding of the systems that are studied, and is far superior to mere verbal reasoning. It is therefore worth spending time and effort to develop computational representations of systems even when (or precisely when) a mathematical analysis seems hopeless.

How do we know that a system is not amenable to mathematical analysis but can be modeled computationally? Maybe, it is easier first to understand what it is that makes a system suitable for mathematical analysis. Possibly the most successful and influential mathematical model in science is that formulated by Newton's laws of motion. These laws can be applied to a wide variety of phenomena, ranging from the trajectory of a stone thrown into a lake, to the motion of planets around their stars. One feature that Newton's laws share with many models in physics is that they are *deterministic*. The defining feature of a deterministic system is that, once its initial conditions are fixed, its entire future can be calculated and predicted. The word "initial" implies somehow that we are seeking to identify the conditions that apply at the "beginning" of a system. In fact, initial conditions are often associated with the condition at the time $t = 0$; in truth, this is only a convenient label for the time at which we have a complete specification of the system and does not, of course, denote the origin of time. All that matters is that we have, for a single time point, a complete specification of the system in terms of all the positions and velocities of its components. Given those, we can compute the positions and velocities of the system for all future (and, indeed, past) times, if the system is deterministic.

A good example of such deterministic behavior is that of the planets within our solar system. Since we know the laws of motion of the planets, we can predict their positions for all time, if we only know their positions at one specific time (that we would arbitrarily label as time $t = 0$). The positions of the planets are relatively easy to measure, so determining the initial conditions is not a problem. Equipped with this knowledge, astronomers can make very accurate descriptions of phenomena such as eclipses or the reappearance of comets and meteors (that might or might not collide with the Earth). Applying the same laws further, we can predict when a certain beach will reach high tide or how long it will take before a tsunami hits the nearest coast.

Determinism is also the essential requirement for our ability to engineer electrical and electronic circuits. Their behaviors do not follow from Newton's laws, but they are also deterministic, as are many of the phenomena described by classical physics.

2.1.1 Limits to Modeling

2.1.1.1 Randomness

In the intellectual history of science, determinism was dominant for a long time but was eventually replaced by a statistical view of the world—at least in physics. Firstly thermodynamics and then quantum mechanics led to the realization that there is inherent randomness in the world. The course of the world is, after all, not determined once and forever by its initial conditions. This insight was conceptually absorbed into the scientific *weltanschauung* and, to some extent, into the body of physical theory. Despite the conceptual shift from a deterministic world-view to a statistical one, mathematical modeling, even in the most statistical branches of science, continues *not* to represent the randomness in nature, at least not directly. Take as evidence that the most basic equation in quantum mechanics (the so-called Schrödinger equation) is a deterministic equation, even though it represents fundamentally stochastic, indeterminate, physical phenomena. Randomness in quantum mechanics only enters the picture through the interpretation of the deterministic equation, but the very random behavior itself is not modeled.

To be fair, in physics and physical chemistry there are models that attempt to capture randomness, for example in the theory of diffusion. However, the models themselves only describe some deterministic features of the random system, but not the randomness itself. Mathematical models tell us things such as the expected (or mean) behavior of a system, the probability of a specific event taking place at a specific time, or the average deviation of the actual behavior from the mean behavior. All these quantities are interesting, but they are also deterministic. They can be formulated in equations and, once their initial conditions are fixed, we can calculate them for all times. Mathematics allows us to extract deterministic features from random events in nature. True randomness is rarely seen in mathematical models.

An example might illustrate how inherently stochastic systems can be described in a deterministic way. Think of a small but macroscopic particle (for example, a pollen grain) suspended in a liquid. Observing this particle through a microscope will show that it receives random hits from time to time, resulting in its moving about in the liquid in a seemingly random fashion. This so-called *Brownian motion* cannot be described in detail by mathematical modeling precisely because it is random. Nevertheless, very sophisticated models can give an idea of how far the particle will travel on average in a given time (mean), by how much this average distance will be different from the typical distance (standard deviation), how the behavior changes when force fields are introduced, and so on. Note that the results of this mathematical modeling, as important as they are to characterize the motion of the particle, are themselves deterministic. This does not make them irrelevant; quite the opposite. The point to take from this discussion is that the description of the random particle is indeed a description of the deterministic aspects of the random motion, and not of the randomness itself.

One could argue that there is not much more we could possibly want to know about the pollen grain beside the deterministic regularities of the system. What good is it to know about the idiosyncracies of the random drift of a particular particle? And yes, perhaps in this case we really only want to know about the statistical regularities of the system, which are deterministic. In that respect, the randomness of Brownian motion is quite *reducible* and we are well served with our deterministic models of the random phenomenon.

Another example of systems where stochasticity is reducible are gases in statistical mechanics. A gas (even an ideal one) consists of an extremely large number of particles, each of which is characterized, at any particular time, by its position and momentum. Keeping track of all these individual particles and their time evolution is hopeless. Collisions between the particles lead to constant re-assignments of velocities and directions. In principle, one could calculate the entire time-evolution of the system if given the initial conditions—ideal gases are deterministic systems. In reality, of course, this would be an intractable problem. Moreover, the initial conditions of the system are unknown and unmeasurable.

As it turns out, however, this impossibility of describing the underlying motion of particles in gases is not a real limitation. All we really need to care about are some stochastic regularities emerging from the aggregate behavior of the colliding molecules. Macroscopically, the behaviors of gases are quite insensitive to the details of the underlying properties of the individual particles and can be described in sufficient detail by a few variables. It is well known that an ideal gas in thermal equilibrium can be described simply in terms of its pressure, volume and temperature:

$$PV \propto T$$

There is no need to worry about all the billions of individual molecules and their interactions. We do not need to know where every molecule is at any given time. All we care about is how fast they are on average, the probability distribution of energies over all molecules, and the expected local densities of the gas. Once we have those details we have reduced its random features to deterministic equations.

This approach of reducing randomness really only works if the individual random behavior of a specific particle in our system is unimportant. In physics, this will often be the case, but there are systems that are *irreducibly random*; systems where the random path of one of its components *does* matter and a deterministic description of the system is not enough. One example is natural evolution in biological systems. Random and unpredictable mutations at the level of genes cause changes at the level of the phenotype. Over evolutionary time scales, there will be many such mutations, striking randomly, and often leading to the demise of their bearer. Sometimes a mutation will have no noticeable effect at all but, on rare occasions, a mutation will be beneficial and lead to an increase in fitness.

Imagine now that we wish to attempt to model this process. If we tried to come up with a deterministic model of mutations and their effects, after much effort and calculation we would perhaps know how beneficial mutations are distributed, for how long we need to wait before we observe one, and so on. This might be what we want to know, but maybe we want to know more. Imagine that we would like to model the actual evolution of a species; that is, we are interested in creating a model that allows us to study how and when traits evolve and what these traits are. Imagine that we want to model the evolutionary arms race between a predator and its prey. Imagine that we want to have a model that allows us to actually see evolution in action. In this case, we do not care about all the mutations that led nowhere. All we are interested in is those few, statistically insignificant events that resulted in a qualitative change in our system: a new trait or defense mechanism, for instance. What we would be interested in are *the particulars* of the system, not its general features.

Mathematical models are normally not very good at representing particulars of random systems; different approaches are required if we are interested in those. We can say, therefore, that if our system is irreducibly random then mathematical approaches will be limited in their usefulness as models.

2.1.1.2 Heterogeneity

Another aspect of natural systems that limits mathematical tractability is system *heterogeneity*. A system is heterogeneous if: (i) it consists of different parts, and (ii) these parts do not necessarily behave according to the same rules/laws when they are in different states. As a simple example, one can think of structured populations of animals, say lions. Normally lions live in packs, and each pack has an intrinsic order that determines how individuals act in the context of the entire group. This group structure has evolved over time and is arguably significant for the survival of lions. It is also difficult to model mathematically.

One could try to circumvent this and ignore the detail. Often such an approach will be successful. If one wanted to model the population dynamics of lions in the Serengeti, it may be sufficient to look at the number of prey, the efficiency with which prey is converted into offspring by lions, and the competition (in the form of leopards, cheetahs, etc.). With this information, we might then formulate a reasonable model of how the lion population will develop over time in response to various environmental changes, despite the model having ignored much of the structure of lion populations.

The interactions between the individual animals would be difficult to model in a mathematical model. Lions behave differently depending on their age, their rank within the group and their gender. Keeping track of this in a set of equations would quickly test the patience and skill of the modeler. Moreover, lions reproduce and die.

Equations don't seem to be able to offer a good approach in this case. At the same time, it is not unthinkable that one may want to model the behavior of packs of lions formally. One might, for example, be interested in the life strategies of lions and complement empirical observations with computer models. The heterogeneity of the pack is irreducible in such a model. It would make no sense to assume an "average" lion with some "mean" behavior. The dynamics of the packs and how the behavior patterns contribute to the evolutionary adaptability of lions rest crucially on the lions being a structured population. We are not aware of attempts to model life histories of lions, but in the context of social insects and fish there have been many attempts to use computer models to understand group dynamics (see [1,2]); but never have these studies used purely equation-based approaches.

While we can clearly see that most systems are heterogeneous, often they are *reducibly* so. The difference between the component parts can often be ignored and be reduced to a mean behavior while still generating good and consistent results. Whether or not a system is reducibly or irreducibly heterogeneous depends on the particular goals and interests of the modeler and is not a property of the system *per se*.

In physics, the heterogeneity of most problems is reducible, which has to do with the type of questions physicists tend to ask. In biology things are different. Many of the phenomena bioscientists are interested in are essentially about heterogeneity. Reducing this heterogeneity is often meaningless. This is one of the reasons why it has been so difficult to base biology on a mathematical and formal theoretical basis, whereas it has been so successful in physics. Irreducible heterogeneity makes mathematical modeling very difficult and life consequently harder.

2.1.1.3 Interactions

Finally, a third complicating property of systems is component interaction. Unlike heterogeneity and randomness, interactions have been acknowledged as a problem in physics for a long time. In its most famous incarnation this is known as, "the n-body problem". Theoretical physics has the tools to find general solutions for the trajectories of 2 gravitating bodies. This could be two stars that are close enough for their respective gravitational fields to influence each others' motions. What about three bodies? As it turns out, there is no nice (or even ugly) formula to describe this problem; and the same is true for more than three bodies.

Interaction between bodies poses a problem for mathematical modeling. How do physicists deal with it? The answer is that they do not! In the case of complex multi-body interactions it is necessary to solve such problems numerically. Furthermore, there are cases of systems that are so large that one can only approximate the inter-action between parts with a "mean field", which essentially removes any individual interactions and makes the system solvable. There are many cases of such *reducible interactions* in physics, where the mean-field approximation still yields reasonable results. Before the advent of computers, irreducible interactions were simply ignored because they are not tractable using clean mathematical approaches.

In biology, there are few systems of interest where interactions are reducible. Nearly everything in the biological world, at all scales of magnification, is interaction: be it the interactions between proteins at a molecular level, inter-cell communications at a cellular level, the web of interactions between organisms at the scale of ecology and, of course, the interaction of the biosphere with the inanimate part of our world. If we want to understand how the motions of swarms of fish are generated through the behavior of individual fish, or how cells interact to form an embryo from an unstructured mass of cell, then this is irreducibly about interactions between parts.

The beauty of mathematical modeling is that it enables the modeler to derive very general relationships between variables. Often, these relationships elucidate the behavior of the system over the entire parameter space. This beauty comes at a price, however: the price of simplicity. If we want to use mathematics then we need to limit our inquiry to the simplest systems—or at least to those systems that can be reduced to the very simple. If this fails then we need to resort to other methods, for example computer models. While computational models tend not to satisfy our craving for the pure and general truths that mathematical formulas offer, they do give us access to representations of reality that we can manipulate to our pleasure.

In a sense, computational models are a half-way house between the pure mathematical models as they are predominantly used in theoretical physics, and the world of laboratory experimentation. Computer models are formal systems, and thus rigorous, with all the assumptions and conditions going into the models being perfectly controllable. Experimentation with real systems often does not provide this luxury. On the other hand, in a strict sense, computer models can only give outcomes for a particular set of parameters, and make no statement about the parameter space as a whole. In this sense, they are inferior to mathematical models that can provide very general insights into the behavior of the system across the full parameter space. In practice, the lack of generality is often a problem, and it makes computer models a second-best choice—for when a mathematical analysis would be intractable.

2.2 Agent-Based Models

The remainder of this chapter focuses on a particular computer modeling technique— *agent-based modeling* (ABM)—which can be very useful in modeling systems that are irreducibly heterogeneous, irreducibly random and contain irreducible interactions. The principle of an ABM is to represent *explicitly* the heterogeneous parts of a system in the computer model, rather than attempting to "coarse grain" it. In essence, this is achieved by building a virtual copy of the real system [3]—the model represents components of the real system explicitly and keeps track of the behavior of individuals over time. So, in a sense, each individual lion of the pack would have a virtual counterpart in the computer model. As such, ABMs are quite unlike mathematical models, which represent components by the values of variables rather than by behaviors. In an ABM, the different components (the "agents") represent entities in the real world system to be modeled. As well as the individual entities, an ABM also represents the environment inhabited by these entities. Each of these modeled entities has a *state* and exhibits an explicit *behavior*. An agent can interact with its environment and with other entities.

The behavior of agents is often "rule-based." This means that the instructions are formulated as *if-then* statements, rather than as mathematical formulas. For example, in a hypothetical agent-based model of a lion, one rule might be:

If in hunting mode and there is a prey animal closer than 10 meters, then attack.

A second rule would likely determine whether the lion should enter hunting mode:

If not in hunting mode, and T is the time since the last meal, then switch into hunting mode with probability $(T_{max} - T)/T_{max}$.

This second rule clearly contains a mathematical formula, which illustrates that the distinction between rules and mathematical formulas is somewhat fuzzy. Rules are not always purely verbal statements but often contain mathematical expressions. However, what is central to the idea of rule-based approaches is that ABMs never contain a mathematical expression for the behavior of the system as a whole. Mathematical expressions are a convenient means to determine the behavior of individual constituent parts and their interactions. The behavior of the system as a whole is, therefore, *emergent* on the interaction of the individual parts.

The underlying principle of an ABM is that the behavior of agents is traced over time and observed. The approach is thus very different from equation-based mathematical models that try to capture directly the higher-level behavior of systems. ABMs can be thought of as *in silico* mock-ups of the real system. This approach solves the representational problems of standard mathematical models with respect to irreducible interaction, randomness and heterogeneity. An explicit representation of random behavior is unproblematic in computer models: (pseudo) random number generators can be used to implement particular instances of random behaviors and to create so-called sample trajectories of stochastic systems, without the need to reduce the stochasticity to its deterministic aspects. Similarly, while it is often difficult to represent irreducible heterogeneity in mathematical models, in ABMs this aspect tends to arise naturally from the behavioral rules with which the agents are described—distinct individuals of the same agent type naturally end up in distinct states at the same time. There is a similar effect with respect to interactions.

The major drawback of ABMs is their computational cost. Depending on the intricacy of the model, ABMs can often take a very long time to run. Even when run on fast computers, large models might take days or weeks to complete. What exacerbates this feature is that the result of an individual model run is often shaped by stochastic effects. This means that the outcome of two simulation experiments, even if they use the same configuration parameters, may be quite different. It is necessary, therefore, to repeat experiments multiple times in order to gain (statistical) confidence in the significance of the results obtained.

Altogether, ABMs are a mixed blessing. They can deal with much detail in the system, and they can represent heterogeneity, randomness and interactions with ease. On the other hand, they are computationally costly and do not provide the general insights that mathematical models often do.

2.2.1 The Structure of ABMs

Let us now get to the business of ABM modeling in detail. ABMs are best thought of in terms of the three main ingredients that are the core of every such model:

- the agents;
- the environment inhabited by the agents;
- the rules defining how agents interact with one another and with their environment.

2.2.1.1 Agents

Agents are the *raison-d'être* of any ABM, representing the entities that act in the world being modeled. These agents are the central units of the model and their aggregate behavior will determine its outcome. In an ABM of a pack of lions, we would most likely choose the individual lions as agents. Models will often have more than one type of agent with, typically, many instantiations of each particular type of agent. For example our model may have the agent types: lion, springbok, oryx, etc., and there will exist many instances of each in the running model. A particular agent type is characterized by the set of internal states it can take, the ways it can impact its environment, and the way it interacts with other agents of all types, including its own. A lion's state, for example, might include its gender, age, hunger level and whether it is currently hunting; its interactions would likely include the fact that it can eat oryxes and springboks. On the other hand, a springbok's interactions would not involve it preying on either of the other two species. In a model of a biochemical system the agents of interest might be proteins, whose internal state values could represent different conformations. Depending on the model, the number of internal states for a particular type of agent could be very large. While all agents of a specific type share the same possible behaviors and internal states, at any particular time agents in a population may differ in their actual behaviors. So, when hunted by a lion, one oryx may get eaten lion while another escapes; one lion is older than another, and so on.

In the context of ABMs in biology, it is often useful to limit the life-time of agents. This requires some rules specifying the conditions under which agents die. In order to avoid the population of agents shrinking to zero, death processes need to be counterbalanced by birth or reproduction processes. Normally the reproduction of agents is tied to some criterion, typically the collection of sufficient amounts of "food" or some other source of energy. Models with birth and death (more generally: creation and destruction) processes allow a particularly interesting kind of effect: If the agents can have hereditary variations of their behavior (and possibly their internal states) then this could make it possible to model evolutionary processes. In order for the evolutionary process to be efficient, one would need one more feature—competition for resources. Resources need not necessarily be nutrient, but could be space, or even computational time. An example of the latter is the celebrated simulation program Tierra [4], where evolving and self-reproducing computer programs compete against one another. Each program is assigned a certain amount of CPU-time and needs to reproduce as often as possible within the allotted time. The faster a program reproduces, the more offspring it has. This, together with mutations—i.e., random changes of the program code—leads to very efficient reproducers over time.

There have been many quite successful attempts to model evolutionary processes mathematically. These models usually make some predictions about how genes spread in a well defined population. Such mathematical predictions of the behavior of certain variables in an evolutionary process are very different from the type of evolutionary models that are possible in ABMs. In a concrete sense, evolution always depends on competition between variants. The variants themselves are the result of random events, i.e., mutations. The creation of variants in evolutionary models requires an explicit representation of randomness (rather than a summary of its statistical properties); and the bookkeeping of the actual differences between the variants requires the model to represent heterogeneity. Agent-based modeling is the ideal tool for this. If we think again of lions, we could imagine a model that allows each simulated lion to have its own set of hunting strategies. Some lions will have more effective strategies than others, an effect which could drive evolution if lions that are more successful have more offspring.

2.2.1.2 Environment

Agents must be embedded in some type of environment, that is, a space in which they exist. The choice of the environment can have important effects on the results of the simulation runs, but also on the computational requirements of the model. How to represent the environment and how much detail to include will always be a case-specific issue that requires a lot of pragmatism. In the simplest case, the environment would be simply an empty, featureless container with no inherent geometry. Such featureless environments are often useful in models that assume so-called "perfect mixing" of agents. More on that later.

A simple, featureless space can be enhanced by introducing a measure of distance between agents. The distance measure could be discrete, which essentially means that there are compartments in the space. In ABMs it is quite common to use 2-dimensional grid layouts. Agents within the same compartment would be considered to be in the same real-world location. A further progression would be to introduce a continuous space, which defines a real-valued distance between any pair of agents. For computational reasons, continuous spaces used in practice are often 2-dimensional, but there is nothing preventing a modeler from using a 3-dimensional continuous space if required, and computational resources permit.

In ABMs, spatially-structured models are usually more interesting than completely featureless environments. The behavior of many real systems allows interactions only between agents that are (in some sense) close to one another. This then introduces the notion of the *neighborhood* of an agent. In general an agent A tends to interact with only a subset of all agents in the system at any particular time. This subset is the set of its neighboring agents. In ABMs "neighborhood" does not necessarily refer simply to physical proximity; it can mean some form of relational connectedness, for instance. An agent's neighborhood need not be fixed but could (and normally will) change over time. In many ABMs, the only function of the environment is to provide a proximity metric for agents, in which case it is little more than a containing space rather than an active component. There are environments that are more complex and, in addition to defining the agent-topology, also engage in interactions with agents. A common, albeit simple example, is an environment that provides nutrients. Sometimes, very detailed and complicated environments will be necessary. For, example, if one wanted to model how people evacuate a building in an emergency, then it would be necessary to represent the floor-plans, stairs, doors and obstructions. There have been attempts to model entire city road systems in ABMs, in order to be able to study realistic traffic-flow patterns. A description of such large modeling attempts would go beyond the purpose of this book, and the interested reader is encouraged to read Casti's "Would-be Worlds" [3].

2.2.1.3 Interactions

Finally, the third ingredient of an ABM is the rules of action and interactions of its agents. In each model, agents would normally show some type of activity. Conceptually, there are two types of possible activities of agents: (i) taking actions independent of other agents, or (ii) interacting with other agents. In the latter case, agents are generally restricted to interacting with only their neighbors. The behavior rules of agents are often rather simple and minimalistic. This is as much a tradition in the field as it is a virtue of good modeling. Complicated interactions between agents make it hard to understand and analyze the model, and require many configuration parameters to be set. In most applications it is a good idea to try to keep interactions to the simplest possible case—certainly within the earliest stages of model development.

2.2.2 Algorithms

Once the three crucial elements of an ABM (the agents, the environment, and the interactions) have been determined, it still needs to be specified how and under which conditions, the possible actions of the agents are invoked. There are two related issues that we consider when modeling real systems: the passage of time and concurrency.[1]

In real-world systems, different parts usually work concurrently. This means that changes to the environment or the agents happen in parallel. For example, in a swarm of fish, every individual will constantly be adjusting its swimming speed and direction to avoid collisions with other fish in the shoal; lions hunting as a pack attack together; in a cell some molecules will collide, and possibly engage in a chemical reaction,

---

[1]Concurrency denotes a situation whereby two or more processes are taking place at the same time. It is commonly used as a technical term in computer science for independent, quasi-simultaneous executions of instructions within a single program.

while, simultaneously, others disintegrate or simply move randomly within the space of the cell.

Representing concurrency of this sort in a computer simulation is not trivial. Computer programs written in most commonly-used programming languages can only execute one action at a time. For example, if we have 100 agents to be updated (or to perform an action), this would have to be done sequentially as follows:

1. Update agent number 1.
2. Update agent number 2.
3. ...
100. Update agent number 100.

Clearly, if the action of one agent depends on the current state of one or several other agents, then sequential implementation of what should be a concurrent update step may mean that the outcome at the end of the step might be affected by the order in which the agents are updated. This should not be the case. In order to simulate concurrent update it is therefore necessary that the state of agents at the beginning of an update cycle is remembered, and all state-dependent update rules should refer to this. Once all agents have been updated the saved state can be safely discarded.

For managing the passage of time, there are essentially two choices: time-driven and event-driven. The simplest is time-driven, which assumes that time progresses in discrete and fixed-length time steps, or "ticks", in an effort to approximate the passage of continuous time. On every tick, each agent is considered for update according to the changes likely to have happened since the previous time step. For instance, if the time step is one day in the lion model, then the hunger-level of each lion might be increased by one unit, possibly leading to some lions switching into hunting mode. The real-world length of the time step is obviously highly model dependent; it could be nanoseconds for a biochemical reaction system or thousands of years for an astrophysics model. One of the skills of the modeler will be choosing an appropriate time-step size that is small enough to capture all significant effects of interests, but not so small that the model's runtime is prohibitively long.

The event-driven approach also models a continuous flow of time but the elapsed time between consecutive events is variable. For instance, one event might take place at time $t = 1$, the next at $t = 1.274826$, and the next at $t = 1.278913$—or at any time in-between. *Event-driven algorithms* are appropriate in models of scenarios where events occur naturally from time to time rather than periodically. This is the case, for example, in chemical-reaction systems. Individual reactions occur at system-dependent points in continuous time, and between two reactions nothing of interest is considered to take place.

Both time-driven and event-driven approaches naturally lead to their own distinctive implementations in computer models and we illustrate both, along with further detail, in the following sections.

2.2.3 Time-Driven Algorithms

Time-driven algorithms (Algorithm 1) are also often referred to as *synchronous* update models, in contrast to event-driven algorithms being referred to as *asynchronous*. A synchronous updating algorithm approximates continuous time by a succession of discrete time steps in which all agents are considered for update. Synchronous updating is inexact and can only approximate real-world time. Depending on the specific application, this could make a significant difference to the result, although in many cases the approximation is very good. It is also often the case that the modeler does not care about quantitative correctness and synchronous updating becomes a choice of convenience.

Algorithm 1 Update scheme for synchronous/time-driven ABM.

Set initial time.
Set initial conditions for all agents and the environment.
loop
    for All agents in the model **do**
        Invoke update rule.
    end for
    Increment time to next time step.
end loop

In those cases where quantitative accuracy is important, the size of the chosen time step determines the temporal granularity of the simulation, and hence the accuracy of the simulation. In essence, synchronous models chop time up into a sequence of discrete chunks; the finer the chunks the better the resolution of events. In the limiting case, where each time step corresponds to an infinitesimal amount of time, synchronous algorithms would be precise. In each time step the updating rule would then simulate what happens in the real system during an infinitely short moment of time. As one can easily see, in this limiting case it would take an infinite number of time steps to simulate even the shortest moment of real time. In practice, this limit is therefore unattainable. We cannot reach it, but we can choose to make our time intervals shorter or longer depending on the desired accuracy of the simulation. The trade-off is usually between numerical accuracy of the model and the speed of simulation. The shorter the time interval, the longer it takes to simulate a given unit of real time.

Take as an example a fish swarm. If we simulate a shoal of fish using synchronous update, then each time step would correspond to a given amount of physical time (i.e., fish time). During each of these intervals the fish can swim a certain distance. Given that we know how fast real fish swim, the distance we allow them to travel per time step defines the length of an update-step in the model. The shorter the distance, the shorter the real-world equivalent of a single update step. In practice, the choice of the length of an update step can materially impact on the behavior of the model.

Real fish in a shoal continually assess their distance to their neighbors and match speed and direction to avoid collisions and to avoid letting the distance to the neighboring fish become too large. If we simulate a swarm of fish using a synchronous updating algorithm, then each fish will assess its relationship to its neighbor at discrete times only. If we now choose our temporal granularity such that a fish swims about a meter between update-steps, then this will result in very inaccurate models. A meter is very long compared to typical distance between fish. It is therefore quite likely that after an update step two animals may overlap in space. Between two time steps the nearest neighbors may change frequently. On the whole, the behavior of the simulated swarm is likely to be very different from the behavior of a real school of fish and the model a bad indicator of real behavior. The temporal resolution is too crude.

The accuracy of the overall movements will be increased if we decrease the "size" of each update step. If we halve the time step, say, then the forward distance will then be only 0.5 m per time step. The cost would be a corresponding doubling of the number of update steps, in the sense that for every meter of swarm movement we now need to update two cycles rather than just one. One could, of course, make the model even more accurate with respect to the real system if each fish only swam 1 cm per time step at the cost of a corresponding increase in update steps and run time. The "right" level of granularity ultimately depends upon the level of accuracy required and the computing resources available.

It will not be the case with all models that every agent is updated similarly on each time step. The heterogeneous nature of some systems means that differential updating is quite likely. Where this is the case, care should also be taken with the time step. It should not be so large that a lot of agents are typically updated at each time step, because that risks missing subtleties of agent changes that might otherwise cause significant effects in the model with smaller steps. A corollary, therefore, is that we should expect many agents to remain unchanged at each individual time step, but that changes gradually accumulate within the population over many time steps.

There is a balance to be struck between the algorithmic cost of considering every agent for possible update at each time step, and the number actually needing updating—which might turn out to be none on some cycles. Where it is possible to identify efficiently only those agents to be updated on a particular cycle then those costs may be mitigated, but the event-driven approach should also be considered in such cases.

2.2.4   Event-Driven Models

The basic principle behind event-driven algorithms is that they are controlled by a *schedule*. A schedule is, in effect, a list of pending future events, typically maintained in sorted order of when each event is due to occur. Such events include the updating of agents, the creation of new agents, and also changes to the environment. All the program does then is to follow the list of updates as specified in the schedule (Algorithm 2). In addition, there needs to be a way for future events to be added to the schedule as, typically, the occurrence of an event leads to the spawning of one or more future events.

Algorithm 2 Update scheme for asynchronous/event-driven ABM.

Set the model time: $t = 0$
Set the initial conditions for all agents and the environment.
loop

Update the model time to that of the next event.
Determine which agent(s) will be updated next and which update rule will be applied.
Apply the selected rule to the selected agent(s).
Update the schedule.

end loop

Let us illustrate this process via a simple example where the focus will primarily be on the association of events and their time rather than the individual agents. Consider a system model consisting of at least two types of molecule, and that these molecules can engage in reactions with one another. Let us assume that reactions between molecules of the two types happen with a rate of 1. This means that, on average, there will be one reaction per time unit. We will assume that the system we are modeling operates in continuous time. However, continuous time is difficult to implement in computers so we use an approximation: We define one update step of the model as representing the passing of one time unit. We can then naively simulate our system as follows:

1. Randomly choose one molecule of each type.
2. Simulate the reaction (this could, for example, be the production of a third type of molecule).
3. Update the time of the model by one.

The system exhibits precisely the behavior we expect, namely that it has one reaction per time unit. We could even, again rather naively, extend this scheme to a reaction rate of 2. In this case we would then execute two reactions per update step.

For some applications, this algorithm would be acceptable, but it will never be an accurate model of the real system. If the reaction rate is 1 per time unit *on average* then this does *not* mean that in any given time unit there will be exactly one reaction. Instead, during a particular time unit of observation there might be two reactions, whereas during the next three time units there might be no reactions. From a simulation point of view, the problem is how to work out how many events actually happen during any given time unit. While it is possible to do this but, in practice, it turns out that there is a better approach.

Instead of calculating how many events take place in any given time interval, there are event-driven algorithms to calculate, for each reaction, at what time it next takes place. In the case of chemical reaction systems, this time can be determined by drawing a random number from an exponential distribution with the mean equal to the reaction rate. Making the example concrete: assume a very simple chemical system consisting initially of $N_A = n_0$ molecules of type A and $N_B = 0$ molecules of type B. Assume, further, that these molecules are embedded in a featureless environment—there is no sense of distance between them—and that the molecules engage in a single chemical reaction:

$$A + A \longrightarrow B$$

This means that two molecules of A are used up to produce one molecule of B. Over time, we would expect that the system approaches a state where there are only B molecules.

Assume that we start at time $t = 0$. The question we have to ask now is: When does the next reaction occur? For this we need to draw a random number, in this case from an exponential distribution.[2] The mean of the distribution will be chosen based on the current value of N_A and the reaction rate. Let us assume that the random number we draw determines $\Delta t = 1.0037$. We now set the new time to $t = t + \Delta t = 1.0037$ and use the update rule which, in this case, is simply to destroy a pair of A molecules and create a new molecule of type B. We would need to choose which pair of A-agents is used up in the reaction and, for the sake of simplicity, let us assume that we simply draw a random pair of A-agents. That was the first step.

The second step is identical. We need to determine a new time, by drawing a random number from an exponential distribution. There is one complication here, namely that the new N_A is $n_0 - 2$ because two of the original A molecules have been used up to form the new B. In real applications, this effect will be important, but for the moment we will ignore it. A detailed treatment of this will be given in Chap. 7. For the moment we will simply note that we need to draw a new random number from an exponential distribution with an appropriately updated mean. This generates a new Δt, say $\Delta t = 0.873$. We set the new time to $t = t + \Delta t = 1.0037 + 0.873 = 1.8767$ and update the relevant agents again. We continue in this fashion until we hit a stopping condition, which will normally be a fixed end time of the model (or the end of the patience of the modeler).

Notice that, here, because there is only one rule and one type of event, we do not even need to maintain a schedule of future events. In general, the situation will be more complicated, because typically there will be more than just two molecular species and more than one possible reaction.

This particular example is somewhat artificial in that ABMs are not the best way to simulate such systems. There is no irreducible heterogeneity in the system, in the sense that all molecules of type A are indistinguishable in their state and identity. It is therefore not necessary to simulate each of the molecules individually. Chapter 7 will provide a detailed discussion of how to simulate chemical systems where there is no heterogeneity between the individual molecules.

---

[2]There are specialized algorithms available to do this, and there will be no need to re-implement them. For programmers of C/C++ the Gnu Scientific Library [5] is one place to look.

In those practical cases where event-driven ABMs are necessary, the details of how the schedule is created and maintained will very much depend on the particular demands and properties of the system to be modelled. It is then essential to think carefully about how to choose the next events and to correctly calculate how much time has passed since the most recent event. If these rules are chosen properly, then event-driven algorithms can simulate continuous time.

2.3 Game of Life

ABMs have been found to be useful over a wide range of application areas. They have been used to study the movement of indigenous tribes in the Americas [6], how sand piles collapse [7], how customers move through supermarkets [8] and even to model the entire traffic system of Albuquerque in the US state of New Mexico [3]. So far, perhaps the greatest following of ABMs is among economists. Agent-based Economics [9] is a field that uses ABMs to study economic systems in a new way.

In this book, we will not re-trace the historical roots of ABMs, although we will look at one of the examples of an ABM that was important in generating initial interest in this modeling technique in the scientific community: *cellular automata* (CA). We will examine a particularly interesting example of a CA that is well known and, to this day, continues to keep researchers busy exploring its properties: the Game of Life. However, we will not present a complete exposition of CAs in all their shapes and varieties, and the reader who would like to explore further is referred to the excellent book by Wolfram [10].

Despite its suggestive name, the Game of Life is not really a model of anything in particular. Rather, it is a playground for computer scientists and mathematicians who keep discovering the many hidden interesting features this model system has to offer. Unlike mathematicians and computer scientists, natural scientists are normally interested in models because they can help them understand or predict something about real systems. The Game of Life fails in this regard, but it is still interesting as an illustration of the power of ABMs, or more generally, how interactions can lead to interesting and complex behaviors. A Java program implementing the game is provided with the materials associated with this book at http://www.cs.kent.ac.uk/imb/.

The Game of Life is played in an environment that is a two-dimensional grid (possibly infinitely large). Agents can be in one of only two possible states, namely '1' or '0' (in the context of the Game of Life these states are often labeled "alive" and "dead", but ultimately it does not matter what we call them). This particular CA is very simple and the range of actions and interactions of agents is very limited: agents have a fixed position in the grid, they do not move, and they have an infinite life-span (despite the use of "dead" as a state name). They have no internal energy level or age, for instance, and time plays no significant role in the model. The most complex aspect of an agent is its interaction rules with other agents. These rules are identical for all agents (as we have only one type of agent) and depend solely on

Fig. 2.1 A configuration in
the Game of Life. The *black
cells* are the Moore
neighborhood of the central
white cell

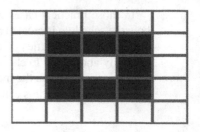

an agent's own state and that of its neighbors. In the Game of Life, neighbors are
defined as those agents in the immediately adjacent areas of the grid. Each agent
has exactly 8 neighbors corresponding to the so-called *Moore neighborhood* of the
grid (Fig. 2.1). The complete set of rules governing the state update of an agent is as
follows:

- Count the number of agents in the neighborhood that are in state 1.
- If the number of neighbors in state 1 is less than two or greater than three then the
  agent's new state will be 0.
- If the agent is in state 1 and it has two or three neighbors in state 1 then it will
  remain in state 1.
- If the agent is in state 0 and it has two neighbors in state 1 then it will remain in
  state 0.
- If the agent is in state 0 and it has three neighbors in state 1 then it will switch to
  state 1.

The global rule is that the state of all agents is updated synchronously.

Note that there is no random element to this behavior—the Game of Life is
completely deterministic. Once it is determined which agents on the grid are in
state 1 and which are in state 0, then the behavior of the simulation is implicitly
determined thereafter. Many, or perhaps most, of the initial conditions will lead
to trivial behavior. For example, if we start with all agents being in state 1 then,
according to the rules above, at the second time step all agents (in an infinite grid)
will have switched to state '0', and they will all remain in that state thereafter. Most
initial conditions will eventually end up in a similarly trivial state. But then again,
given the simple rules of the system, what more do we expect?

It is intriguing, therefore, that there are some configurations whose time evolution
is far from trivial. There are a number of initial conditions that lead to very interesting
and diverse long term behaviors, with new ones continuously being added to the list.
Let us look at a couple of the best known ones. We will here concentrate on local
configurations only; that is, we will assume that at time $t = 0$ all agents are in state
0, except for a few that make up the local configuration.

One of the most famous initial conditions of interest is the so-called "blinker." This is a row (or column) of three agents in state 1. Say, we start with a row configuration:

$$0\ 0\ 0$$
$$1\ 1\ 1$$
$$0\ 0\ 0$$

(remember here we assume that all other agents are in state 0). The middle agents in the top and bottom row are both in state 0 and both have exactly three neighbors in state 1. According to the rules of the Game of Life, these agents now switch to state 1. Similarly, in the middle row, the first and the third agent only have one neighbor in state 1. Following the rules, they will switch to state 0; the central agent in the middle row will not change state.

Synchronous update means that the states of agents are only changed once the new state for each agent has been calculated and the new configuration will be:

$$0\ 1\ 0$$
$$0\ 1\ 0$$
$$0\ 1\ 0$$

The reader can easily convince herself that updating this configuration will lead back to the original horizontal configuration. The blinker will endlessly move back and forth between these two configurations in the absence of intervention from any other surrounding patterns.

The Game of Life illustrates very well the principle of synchronous updating of agents. At every time step the agents should be updated according to the neighborhood they had at the beginning of the update step. Yet the computer program needs to update one agent after the other. The risk is that, as each agent is updated, the environment for its neighbors changes as well, with the result that the outcome of the update step depends on the order in which the agents are updated. This problem has been touched upon in more general terms above. In order to avoid this dependence on the order of update, the full state of the grid must be saved at the beginning of a time step; any application of update rules must be made with respect to this saved copy.

The importance of this can be exemplified using the blinker. Instead of the synchronous update algorithm, let us assume an asynchronous one, that updates agents in order from the top left of the grid to the bottom right. That is, we go from left to right until we hit the end of the grid and then we start again on the left of the next line. The original blinker configuration would lead to the following end state in the next time step:

$$0\ 1\ 1$$
$$1\ 0\ 1$$
$$0\ 0\ 0$$

2.1 Check on paper that failing to keep the old states and new states distinct during the update step leads to the result we have shown.

Fig. 2.2 A partial glider
sequence in the Game of Life.
The four distinct
configurations show elements
of the sequence a glider
would undergo while moving
diagonally north-eastwards
(intermediate stages omitted)

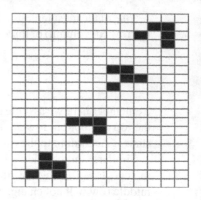

In the standard Game of Life, the blinker configuration oscillating forever is
perhaps not the epitome of an interesting configuration. However, there is more in
store. Another important configuration is the "glider," a structure with this initial
condition[3]:

$$1\ 1\ 1$$
$$1\ 0\ 0$$
$$0\ 1\ 0$$

The glider represents a local pattern of states of agents that propagates through
the environment diagonally (Fig. 2.2). Note that it is not actually the agents that
are moving but the configuration pattern of states. Apart from being fun to watch,
gliders can be thought of as transmitting information from one part of the grid to
another—information that is relayed from one group of agents to another. Gliders are
often used as essential parts in many more complex configurations in the Game of
Life. There have even been successful attempts to built entire computing machines
purely from initial configurations in the Game of Life. In some versions of these
configurations gliders are used as channels to transmit basic units of information
between the parts of the "computer".

2.2 Start with a glider pattern and use the rules to update it by hand for a few time
steps.

2.3 Check, by hand, the effect of different asynchronous update rules on the glider.

2.4 What happens if two gliders collide?

2.5 If Java is available on your system, try using the implementation provided with
the materials associated with this book to experiment with different starting patterns,
including the blinker and glider.

---

[3] Alternative symmetrical configurations would obviously be equivalent.

As far as we are concerned here, the Game of Life is an example of an ABM that allows for only limited actions and interactions between agents, but can still produce complex and interesting behaviors. Unfortunately, it is not really useful beyond being a playground for mathematician. Nevertheless, it does illustrate the principles of agent-based modeling—agents with state and rule-based behavior interacting in a structured environment—and shows that even simple interactions can lead to complex behavior.

The Game of Life also illustrates how ABMs can deal with irreducible heterogeneity. All agents in the system are the same, and only differentiated from one another by their state. Modeling this mathematically would be nearly impossible, but it is easy to implement as a computer model.

2.4 Malaria

The Game of Life is one of the simplest possible non-trivial ABMs; it is an interesting world to explore, but it does not have any immediate use as a model of *something real*. ABMs as models of some aspects of the real world need to be more advanced than the Game of Life. In this section we will discuss a somewhat more advanced (and perhaps practically more useful) example of an ABM, modeling the transmission of malaria. Modeling diseases and their spread is an important thing to be able to do. Models of disease transmission could, at least if they are accurate enough, be used in a variety of contexts, such as in optimizing responses to emerging epidemics, or for efficient planning of immunization programs, particularly when resources are limited.

Particularly in sub-Saharan Africa, Malaria is the single biggest killer, ahead even of HIV/AIDS. The disease is *vector borne*; that is, it is transferred from person to person by a secondary agent, the male *Anopheles* mosquito (the female is not a vector of malaria). An infection is transferred both from human to mosquito and *vice versa* when Anopheles bites a human.

The dynamics of infection depend on the local interactions between individual agents, namely the vectors (mosquitoes) and the humans. Both humans and mosquitoes have a limited range of movement and can only interact when in each other's neighborhood. The exact details of how agents move across their available space—in particular the mosquitoes—could have a major influence on how fast and how far malaria spreads; in this sense there are irreducible interactions. The system is also irreducibly heterogeneous in that some agents are infected while some are susceptible (that is uninfected). There is a random element to the model in that the contact between the agent-types is probabilistic. As will become clear below, this randomness and heterogeneity is, under some conditions, reducible, but irreducible under others. In order to deal with the latter case, an ABM is appropriate.

Let us now leave the preliminaries behind and put into practice what we have discussed in theory. Our aim is to find the simplest non-trivial model of malaria that captures the essence of reality. For the reasons we outlined in Chap. 1, in every modeling enterprise one should always try to find the simplest model first, and only introduce complexities once this simplest model is thoroughly understood. There are many possible choices one can take when modeling the spread of malaria. The shape and form of the final model will depend on the particular goals that are to be achieved. Depending on whether they are predictive, explanatory or exploratory, different requirements will be placed on the model. Here we do not have any particular purpose in mind, other than demonstrating how to design ABMs, and we will therefore design the simplest non-trivial model of the spread of malaria. Generally, it is a good idea to start in this way, even when one has a particular purpose in mind. It is a common mistake for modelers to start designing models that are full of details but are also opaque. Particularly with ABMs, this danger is ever-present.

As a minimal requirement, an ABM of the transmission dynamics of malaria will have to consider at least two types of agent, namely mosquitoes and humans. In a first model, it is best to avoid the complication of distinguishing between female and male mosquitoes, although for some applications this could become relevant. Similarly, humans could be differentiated according to age, as different age-groups will likely have different susceptibilities for the disease and different survival rates. However, for a first attempt it is best to ignore these complicating effects as well. Instead, to keep matters simple for an initial model, we reduce the possible states of the agents to two: either an agent is infected or not. We have to assume that both types of agent could be either infected or not, and these would be the only two states of the agents. The difference between human-agents and mosquito-agents is in the details of the rules that determine how they carry an infection, and for how long they stay infected. To keep the complexity of the model to a minimum, one would initially also ignore that agents die (of malaria or otherwise). This means that there is a fixed number of agents in the environment. Each agent can be either infected or susceptible. Naturally, we have to assume at least one interaction between mosquitoes and humans, namely cross-infection; the model needs to have some mechanism whereby mosquitos can infect humans, and vice-versa. Cross-infection can only happen between neighboring agents (where neighboring refers to some spatial proximity) and between agents of different type.

A feature that we cannot ignore, even in the initial simple model, is the spatial structure of the population. Agents need to be embedded in an environment. In the present case the simplest way to represent an environment is to assume a 2-dimensional space, say a square of size $L \times L$. We will allow the agents to move in their environment. Static agents seem simpler at first, however they are not as we shall see below. Moreover, at least the mosquitoes have to be allowed to move. If all agents were static then no non-trivial disease transmission dynamics could emerge.

Abstracting away all but the most essential details, we end up with a model that has a fixed number of ageless and immortal agents that move about in their environment. We summarize here the essentials of our model:

- There are two types of agent: humans and mosquitoes.
- Agents live in a continuous rectangular environment of size $L \times L$.
- Agents can be in either of two states: infected or susceptible.

- If a mosquito is infected then all humans within a radius b of the mosquito will become infected.
- If a mosquito is susceptible, and there is an infected human within a radius b of the mosquito, then the mosquito will become infected.
- Once infected, humans remain infected for R_h time steps, mosquitoes will remain infected for R_m time steps. The actual numbers chosen will be arbitrary, but should reflect the fact that the life span of a mosquito is very short compared to the likely time a human remains infected.
- All agents move in their world by making a step no larger than s_h (humans) or s_m (mosquitoes) in a random direction; essentially they take a random position within a radius of size s_h or s_m from their present position. This rule introduces randomness into the model; the randomness which will turn out to be irreducible under some conditions. Henceforth we will always keep $s_m = 1$; this is a rather arbitrary choice but it simplifies the investigation of the parameter space by reducing its dimensionality. At the same time, we expect that it will not make a material difference to the behavior of the model as we observe it.

These choices are of course not unique and have a certain degree of arbitrariness to them. Following the same principles of simplicity, one could have come up with a slightly different model. But, for the rest of this section, we accept this choice.

Note that this description includes provision for a number of *parameters* or configuration variables, such as L, b, s_h, etc. Identifying appropriate parameter values is always an important and unavoidable part of the development of an ABM. The problem with parameters is that they are often not known, but their precise value could (and typically does) make a huge difference to the behavior of the model. Finding the correct parameters is, at least within biological modeling, one of the greatest challenges of any modeling project. In practice one often finds that a model has more parameters than one anticipated. (We like to refer to this phenomenon as the *law of mushrooming parameters*.) One strategy to reduce the number of parameters is to leave out interactions, phenomena, and/or details from the model that (i) require additional parameters and (ii) are perhaps not essential to the particular purposes of the investigation. Keeping the mushrooming under control is not something that can be done by following the rules given in a book, but requires the sweat and tears of the modeler who has to go through every aspect of their model and ask, "Do I really need to include this effect right now?" In general it is a good rule to commit model details to the executioner, unless their necessity is proved beyond a reasonable doubt. No mercy should be shown.

The reader will perhaps agree with us that our model of the spread of malaria is a bare-bones representation of the real system. Not much can be stripped away from it without it becoming completely trivial, yet even this simple model has many parameters that need to be determined. Before we start considering these in more detail, we are going to make a seeming digression and explore the way in which the mathematics that we have abandoned in favor of computer modeling still has something left to offer us.

2.4.1 A Digression

One of the problems of computer simulation models in general, and ABMs in particular, is their lack of transparency. Even when a model is relatively simple, it is often hard to understand exactly how one would expect the model to behave. This is particularly true during the early stages of the modeling cycle. Even more so, one can never be quite sure that the model is actually behaving the way it is meant to. One source of problems is simply programming bugs. Subtle errors in the code of the program are often hard to detect, but could make a big difference to the simulation results. In the case of the Game of Life, testing the model is not such a problem. Once one has tried out a few configurations with known behavior, one can be reasonably confident about the correctness of the code. The source of this confidence is that there is a known reference behavior. The user knows how the model should behave and can compare the expected behavior with the actually observed one. Unfortunately, in most realistic modeling ventures, this will not be true, at least not across the whole parameter space. However, it is often possible to find parts of the parameter space where the dynamics of the model simplifies sufficiently to make it amenable to mathematical modeling. The mathematical model will then provide the reference behavior against which the correctness of the ABM can be checked.

Apart from the correctness aspect, there is another advantage in making a mathematical model: A mathematical formulation will often provide some fundamental insights into how changes of parameter values impact the overall behavior of the model. These insights might not carry over precisely to more general situations in the parameter space, but they often still provide basic insights that might help shape the intuition of the modeler.

Let us consider a few key parameters of our Malaria model—those affecting the ranges of movement of both the humans and the mosquitoes. The simplest case to consider is that humans move by taking a random position in their world; this corresponds to $s_h = L$. This particular choice of parameter is perhaps not very realistic with respect to the real system, but it does have the significant advantage of making the model amenable to a mathematical analysis. When all humans choose a random position in their world then the spatial configurations of the agents become unimportant (we keep $s_m = 1$). This limiting case then corresponds to what we call a *perfectly mixed* model. In what follows, we will walk through the process of comparing the model with a mathematical prediction of its behavior.

In the Malaria model with perfect mixing, a mathematical model can be easily formulated (see [11] for details). If M is the total number of mosquitoes, m the number of infected vectors, H and h the corresponding variables for the numbers of humans, R_m and R_h the times for vectors and humans to recover from an infection, b the area over which a vector can infect a human during a single time step and W the total area of the world; then the proportion of infected vectors and humans can be written as follows:

$$\frac{m}{M} = 1 - (1 - Q(h))^{R_m}$$

$$\frac{h}{H} = 1 - (1 - Q(m))^{R_h} \tag{2.1}$$

Here the function $Q(x)$ is given by:

$$Q(x) = 1 - \exp\left(-\frac{bx}{W}\right)$$

Note that these equations depend only on the density of the agents (the number of agents over the size of the environment) and not on the absolute numbers.

Considering these equations, the astute reader will immediately object that the variables m and h are not dependent on time. This appears to be counter-intuitive. There surely are parameters that allow overall very high infection levels? Given such parameters, if we introduced a single infected agent into a healthy population, it will take some time before high infection levels are reached. It is not clear how the mathematical model could describe the transient infection levels, given that it has no representation of time.

This objection is correct, of course. The model does not describe initial conditions at this level of detail and the model does not contain time. Instead, it describes the long-term or steady state behavior of the model. The steady state of a system, if it exists at all, is the behavior that is reached when the transient dynamics of the model has died out. Often, but not always, the steady state behavior is independent of the initial state. Certainly in the case of the Malaria model, in the long run the behavior will be the same, whether we start with a single agent infected or all of them infected, It is this steady state behavior that the mathematical model (2.1) describes.

There is one complication: What if we start with no infection at all? According to the rules of the ABM, independent of the values of b and W an infection can never establish itself unless there is at least one infected agent in the system. This seems to indicate that, for the special case of $m = 0$ and $h = 0$, initial conditions are important after all. A second glance at (2.1) reveals that the mathematical model can indeed represent this case as well. The reader can easily convince herself that the case $h = 0$ and $m = 0$ always solves the set of equations, independent of the values of b and W. This tells us that, for some parameter sets, the model has two solutions:

- The case where a certain fraction of the mosquitoes and the humans are infected. This solution, if it exists, will always be observed in the model, unless we start with no infected agents at all.
- The other case is possible over the entire space of parameters, and describes the situation where there is no infected agent in the model.

Whether or not there are two or just one solution to the model depends on the parameters. If there are two solutions then the solution corresponding to no infection at all (the "trivial solution") is, in a technical sense, unstable. In practice this means that the introduction of a single infected agent (human or mosquito) is sufficient for the system to approach the non-trivial solution. In this case, the disease would establish itself once introduced.

In the part of the parameter space that admits only one solution, the long-term behavior will always be the trivial behavior. No matter how the initial state is chosen—whether there are many infected agents, or no infected agents—malaria will never be sustained in the system. For the particular mathematical model in (2.1) it is possible to derive the range of parameters for which there is only one solution. We will not derive the formula here, but be content to state the result. The reader interested in exploring the details of the model is referred to the primary literature [11]. The model in (2.1) has two solutions only if the following condition is fulfilled:

$$\frac{b}{W} > \frac{1}{\sqrt{MHR_mR_h}} \tag{2.2}$$

Essentially, this means that the area b over which the mosquitoes can infect humans within a single time step needs to be sufficiently large in relation to the overall size of the system (W) so that an infection can establish itself. What counts as "sufficiently large" depends on the total numbers of mosquitoes and humans and for how long they remain infected once they are bitten by a mosquito.

2.4.2 Stochastic Systems

There is one additional qualification to the validity of (2.2) that is of more general importance in modeling. Strictly speaking, the condition is only valid in the case of a *deterministic system*. The concept of a deterministic system is perhaps best understood by explaining what is *not* a deterministic system.

Assume in our Malaria model a set of parameters that allows a non-trivial solution. Assume further that we choose our initial conditions such that all human agents are susceptible, and only a single mosquito is initially infected. According to the preceding discussion we would now expect the infection to take hold in the system and, after some time, we should observe the non-trivial steady state infection levels as predicted by the model (2.1).

In the model, this is not necessarily what we will observe. Remember that, in the case of perfect mixing, both mosquitoes and humans take up random positions at every time step. It is not impossible that, by some fluke event, at the beginning of the simulation the single infected agent is in an area where there are no other human agents at all. This probability can be easily calculated from the Poisson distribution:

$$P(\text{no human}) = \exp\left(-\frac{Hb}{W}\right) \tag{2.3}$$

Here Hb/W is the average number of human agents one would expect within the bite area of the mosquito. Remember now that the infection of a single agent lasts for R_m or R_h time steps. In the case we are considering here, the probability that an infection is introduced by a single mosquito is not passed on is then given by P^{R_m}. Depending on the values of the parameters, this probability could be very small, but it is never vanishing. Simple stochastic fluctuations can wipe out an infection, particularly when the number of infected agents is low. Hence, the behavior predicted by the mathematical models (2.1) and (2.2) might not always be correct, because of *stochastic fluctuations* in the system.

There is another—probably more important—sense in which our Malaria model is non-deterministic. When simulating the model, one will typically find that the actually observed number of infected agents is not exactly equal to the predicted number, but will be a bit higher or lower than the predicted number. The reason for this is that the mathematical model only tells us about the *mean* number of agents in the long run. Stochastic fluctuations ensure that actual infection levels will vary somewhat over time, and either be a bit higher or a bit lower than the mean. The source of these fluctuations is the randomness in the rules of the system, particularly the randomness in contact between humans and mosquitos. A good comparison is a coin flipping experiment. If one flips a coin 10 times then, on average, one would expect to see a "head" 5 times. In an actual experiment this could happen, but one will often observe that the number of heads deviates from this "mean." In essence, this is the same type of stochastic fluctuation that leads to the deviation from the calculated mean behavior in the Malaria model.

Figure 2.3 shows an example of a simulation of the Malaria model. The particular graph plots the proportion of infected agents at every time step. The simulation shown in the figure starts with no humans being infected. In order to avoid the trivial solution, we start with half of the mosquito population being infected. In the perfect-mixing

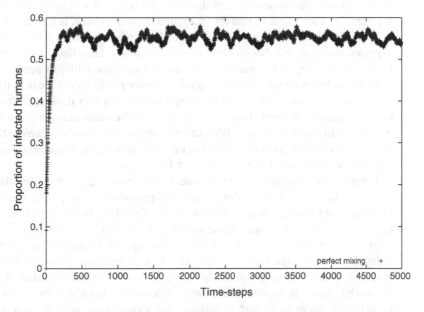

Fig. 2.3 An example of the time-evolution of the proportion of infected humans in the Malaria model when perfect mixing is assumed. We start with half of the mosquitos infected. After a rapid initial increase in the infection levels, the system finds its steady-state and the infection level fluctuates around some long-term mean value

case, the population of infected agents rises initially, until it reaches its expected steady-state behavior. After this transient period, the infection level of the humans stays close to the theoretical steady state value, as calculated from the model (2.1). The level of infection never quite settles on the expected behavior, which is another manifestation of the stochastic fluctuations.

Stochastic fluctuations are often referred to as *noise*. Just how much noise there is in this system depends on the size of the population. In practical modeling applications it is often necessary to reduce the noise. In these situations it will be useful to *scale* the model. By this we mean, a change of the parameters that leaves the fundamental properties of the model unchanged but, in this particular case, increases the number of agents in the system. What counts as a "fundamental property" is problem dependent. In our case here, the fundamental variable we are interested in is infection level, so the fraction of agents that carry the infection.

If we attempted to reduce the noise by simply changing the number of humans in the model, we might reduce the noise, but we would also change the expected levels of infection by changing the density of humans in their "world" and the number of humans in relation to mosquitoes; this would not be scaling. In order for the properties of the model to remain the same, we would also need to change the size of the world and the number of mosquitoes, such that the scaled up model has the same density of mosquitoes and agents as the original model.

The expected behavior, as calculated from (2.1), will be the same at all scales, but the relative fluctuations around the expected behavior will be smaller the greater the number of agents; note that the mathematical model itself is formulated in terms of agent densities only. Only in the limit of an infinitely large population will the expected steady state behavior equal the observed one. The effect of scaling on noise is illustrated in Fig. 2.4. Particularly for smaller systems, with few agents, the mean or expected behaviors of a system can be very poor predictors of the actual behavior.

In general, stochastic fluctuations are mathematically very difficult to describe and model, unless the system is extremely simple. At the same time, they are important to consider in many contexts. ABMs can be very powerful tools to explore the effects of stochastic effects in biological systems. In addition, there are other approaches that will be discussed in both Chaps. 6 and 7.

Let us summarize before we continue. If we set our parameters such that at each time step the agents take random points in their environment, then we can predict the long term (or steady state) infection levels of the system using a relatively simple mathematical model. However, the accuracy is limited by the level of noise in the model that, in turn, depends on the number of agents. Noise can be systematically reduced by scaling the model up. The steady state behavior predicted by the mathematical model will be correct only in the case of an infinite number of agents in the model. Also, the mathematical model tells us nothing about the behavior of the model when there is no perfect mixing. For those cases, we will have to rely on simulation.

As we reduce the scope for agent motion in the model, the spatial structure of the population becomes more important and the perfect-mixing approximation, as expressed by the mathematical model (2.1), becomes more inaccurate.

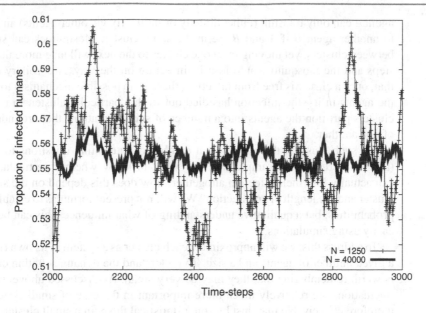

Fig. 2.4 Noise in agent-based simulations depends on the population size. These are simulation runs for perfectly-mixed populations using the same parameters as in Fig. 2.3, with two different population levels. The stochastic fluctuations are clearly larger for the smaller population size

2.4.3 Immobile Agents

Let us now consider the opposite extreme of perfect mixing, namely that humans do not move at all[4] ($s_h = 0$). Humans are initially dropped randomly into the world, and remain in their place throughout the simulation. Remember that humans cannot infect each other directly, but any transfer of the disease must be mediated by a mosquito.

Mosquitoes lose their infectivity after a relatively short time—2 time steps, say. Since we set their maximum step size $s_m = 1$, it will not be possible to transmit an infection over distances greater than 2. This means, in effect, that an infection can only be carried from one agent to another if these agents are within a distance of 2 of each other.

To help in the following discussion, let us define *clusters* of human-agents that allow cross-infection: Two agents are within the same cluster if the distance between them is 2 or smaller. Depending on the size of the system and the number of humans in it, there could be only a single cluster encompassing all agents, or there could be as many clusters as there are agents. In the latter cases we would have a very sparsely populated environment. The idea of the definition of a cluster is that an

---

[4]We assume that the mosquitoes continue to move; if all agents were immobile, this would be a meaningless model.

agent A can only transmit (either directly or indirectly via other agents) an infection to another agent B if A and B are in the same cluster. Mosquitoes can still move between clusters; yet moving from one cluster to the next will take more than 2 time steps and the mosquito would lose its infection on the way. A corollary of this is that, once a cluster is free from infection, there is no possible mechanism to re-infect the agents in it—the infection has died out in that particular cluster. In this sense, clusters partition the agents into a number of sub-populations that are independent of one another.

For such clusters, there are now a number of interesting questions to be asked (and answered). For example, we may want to know whether particular clusters can actually lose their infection altogether; how does this depend on the size of the cluster and the length of an infection? We will not present formulas to calculate these probabilities, but a qualitative understanding of what influences this can be reached easily using simulations.

To address this, we will approximate each cluster as a system in its own right, with a given number of agents and a size. To understand the dynamics within clusters, it is worth remembering that they may be very small. As discussed above, stochastic fluctuations are relatively much more important in the case of small systems. It is therefore conceivable that, just by some statistical fluke, in a small cluster infection dies out altogether, sooner or later.

Figure 2.5 addresses this by simulation. It shows three simulations of the Malaria model, each with very few agents (and a correspondingly small world size). To make things a bit simpler, we assume perfect mixing in the clusters. In these simulations we are interested in understanding how long it takes for small populations to lose infection levels altogether by chance. To do this we assume a certain infection level as an initial condition and let the model run until there are no more infected agents in it. We then record the time it took for this to happen. From the figure it is clear that the population dies out relatively quickly for the particular parameters chosen.

Clearly, in this world infections are not sustainable. Note, however, that this result of the computer model contradicts the predictions of the mathematical model. For the parameters used in Fig. 2.5 the stability condition (2.2) predicts stability of the non-trivial solution. The mismatch between the mathematics and the simulation is purely due to the stochastic fluctuations around the mean, that are not accounted for in the mathematical model. Even though the simulations in Fig. 2.5 assume perfect mixing, they still do not behave as the mathematical formula predicts, because there are so few agents in the system and the stochastic effects become large.

2.6 Confirm that the mathematical model predicts stability for the parameters in Fig. 2.5.

Returning now to the full models with immobile humans. The individual clusters do not quite behave like perfectly mixed sub-populations, although they are perhaps not far away. Nevertheless, the clusters of immobile agents are subject to the same random fluctuations that led to a loss of infection in the small models with perfect mixing. Hence, the simulations in Fig. 2.5 explain why the infections in the small

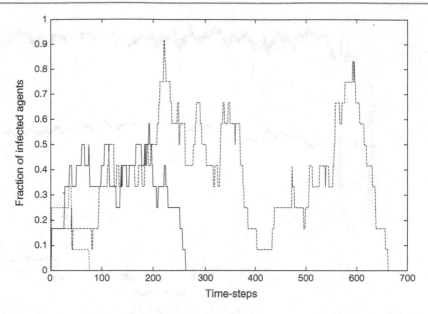

Fig. 2.5 Three runs of a system consisting of 25 agents in a world of size 2×2. The infection levels fluctuate wildly. The times required for the infection to disappear vary between 100 and nearly 700 time steps. For systems of this small size, the mean steady-state infection levels are no longer a good description of the system

clusters are lost over time. Overall we can predict, therefore, that in the case of immobile human-agents it takes much higher population densities to sustain an infection than predicted by the steady state solution for perfect mixing cases (as formulated by the condition in (2.2)). We still do not know exactly how dense the population must be to sustain the infection, but at least we can make a qualitative statement about the behavior of the system: Infection levels in the case of immobile agents are lower than in the case of perfect mixing.

What now for intermediate values of s_h? After all, we would expect that realistic cases are somewhere in between perfect mixing and immobile humans; we would also expect that long-term infection levels are somewhere in between the two extreme cases, but we really need to simulate the system to find out. Figure 2.6 shows a comparison of the Malaria model with two different levels of agent mobility. The parameters of these example simulations are the same as in Fig. 2.4; the difference between the simulations is only the level of movement of the humans, s_h. The perfect-mixing case settles around a steady state of between 0.5 and 0.6. For $s_h = 1$, the infection level in the population is markedly lower and only about $1/3$ of the population is infected. In simulation runs where s_h is set to zero, the infection dies out quickly (not shown).

This dependence of the model result on the mobility of agents is a recurrent theme in ABMs. Often it is even the main purpose of a particular ABM to explore the dependence of system behavior on the range of allowed motion of agents. It is certainly one of the strengths of ABMs to be able to address this kind of question.

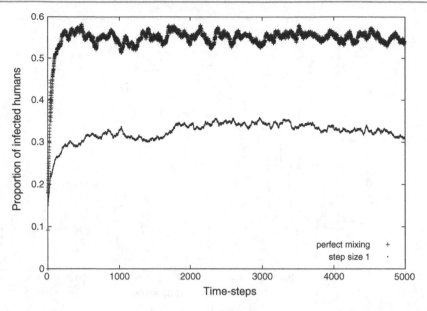

Fig. 2.6 The perfect mixing case is compared with a step size of 1. If agents do not move at all, the infection will die out. Moderate levels of movement lead to relatively low infection levels compared to perfect mixing

2.5   General Considerations When Analyzing a Model

While this particular, simplified model of malaria is unlikely to be useful in any real-world scenario, it is a very useful model for understanding how various assumptions about the nature of agents impact on the results one can expect. The Game of Life and Malaria models highlighted a number of features that suggest a modeling approach based on ABMs, which we summarize as follows.

- **Simplicity and complexity**: Systems consisting of very simple components can often display complex behavior at the aggregate level. The Game of Life was an example of such a system. Agent-based models are useful in modeling how interaction between many types of agent leads to interesting behavior.
- **Stochasticity**: The Game of Life is an example of a system where agents are strictly deterministic. This is rather unusual because the behavior of agents in most models will have stochastic components. When this is the case, the aggregate behavior will often (but not always) show stochastic fluctuations. In general, the fewer agents there are in the system, the larger the relative fluctuations. For some systems, these fluctuations can be so large that they become an important feature of the system in which case the randomness of the system is irreducible. The Malaria model with static human agents is a case in point. More generally, in many biological systems noise is increasingly becoming accepted as an important effect. As discussed

above, sometimes stochastic behaviors can be described deterministically, often they are irreducible.

- **Interactions**: The transmission of diseases over geographical areas is a result of many interactions of agents, as are the patterns produced in the Game of Life. When these interactions become irreducible to a mean-field approximation then ABMs are the method of choice. A case in point is the Game of Life where local interactions are essential for the large scale emergent patterns.
- **Heterogeneity**: Systems consisting of agents that assume different internal states over time are often not reducible to their average dynamics, and a mathematical representation of the system becomes hard or impossible. In the Malaria model, an example of heterogeneity is the infection state of the agents and their position and movement properties. More generally, two of the most important sources of heterogeneity in systems are the spatial distribution patterns of agents, and the internal states which distinguish one agent from another one. In the case of the Malaria model, perfect mixing renders the heterogeneity reducible, which makes mathematical approaches feasible. If one relaxes the assumption of perfect mixing then the heterogeneity of agents becomes irreducible and an analytic treatment very difficult.
- **Steady state**: ABMs often, but not always, approach a steady state behavior after a certain time. In a sense, this is good from a modeling perspective because it means that the initial conditions of the model are not important. Yet, natural systems do not always reach a steady state and sometimes the transient state is crucial to the understanding of the system. Mathematically, transient behavior is much more difficult to model, whereas this is not difficult in ABMs.

2.5.1 Testing ABMs

Sooner or later, every modeler using ABMs faces the difficulty of establishing whether or not her models are correct. By "correct" we do not mean that the model reproduces reality sufficiently well. That is a concern too, indeed philosophically a very deep one. Philosophically less difficult, yet practically of at least the same importance, is a different sense of correctness: Does the model actually show the behavior the modeler intended?

Larger ABMs can be quite complex pieces of software. Even the most experienced programmers will find that their programs contain programming errors. Some of these bugs lead to quite obviously wrong behavior but the majority, by far, do not and can be very hard to detect. In principle, one can never be sure that one's model behaves as intended. There are some strategies, however, to reduce the probability of falling prey to such bugs.

As a general rule, the modeler should always keep a healthy scepticism about the correctness of her software. Both experience and logic tell us that exhaustive testing of even very simple software is impossible, hence it is never possible to prove the absence of bugs [12]. Therefore, the basic assumption must be that the model is incorrect rather than correct. While such a standpoint might appear very depressing at first, there are a few strategies that can be used to detect errors in a model:

1. Mathematical modeling;
2. Extreme parameter tests;
3. Hypothesis-test cycles;
4. Reproducibility.

Mathematical modeling is the best method to create confidence in agent-based models. Unfortunately, the cases where agent-based models can be treated mathematically over the entire parameter space are rare. Of course, if there were an efficient way to use mathematics to describe the system then there would be no need for the ABM in the first place! Mathematical models provide much more general and powerful results than computer models. When it is possible to approximate the behavior of the ABM mathematically for some parameters (as in the case of the Malaria model) then this can provide the reference behavior against which to compare the computer model. This will not always be possible and, even if it is, one should never take this partial consensus between model and mathematics as a confirmation of the correctness of the model. The bug may well only strike in the area of parameter space that is not accessible to equations.

Another method to test the model is to perform extreme parameter tests. The behavior of models of interacting agents is often hard to predict in general, but sometimes easier to understand when some of the parameters are extreme. For example, in a model of agents that take up food, decreasing the supply of food to zero should lead to the entire population dying within a time that is defined by the life-time of the agent. Running the simulation with this extreme parameter will immediately test whether or not the part of the program relating to the death of agents works as intended. This is just an example. Which extreme parameter tests are useful and which ones are not very much depends on the model. In practice, these test are crude but they are a time-efficient way to catch some basic mistakes early on in the development cycle.

A more refined way to find errors in simulation models is to use hypothesis/test cycles. This is a variation on the extreme parameter methods. It is usually very difficult to understand how a given model behaves for a specific set of parameters, but one can normally predict how a *change* of parameters affects the behavior of the model qualitatively. A simple example would be that the population size tends to increase with the food supply, or that an increase in the length of a malaria incubation leads to a higher proportion of sick agents, and so on. During testing of a model, the modeler should form as many hypotheses of this sort as possible and systematically test them. Sometimes the prediction will not be confirmed by the model. If that is the case then it is necessary to understand why exactly the behavior differs from the expectation. The iterative process of forming hypotheses and trying to predict the effect of this change does not only sharpen the intuition of the modeler but, most importantly, can uncover internal inconsistencies in the model. So it is an important part in the process of the modeler gaining confidence about the model and eradicating errors.

Finally, the results of a particular simulation run should be reproducible. Because ABMs are usually stochastic, two simulations will not normally reproduce exactly the same data (if they do then this usually indicates a problem with the random number generator). However, when repeating many runs, the qualitative behavior should be the same in most of them.

In the case of evolutionary simulations (as they will be discussed in the next section), it might be that a specific behavior only evolves with a certain probability, rather than with certainty. In those cases, the model should still be stochastically reproducible in the sense that this probability can be reliably assessed by repeated experimentation.

2.6   Case Study: The Evolution of Fimbriation

The popular account of biological evolution is phrased in terms of individual selection, prominently featuring the "survival of the fittest." The basic idea of the individual selection narrative is that individual organisms are not equal in their ability to collect food and produce offspring. Those that are slightly more efficient will tend to have more offspring. If the more efficient organisms have some traits that the less efficient do not have then, over time, these traits will spread. To illustrate this, imagine a population of animals of species X. Suppose that, over a long period of time, those members of X that had bigger ears had, on average, more offspring. Over evolutionary time periods one would then likely observe a gradual increase in the mean size of ears among all members of X.

Of course, this very brief account of evolution is only a cartoon of the reality of evolutionary theory, and it would go far beyond the scope of this book to explore this topic in any depth. However, the simplified account of the evolution of the hypothetical species X reflects the basic idea of evolution by natural selection. Phenotypical changes are driven by heritable genotypical variations. Typically, one thinks of evolution as acting at the level of the individual (which in the example above would be the owner of the pair of ears).

2.6.1   Group Selection

It is not necessary to dig very deep into biological examples before one finds that evolutionary change is not always driven by variations between individuals, but could also be thought of as acting on entire groups of individuals. To illustrate this, imagine a group of birds sitting on a tree doing whatever birds do. Assume now that a predator approaches. One of the birds of the flock spots the danger and produces warning noises. This alerts the entire group, enabling its members to take-off and flee to safety. In this example, the action of an individual warning-bird has saved multiple members of the flock from potential death through predation. At the same time, however, the warning-bird itself has perhaps directed the attention of the predator to it, thus increasing its chances of ending its days as a dinner. As long as one assumes that evolutionary change is driven strictly by differences between individuals (e.g., size of ears), it is difficult to explain how the "warning bird gene" ever evolved.[5]

The warning-bird example seems to suggest a picture of evolution that acts on the entire group; a narrative that is different from individual-selection scenarios. A higher fitness of the group as a whole need not mean that each of its members has a higher fitness, too. In fact, as the bird-example shows, some of its members could be worse off, even if the fitness of the group as a whole increases.

Whether or not evolution does indeed act at a level of the group, as well as on differences between individual organisms, is still the subject of a heated debate between evolutionary theorists. Much work has been done on "group-selection" theory (for example, [13–17]) in an attempt to understand how it could evolve. The topic also receives significant attention from evolutionary biologists (see, for example, [18,19]). Debates and models in group selection are often centered on the idea of cooperation in social groups. In this context, "social group" does not necessarily mean a group of higher animals that have a group structure of some complexity; instead, it simply means that there exists a population with some form of interaction between its members.

As an example, one might think of *E. coli* bacteria. Under starvation conditions, some strains of these produce *siderophores*. These are a class of chemicals that make iron soluble, which enables the bacteria to scavenge the metal from the environment in which they live—normally the gut of a host. Siderophores are simply excreted into the host environment by the bacteria. The concerted action of the group facilitates the release of significant amounts of iron to the benefit of all group members. The simultaneous release of siderophores by all bacteria can be seen as a kind of cooperation. Each cell participates in constructing a small amount of the chemical, but only the collective effort leads to a reasonable and useful amount of dissolved iron.

So far so good. The problem starts when one considers that a cooperation of this sort comes at a cost. In order to synthesize the siderophores, the bacteria have to use energy. They do this, of course, in the "expectation" that their "investment" will reap an adequate return. The energy and resources the bacteria exert on producing siderophores could alternatively have been invested in growth and production of offspring. However, if an individual cell stops producing the siderophores it will still be able to benefit from those released into the shared environment by other cells and continue capturing iron. At the same time, unburdened by the cost of producing siderophores it would grow faster and produce more offspring. Evolutionarily, it would be better off.

The situation here is similar to that of a shared coffee-making facility as is common in some university departments. Assume that the purchase of new coffee beans is funded through the contributions of coffee-drinkers. Each time somebody takes a cup of coffee she is expected to contribute a small amount of money to the communal

---

[5]We do not really assume that a single warning bird gene exists, and this expression should therefore be understood as a label for a number of genetic modifications that impact on the said behavior.

coffee fund. Normally, such an arrangement is unenforced, in the sense that it is fully possible to drink the coffee without paying for it. A coffee scheme run in this way relies on the honesty of the coffee drinkers for its continued functioning. If a large number of drinkers suddenly failed to pay their expected fees, there would be no money left to purchase new coffee once supplies run out. In this case, anybody who wanted a coffee would have to go to the campus café, where a coffee costs considerably more (furthermore, for busy academics the time investment would be even more punishing). In this sense, the communal coffee scheme relies on the coffee drinkers cooperating; all drinkers are better off as long as everybody contributes their share.

The crucial point is that, as long as most members are honest, an individual is better off not paying. Using such a "cheating" strategy means the get to drink coffee for zero cost.

In the realm of bacteria and siderophores, the situation is similar but worse. Unlike dishonest coffee-drinkers, bacteria that fail to contribute their share of siderophores will tend to reproduce faster than their honest conspecifics. Assuming that the dishonesty in bacteria is an hereditary trait, non-cooperators will tend to increase over time, ultimately resulting in a breakdown of the cooperation between the cells. That is the naive expectation, anyway.

We will not go any deeper into the details of siderophores and communal coffee facilities. Suffice to say that the problems of cooperation in those realms of life seem to be relevant for many other aspects of social life on earth. Cooperation between individuals (be it coffee-drinkers, bacteria, or birds in a flock) seems to exist in real systems. Classical "survival of the fittest" accounts struggle to explain how such cooperation can be sustained unless there is some type of selection at the level of the group. Group-level selection would have to function somehow so as to restrain the individual organisms from following their selfish path to defection (which would ultimately leave everybody worse off).

The problem with group selection is that it lacks a convincing mechanism that could explain it. While the core narrative of the survival of the fittest individual is intuitive and convincing, there is no corresponding story telling us how evolutionary forces could work at higher levels. Groups seem to lack the features of classical Darwinian individuals, in the sense that they are often not clearly defined; they do not reproduce/leave offspring in any defined sense, and it is not immediately clear in what sense they have fixed traits that can be reliably inherited by their offspring and lead to fitness differentials between groups. For this reason, group-selection theories have been eyed with suspicion by many evolutionary theorists.

Computer models are ideal tools for studying this. One can use ABMs to implement evolutionary processes and test explicitly under which conditions group-selection works. The advantage of computer models over observation and interpretation of real systems is that ABMs do provide a fully specified system. Every underlying assumption is, by necessity, spelled out in the code of the implementation. In the remainder of this chapter we will discuss in detail a case study that uses ABMs in this context. This will illustrate how these computer models can be used to model evolution. However, as the primary aim of the following discussion is to demonstrate in detail how a modeling idea is translated into a design for an ABM, the case study should also be of interest for readers who do not specialize in evolutionary modeling.

2.6.2 A Model of Martian Mice

2.6.2.1 Group Selection Without Dilemmas

Group-selection is often seen as an issue only when there is the potential for cheating—when there is a social dilemma. A group of individuals and organisms contribute to and take from a common pool. This tempts some into defection, or cheating, meaning that they draw from the pool without contribution—very much like the coffee drinkers who do not pay. Most of the research work on group selection focuses on these kinds of scenarios.

Lesser known are scenarios where there is no social dilemma of this kind, but there still seems to be selection at the level of the group and it is unclear how this group-level selection could evolve and under which conditions. This is best illustrated using an example. Imagine a population of entities, say Martian mice. These live in caves and exist in two main forms: those having a green tail and those with a red tail. They are otherwise identical. Throughout the life of a mouse, the color of its tail may change from red to green, or vice versa. These color-changing events are essentially random, but the rate of change is somewhat influenced by environmental conditions. Since there is not much vegetation on Mars, the mice depend on a rather peculiar mechanism to acquire food: at the back of each cave, Martian cheese mushrooms grow. These mushrooms are an important source of food for the mice, although they do not entirely rely on this food source. What makes these cheese mushrooms and their interaction with the mice scientifically interesting is the fact that the growth rate of the mushroom depends on the color of the tail of the mice (by a mechanism that, to this day, puzzles Martian—and indeed all other—scientists). Up to a certain point the growth rate of the mushrooms increases with the number of mice with red tails in the cave. However, once there are more than 20 mice with red tails, the mushrooms turn toxic (because of an excessively high growth rate) and kill any mice that ingest them.

Interestingly, the probability of a color change depends on the amount of mushrooms the mice eat. The more mushrooms they eat, the more likely it is that red tailed mice change into green tailed mice. The switch from green to red, however, has been found to be independent of the mushroom consumption (or any other environmental condition). Experimental work by Martian biologists has led to a rather detailed understanding of the mechanisms that lead to the color change. As it turns out, the switching rate from one color to another is genetically controlled, whereas the individual color changing events are purely stochastic. This means that the tail color of an individual mouse cannot be predicted with certainty based on knowledge of the environmental conditions alone. It is, however, possible to predict the probability of a particular mouse being red- or green-tailed given the amount of mushrooms available in a particular cave. Observations have also led to the discovery that there are usually between 10 and 15 red-tailed mice in a cave.

Martian mice and their tail color are a typical example of a group selection effect without a social dilemma. Maintaining either a red or a green tail is energetically equivalent, that is having either color is not an advantage or burden for the mice in an evolutionary sense. Both red-tailed and green-tailed ones will have the same amount of offspring, on average, if they have the same amount of food. For an individual mouse, having a red tail or a green tail is energetically equivalent. On the other hand, at the level of the group the number of red-tailed mice is very important; it determines the growth rate of the mushrooms in a cave. Since the mushrooms are shared equally between all mice in a cave, the growth rate in turn determines the size of the mouse population in the cave. This tells us that the genes that determine the rate of the color changing events are significant for the group, but do not provide differential fitness advantages to the individuals.

There is, quite apparently, an optimum state in which a cave has exactly 20 red-tailed mice. This would be the best state for the population as a whole, because at this point mushroom growth is maximal, but mushrooms have not yet turned toxic. It is, however, not entirely clear how the system could evolve to that state. We have to assume that, initially (or at least at some time in the past), the gene networks that implement the change of tail color were not optimally adapted. Either there were too many red-tailed mice around, leading to toxic mushrooms, or there were too few, and the population did not grow as fast as it could have. The question is now: By what mechanisms can evolution steer the group as a whole to the optimum?

The standard individual-selection process that relies on the fitness differential between individuals in a population would not work here. To see this, consider the following: Assume the parameters of the genetic networks regulating the color switching rate are such that the population is not getting the best mushroom yield because the switching rate from a green tail to a red tail is too low. In this case, the mice have to wait for a mutation to improve their lot. Note that such a mutation will happen at the level of an individual mouse, so will change the genetic make-up of one individual, not the entire group. What is needed is a mutation that either reduces the switching rate from red to green or increases the opposite. As long as the mice are patient enough, such a mutation will eventually happen. In fact, it is not that unlikely. Any mutation affecting the switching probability can lead to either an increase or a decrease of the probability of the affected mouse being red-tailed. Hence, the chances of a beneficial mutation are quite good.

Let us assume now that, at some point, such a beneficial mutation has occurred, leaving one of the mice more likely to be red-tailed. At the level of the group, this will have the effect that on (time-)average there will be more red-tailed mice than before the mutation took place. Naturally, the effect will only be very small because the mutation affected a single mouse only. This is not a problem *per se*. The question is, rather, whether there is a mechanism that allows this small improvement to be amplified: Will the mutation spread?

This question is not straightforward to answer. For one, in finite populations simple stochastic fluctuations can prevent the spread of even the most beneficial mutation, even in classical individual-based systems. The effect is very similar to the stochastic extinctions discussed in the case of the Malaria model above. The existence and importance of such effects have been well studied in theoretical evolutionary biology. Stochastic fluctuations of this sort are certainly a relevant effect in the case of the Martian mice because the number of mice in a cave is very low—small population numbers tend to re-enforce the importance of stochastic fluctuations. However, in the present case this is not the main problem and we will ignore this complication for the moment.

More related to the question of group selection is another complication; assuming that the new mutant mouse has only a slightly higher probability of being red- than green-tailed then, depending on the population size, the effect on the population could be miniscule. The increase of the (time-)average number of red-tailed mice might be tiny and, given the stochastic nature of the switching between the tail colors, the effect might completely disappear in the normal background noise of stochastic switching. A small mutation in one cave could, therefore, result in no significant change in the growth rate of the mushrooms. Admittedly, whether or not this is true depends on the population size and the specifics of the relation between the growth rate of the mushrooms and the number of red-tailed mice.

The larger the mutational change in a particular mouse, the larger the potential impact on the population as a whole. A mutation in a mouse could result in a dramatically increased probability of being red-tailed—say it is always red-tailed. In this case, the effect on the (time-)average number of red-tailed mice in a cave may be sufficiently strong to make a material difference to mushroom growth.

Let us now consider what happens if, by some mechanism that we still do not understand, this initially beneficial mutation spreads in the population; that is, all the mice become red-tailed. By the rules of the mushroom, this would be disastrous for the mouse population. If all mice were red-tailed all the time, then all mushrooms would turn poisonous, spelling the end of the mice in this cave. They would all die. These *gedankenexperiments* highlight an inherent tension in the evolutionary dynamics of the red-green switch in Martian mice. Small mutations will not have any effect and too large mutations will lead to the extinction of the colony, at least if they spread.

So far we have only considered mutations that are (at least in the short term) beneficial in the sense that they lead to an increase of the mushroom yield in a cave. What we have ignored are all those mutations that have the opposite effect—those leading to a short term decrease of the mushroom yield. Since mutations are random events, "good" and "bad" mutations happen at roughly the same frequency and will therefore tend to cancel each other out. The net effect will be zero. On average, the population moves nowhere. Note, however, that this does not mean that there cannot be significant biases in particular populations. As discussed above, stochastic fluctuations can take systems very far away from the expected mean behavior.

Note that in the case of a classical individual-based Darwinian evolution the balance between beneficial and detrimental mutations would be no problem. The bearers of the beneficial mutations will tend to have more offspring, whereas the sufferers of detrimental mutations will leave few, if any, offspring. In the case of the

Martian mice, classical Darwinian arguments cannot be invoked to explain the spread of "good" mutations because the mushrooms are shared equally among all members of the cave. There is no fitness-feedback to bearers of individual mutations, always only to the group as a whole. If, by some stroke of luck, there are two beneficial mutations then everybody in the group will equally enjoy the increased mushroom yield, and the population as a whole will expand. Even those who do not have the beneficial mutation will have more offspring (on average). Similarly, there might be times when, by some statistical flukes, there are mutations leading to fewer red-tailed mice on (time)-average. Again, the population will contract—both the bearers of the detrimental mutations and all others. Over evolutionary time, the average number of mice will perform a random walk. One unfortunate consequence of this is that, sooner or later, the number of red-tailed mice in a cave will go beyond the crucial threshold of 20, the mushrooms will go sour, and the population become extinct. This means that, as far as the evolution of tail-color-switching in Martian mice is concerned, we need to explain two things: (i) how can an optimal switching frequency evolve and (ii) how can random extinctions be prevented?

As it turns out, there is a model that allows the spread of beneficial mutations. A *sine qua non* for this is that individual caves are not isolated. In other words, there needs to be more than one cave and it must be possible for mice to migrate between the caves. There is no mechanism by which beneficial adaptations can spread within a single cave (except chance). In fact, in every particular cave the mouse population will eventually die out, as we have just seen. However, if there is occasional migration between caves, then empty caves can be re-colonized by migrants and new colonies can be established, and a total population crash can be avoided.

To simplify the detailed explanation of the evolutionary mechanisms, let us introduce some shorthand notation. When we say that a mouse is *fitter* than another, then we mean that its switching rates from red to green and vice versa are such that, if an entire population had these switching probabilities then this population would be closer to the *achievable optimum*. Rather than referring to the individual, the notion of fitness here implicitly always refers to a group. By "achievable optimum" we mean the following: The more red-tailed mice there are (up to a certain point), the more mushrooms grow and the larger the population of mice will grow, since the growth rate of mushrooms depends on the *absolute* number of red-tailed mice. Food is shared between the mice, and more red-tailed mice means a higher population growth of all mice. Higher growth means that the population increases, which means that there will be more red-tailed mice, which further increases growth until there are too many red-tailed mice and the mushrooms become toxic. The optimal point, or the point of highest fitness, is just before mushrooms turn toxic.

We can now explain how migration between caves can serve as a mechanism for adaptation of the switching between red and green tails in a population of Martian mice. As before, we assume that the system starts in an unadapted state, i.e., an average number of red-tailed mice that is too low—we do not need to consider the case of too high a number of red-tailed mice, because such populations would immediately go extinct. Let us further assume that there are many caves with mice. Each sub-population is genetically diverse at first, in the sense that the switching rates between red and green within a cave may be very different. Starting from such an initial state, after some time has passed sub-populations will start to become extinct—then statistical flukes will lead to this. The key point is that it is extremely unlikely that all caves become extinct simultaneously, to the extent that we can ignore the possibility. As long as some caves are populated then a total collapse of the population can be avoided by occasional migration of mice from one cave to another. Empty caves will be discovered by migrating mice who then establish new populations. If we assume that migration is relatively rare, then we would expect a strong *founder-effect*. As a result, the newly established cave-populations will be genetically rather homogeneous. All offspring of the founder will have the same, or at least similar, color switching probabilities as the original ancestor. Some variation enters through occasional mutations that adjust the probability of some mice to be red-tailed. Another source of variation is influx from relocating mice, i.e., mice that come from other caves. While there will always be some diversity within the caves, we can assume for the moment that this diversity is relatively low.

Once the population in each cave has become extinct at least once, we would expect a high heterogeneity between different caves, but relative genetic homogeneity within each cave. The differences between the sub-populations will manifest themselves in different population sizes in the caves and different expected times before the inevitable extinction event happens. This difference is a key element for the adaptation mechanism. The longer it takes before a population becomes extinct, and the larger a population in a specific cave, the more mice will, on average, leave it by migration. This translates into an increased rate of colonization of empty caves. Hence, bigger and longer-lasting sub-populations are more likely to be the source for new founder mice in new caves. Given the homogeneity within each cave, we can also assume that a newly established population of a cave, will normally be very similar to the parent population (although this will not always be true).

This provides a direct mechanism for competition between caves. A cave is fitter if it is less likely to become extinct and if it has a larger population. Re-colonization is the group-level equivalent of reproduction. Group-level mutations are achieved by a constant influx of outside mice and mutations within a cave. Any newly established cave will be similar to its "parent-cave," but not necessarily equal.

Altogether, this suggests a possible mechanism for selection at the group level. Unfortunately, verbal reasoning is limited in its deductive powers. While this scenario for how group selection could work seems plausible, one needs more precise reasoning to be sure, and to be able to determine under which conditions it can work.

As a modeling problem, this scenario has all the ingredients that make it suitable' for an ABM and hard for many other techniques (specifically mathematical models). There is irreducible randomness in the system. What we are interested in in evolutionary systems is to see how the action of rare events (positive mutations) can shape the fate of a population. A statistical description of the system is in this context not satisfactory. With the help of a random number generator, ABMs can generate instances of random events and replay evolution. Also crucial to the model is the heterogeneity of the system. All the mice are different, or at least potentially different to one another in the sense that they differ in their switching rates. This difference cannot be reduced to a description of the "mean" switching rate, or some other macroscopic variables, whence the system is irreducible heterogeneous. Moreover, crucial to the system is the interaction between the agents (mice) and their environment (cave). Again, this interaction is irreducible. We conclude that an ABM is the ideal method to model this system.

Before we commit to the expense of setting up a model and simulating it, we should convince ourselves that the problem is indeed worthy of our attention. Martian mice are perhaps too remote to earthly concerns to justify the investment in time and effort that a model requires. However, there is a system here on earth that behaves in a similar way.

2.6.2.2 Fimbriation in *E. coli*

Fimbriae are hair-like structures that grow on the surface of some bacteria, such a *E. coli*. They are so-called *adhesins*, which essentially means that they are used to attach to host cells. Fimbriae are what biologists call a *virulence factor*. This means that they directly cause an immune reaction and make the host ill. *E. coli* comes in many genetic variants and not all of them are virulent. In fact, some *commensal* (that is non-disease causing) strains of *E. coli* permanently colonize human gastro-intestinal tracts. One thing that makes them interesting from a health perspective is that even these commensal strains still express some of the virulence factors, such as fimbriae, but only do so at a low level. In the current context, "low level" means that, at any one time, only a small proportion of the population actually expresses them. Whether or not a specific cell expresses fimbriae is a random decision that is genetically coded— by the so-called *fim* operon. Within the *fim* operon is an invertible element called *fimS*. Two proteins, FimB and FimE can (by themselves) catalyse an inversion of *fimS*. They do so by binding on each side of the invertible element, then literally cutting out the element and re-inserting it in the opposite orientation. The expression level of fimbriae depends on the orientation of *fimS*: If it is inserted in the "on-orientation" then it is expressed; if it is in the "off-orientation" it is not expressed (Fig. 2.7).

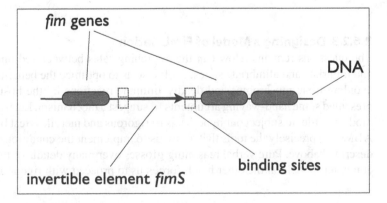

Fig. 2.7 The organization of the *fim* operon and the fim switch

There is a subtle difference in the way FimB and FimE function. FimB switches at about the same rate in both directions. Its overall switching rate, however, is very low. FimE, on the other hand only switches efficiently from on to off and it does so with a much higher efficiency than FimB. Yet FimE is not expressed when the *fim* switch is in the "off" position. In practice, this architecture results in a division of tasks between the two proteins: FimB essentially only switches fimbriae on. Once FimB is expressed it does not play much of a role compared to the much more efficient FimE that switches fimbriation off.

There are a number of factors controlling FimB expression, including temperature and growth rate. The dominant determinant for the FimB concentration is the concentration of sialic acid in the host cell. Sialic acid is a potential nutrient for *E. coli* and is released by host cells in response to low levels of fimbriate bacteria. Up to a certain point, an increase in the number of the bacterial fimbriation levels leads to a further increase in the amount of sialic acid released by the cell. The situation is analogous to the case of the Martian mice, where an increase of the number of red-tailed animals in a cave leads to a higher growth rate of the mushrooms. *E. coli* faces the same fate as the Martian mice once the fimbriation levels become too high. Instead of stimulating further sialic acid release, fimbriation levels above a certain threshold trigger an immune response by the host cells. Such an immune response normally leads to extinction of the bacterial colony, and makes the host sick. In this sense, sialic acid is for earthly *E. coli* what cheese mushrooms are for Martian mice, and fimbriae correspond to red tails.

The Martian population is partitioned into only weakly interacting sub-populations by the caves. In *E. coli* this role is played by different hosts. Ingestion of these bacteria is unavoidable. Hence, even if an immune response leads to the complete elimination of previous populations, the host will be re-colonized by new bacteria. This re-colonization would be the equivalent of migrating Martian mice. Crucially, also equivalent to the Martian mice is that, within each sub-population, we can assume that all bacteria share the benefits of increased release of nutrients. Sialic acid is released by the host into the environment of the bacteria, all of which are free to take it up.

2.6.2.3 Designing a Model of Fimbriation

The questions remains: How can the switching rates between red and green tails, and fimbriate and afimbriate states, evolve so as to optimize the benefit for all, while avoiding toxic mushrooms and deadly immune reactions by the host? Above, we presented some hand-waving arguments to suggest a mechanism. Let us now design a model in order to subject our hypothesis to rigorous and merciless test by a computer. ABMs are precisely the tools that allow us to implement the conditions that we have described above. Pure verbal reasoning glosses over many details of the problem. A computer model, on the other hand, forces us to proceed with rigour and to specify and spell out every single assumption we are making. In what follows, we will discuss in detail how the ideas we have presented on group selection can be converted into an ABM. In particular, this discussion will place emphasis on highlighting how design choices are made. These choices are not definitive, however, in the sense that one could come up with a different model that does the same thing. Nevertheless, the principles are general ones. The overriding aim is to obtain a model that is as simple as possible, while still showing the desired behavior. As always in modeling, before choosing to implement a particular feature we will carefully deliberate whether or not it really is necessary for the particular purposes of the model.

At the beginning of a design process for ABMs we need to ask ourselves three questions:

1. What are the agents?
2. What is the agent's environment?
3. What are the interaction rules: between agents and agents, and between agents and the environment?

2.6.2.4 Agents

Starting with the agents, the entity in the system that is most subject to change is the bacteria, and it is changes in them that are the focus of interest within the model. An important agent type is therefore the *E. coli* bacterium. Since we are (at least for the moment) only interested in a single species of parasite, a single type of agent will suffice in our evolutionary model.

There are a number of minimal requirements that this agent has to fulfill in order to be useful. Firstly, since we are interested in evolutionary change, there should be the ability for agents to die and reproduce. A standard design choice in ABMs is to couple reproduction to the accumulation of nutrients. One way to implement this is to convert nutrients upon uptake into "energy points." Over time, agents accumulate these energy points—the more food they gather, the faster they accumulate energy. In many models one finds that individual agents are required to pay a "maintenance-tax," that is, they lose some part of their accumulated energy at each time step simply in order to stay alive; this simulates the costs of maintaining cell processes. Real bacteria certainly do have ongoing maintenance costs and it would make sense to include such a tax in our model. On the other hand, this introduces an additional parameter to the model. We would have to decide how much an agent pays for maintenance. The additional complication is not very great, but there is no obvious benefit from having it either, at least not in the present case. We are interested in group selection here, and an individual maintenance tax does not seem be very relevant to our model. It seems best, therefore, to simply set the maintenance tax to zero in the present case.

Contrary to popular belief, real bacteria do not live indefinitely. Cell division is asymmetric and one part of the dividing cell inherits the "age" of the parent cell, whereas the other part can be considered the "newly born" cell. We could directly reflect this feature in the model. Yet, again, in order to simplify the problem (without making much of an error) we will assign an age to every agent instead of keeping track of every division event. The age is updated by 1 at every time step. Whenever an agent has reached a certain threshold age it will be eliminated from the model with a given probability per time step. In a simplistic way, this reflects the finite life times of cells. Both the threshold age and the probability to be eliminated need to be set as parameters. One might think that it is simplest to set the probability of death to 1 once the crucial age is reached, as this avoids setting an extra parameter. In practice, this has the undesired side-effect that subsets of the population then tend to act in "lock-step"; for instance, for periods of simulation time no agents die, but at certain intervals a large proportion of the population dies. This effect leads to artificial swings in the population size and complicates the analysis of the model. Making death a probabilistic event (depending on age) de-synchronizes the life-cycles of the agents, albeit at the cost of a new parameter. This cost is not too high, however, because the particular parameter value chosen will not materially influence the outcome of the models and can be kept fixed over all simulations.

Similar arguments apply to agent-reproduction: An agent should only produce offspring if it is successful enough, that is if it has collected sufficient nutrient. Just how much nutrient is "enough" is to some degree an arbitrary decision, as long as we do not insist on quantitative accuracy of the model (and this would be impossible to achieve anyway). So, again, the modeler is free to choose a value that is reasonable. Agents should not be able to collect all nutrients necessary to reproduce in one step because this would mask subtle differences in performance between agents. If finding food once were sufficient, then there are essentially only two categories of agents, those that reproduce and those that do not. Since reproduction is intimately linked with fitness, this would result in a very crude "evaluation" of fitness dominated by noise. On the other hand, if agents have to collect nutrients over many update steps then more computing cycles are spent on evaluating the fitness of agents, which increases the computational cost of the model. A good parameter value for the reproduction threshold strikes a balance between these two opposing demands. Independent of the chosen value of the parameter, reproduction should also be stochastic (like death) in order to avoid simultaneous reproduction of all agents. That is, once the agent has collected enough energy, it will reproduce with a certain probability per time step.

A crucial element of the model is the design of the evolutionary search space the bacteria can explore. This is the feature on which both the credibility and the feasibility of the model depend. One could try and design the artificial cells such that they have a vast space of possible behaviors to explore. This would, perhaps, lead to credible models, but certainly also to infeasible ones. The bigger the search space for evolution the harder it is to program the model and, most of all, the more likely the model contains bugs that are hard to find. Models that are too complex are bad from a practical point of view.

When designing a model, it is paramount to keep focused on the particular research question that motivates the model. Our particular task here is to test a hypothesis about group selection. This hypothesis might ultimately be motivated by an interest in fimbriation or red-tailed mice. Yet not all details of these systems are necessarily relevant for the particular evolutionary scenario we are interested in. Clearly, Martian mice and *E. coli* are very different. Yet if we have understood the evolution of the switch from red to green tails, we have also understood how the control of fimbriation evolves. When modeling the evolution of either system, we can therefore ignore the features that are not shared by them. Considering the ultimate goal and motivation of a model is usually helpful to make key-simplifications. In modeling, stripping away detail often results in better models rather than worse ones.

The genetic network that controls the *fim* switch is reasonably well understood and could be explicitly implemented in biochemical detail. However, in practice this would not work very well. Firstly, biochemical interactions are slow to simulate. This means that we would spend much of the simulation time calculating the interactions of molecules, which is not the focus of the model. Secondly, the time scale of biochemical interactions is very much shorter than the time scale over which evolutionary change happens. Hence, taking into account biochemical interactions would transform our model into a multi-scale simulation problem, which requires significantly higher technical skills than single-scale models. Thirdly, and most importantly, there is no need to implement detailed biochemical models. The detailed genetic processes in Martian mice and *E. coli* are certainly different, which should tell us that these differences are irrelevant for our particular research question. In the context of our model, we are interested in the *strategy* cells use. The specifics of the *fim* switch are only of parochial, earth-centered importance, and should be left out. What we are interested in is the rate with which bacteria switch from fimbriate to afimbriate, which constitutes their evolutionary strategy. Any such strategy could be implemented in a number of ways biochemically, and any conclusions reached about the strategy are likely to be of inter-planetary importance, rather than limited only to a particular earth species.

Instead of worrying about the underlying molecular mechanisms that implement the *fim* switch in reality, we simply model it by two numbers, representing the rate of on-to-off and off-to-on switching. In the particular case of *fim*, the on-to-off can be thought of as essentially fixed, albeit with an unknown value. The switching rate is a core issue in the evolution of the system and we leave it therefore to adaptive forces to determine this value; the additional benefit is that this relieves us of the burden of choosing a value for it. The off-to-on switch is somewhat more complicated because there is a coupling between the amount of sialic acid that bacteria find in their environment, and their probability of switching from off-to-on.

In order to find a plausible representation of the relationship between nutrient/food and the tail/fimbriae switching rates, we choose to be inspired by the *fim* genetic networks of *E. coli*. FimB needs to bind to four binding sites to effect the switch (see Fig. 2.7 on p. 57). Furthermore, there could be cooperativity between the binding sites, that is the binding is stronger when there are two molecules binding than when only a single site is occupied (see also Sect. 6.3.3 on p. 235). Mathematically, cooperativity is normally modelled using a so-called *Hill function*, which is a step-like function with a parameter h. (For an illustration of the Hill function see Fig. 2.10;

also Sect. 4.5.1 on p. 159.) The higher h, the more the Hill function resembles a true step function. Another parameter of the Hill function, K, determines at which point the function is halfway between the minimum and the maximum value. For our purposes we can write a Hill function as follows:

$$h(x) = C \frac{x^h}{x^h + K^h} \qquad (2.4)$$

The variable of the Hill function would here represent a measure of the availability of nutrient, for example the number of mushrooms in the cave or the concentration of sialic acid. The constant parameter, C, determines the maximum and minimum values of the system. The function $h(x)$ reaches its maximum value for an infinitely high x. This is easy to see. When x is very large then K^h is small compared to x^h and the term containing the fraction will tend towards 1, giving $h(\infty) = C$. On the other hand, for a vanishing x the function approaches zero, i.e., $h(0) = 0$. Given that we are interested in probabilities per time step, and that Hill functions also (crudely) model how molecules bind to the DNA, $h(x)$ seems a good *ansatz* for the switching function we are seeking. It should be stressed here that the main feature that makes Hill functions a good choice is their mathematical flexibility; the fact that they can also be interpreted as a description of how FimB and FimE bind to the DNA is, and should be, secondary.

In bacteria we observe that a higher sialic acid concentration reduces the rate of switching. If we take $h(x)$ as this rate and x as the amount of sialic acid, then we would like $h(x)$ to reach its minimum value for high values of x. One way to achieve this is to use $1 - h(x)$ in the switching function rather than $h(x)$ itself. This works, as long as we constrain C to be between 0 and 1. This leaves us now with the switching function S^+:

$$S^+(x) = 1 - C \frac{x^h}{x^h + K^h} \qquad (2.5)$$

Having determined the functional form of this equation, the next question is, which values to assign to the parameters? The answer is very simple: We do not worry about them, but let evolution do the job. The expression for the switching probability in this equation is a general expression that determines the broad shape of how the probability of switching from off to on depends on the sialic acid concentration x; by changing the parameters (i.e., K, h, C) the function can take a variety of different responses. In this model, we are precisely interested in finding out how evolution shapes these parameters to find a suitable response function that enables stable and well adapted populations of *E. coli* (or Martian mice). The only thing we have to do, as a modeler, is to set an initial value for these parameters, to give evolution some starting point. The presumption is that this initial value does not matter, and the simplest thing is therefore to assign random numbers to them. After that we will leave it to evolutionary change to find more suitable values.

2.6.2.5 Environment

A potential second category of agents is the hosts (i.e., the equivalent of the caves on Mars). It might be a good choice to introduce hosts as a type of agent if we plan on expanding the model. At the moment it is simpler not to have host agents. We assume that there are several hosts, but the hosts themselves are all the same. We also assume them to be static. Instead of being agents, hosts are modelled as compartments in the environment. These compartments each contain a sub-population of agents. There is limited migration between them. To simplify the model further we assume that the agents spend only a negligible amount of time in migration between two hosts compared with the typical residence time within a host. This means that it does not take any time to move from one host to another. In reality this is, of course, not completely correct; yet the error we make by this assumption is very small and the benefit of a simpler model easily outweighs the costs. Most likely, including the time of migration between hosts does not result in a better model at all because: (i) it is unknown how long it takes to migrate from one host to the next; and (ii) we are not primarily interested in numerically-accurate models of Martian mice or the *fim* switch, but in testing an evolutionary hypothesis. This hypothesis can be tested even if we do not know many of the particulars of the system. The reader is again reminded of the mantra: Leaving unnecessary elements out results in better models, not in worse ones.

There is one feature of hosts that does need to be included in the model, namely the release of nutrients. Since we chose to represent hosts as a part of the agent's environment, we need to design our compartments such that they release nutrients in response to the state of the agents within them.

In summary, we choose the environment to be partitioned into a number of compartments, each of which represents a single host. Each compartment contains a number of agents (bacteria) and releases nutrients. Following what we know about the biology of the hosts, we need to couple the nutrient release by the host to the fimbriation levels of the bacterial cells colonizing this particular host (or the growth rate of the mushrooms to the number of red-tailed mice). The precise functional shape of the host response is not known, but some qualitative features of it are: for low levels of fimbriation, only small amounts of nutrient are released; once a threshold level of fimbriation is reached then a full immune response is triggered and the bacterial colony is killed. Since we do not have any quantitative information about the response function, we simply assume it to be of the form of a Hill function again. Other than in the case of agents, we now need to specify the values of the parameters, because the host is assumed to be static. The choice is to some degree arbitrary; we are not interested in quantitative models, only in checking the feasibility of a hypothesis. Let us tentatively assume a response function such as:

$$R(n_f/N) = F \frac{(n_f/N)^h}{(n_f/N)^h + D^h} \tag{2.6}$$

Here we assume that n_f is the number of fimbriate agents in the host, N is the total number of agents, and F is the maximum amount of released nutrient. Essentially, this response function is the same as the preliminary switching function in (2.4); the functional argument here is now the fraction of fimbriate agents in the population, rather than the sialic acid concentration in the environment. There is no real justification for using this function, other than convenience—by varying the parameters h and D, a number of different shapes can be achieved, which in turn allows testing different scenarios.

The response function (2.6) has three independent parameters: the constant D that specifies the point at which the host releases half of its maximum amount of nutrient; a scaling factor F; and the Hill parameter h. The first two parameters, F and D, are arbitrary in the sense that they simply scale the model. A higher F can be counterbalanced by increasing the energy required for bacteria to reproduce. Similarly, D scales the possible population size that is achieved during simulation runs. Even though the values of the individual parameters are arbitrary to some extent, the set as a whole determines the achievable population sizes. This must be chosen with care. Too large a population may slow the simulation down to a level where it becomes infeasible. A very small population, on the other hand, poses the danger that the individual sub-populations in the compartments are dominated by stochastic fluctuations to the extent that a meaningful adaptation becomes impossible. It is crucial for the success of the modeling exercise to find good values for these parameters to balance these conflicting requirements.

The argument to the function in (2.6) is the fraction of fimbriate cells in the given sub-population, n_f/N. This looks reasonable at first, but a comparison with the real world reveals that it is unlikely to be correct. If the response only depends on the *fraction* of cells, then this would allow *E. coli* to grow to any population size without triggering a host response. In the real cell, however, what leads to the host response is the direct interaction between the fimbriate cells and the host. What counts is the *absolute number* of contacts a host-cell makes with the bacterial virulence factors, i.e., the fimbriae. The relevant variable is therefore the absolute number of fimbriate cells and not their proportion. In addition to the sialic acid released in response to fimbriation, we assume that there are also some other food sources in the environment, that are not further specified. We denote this additional user-defined parameter by G; it does nothing to the evolutionary behavior of the model, but it is essential to set $G > 0$ to avoid premature population crashes. The correct response function is therefore:

$$R(n_f) = G + F \frac{n_f^h}{n_f^h + D^h} \tag{2.7}$$

This new host response function also introduces some additional requirements on the parameters. Very small population sizes might have the undesired side-effect that they could never trigger an inflammatory host response. If the population size is $P < D^h$ then, trivially, n_f can never reach the inflammatory threshold D^h. This would make the entire model pointless.

In general, the easiest way to determine the correct parameters in situations where there are no reliable empirical measurements is through trial and error. Doing this is somewhat time-consuming, of course, but has the added advantage that it allows the modeler to obtain a feeling for the run-time requirements of the model under various parameter settings, and choose the parameters accordingly.

2.7 Why do we expect population crashes in the absence of G?

2.6.2.6 Interactions

The final item to be determined is the interaction rules of the model. It is useful to distinguish between interactions between individual agents and the interaction between an agent and its environment. In the current model there is no direct interaction between agents but they do affect each other indirectly via their interactions with the shared environment. At each time step, each host releases an amount of nutrient into the environment. Agents can take up this nutrient. Each agent will obtain approximately the same amount of cheese mushrooms/sialic acid. In the simulation we model this in a simplified way. At every time step we divide the amount of nutrient released by the number of bacteria in the model to determine how much nutrient each bacterium is allocated. This ensures that the entire amount of nutrient is used up at every time step, and shared equally among all agents. This may not accurately reflect real life, but it reflects a key-part of our hypothesis, namely that resources are shared between agents. If one wishes, one could later refine the model and introduce more sophisticated rules of resource sharing. For the moment, the simplest possible solution provides most insight.

The indirect interaction of the agents is mediated through the host response, which is the reaction of the environment to the number of fimbriate agents in the system. At every time step, we assume that the environment releases a certain amount of nutrient that depends on the number of fimbriate agents, according to the corrected response function in (2.7). An inflammatory host response is triggered if the number of fimbriate agents in the host is greater than or equal to D^h, that is, if $n_f \geq D^h$. In this case, the entire sub-population within this host is killed.

2.6.2.7 The Simulation Algorithm

Having clarified the basic structure of the model, we now need to determine the details of the simulation algorithm to use. Evolution clearly takes place in continuous time, but we are not interested in quantitative predictions; there is no need for the additional complexity of an event-driven structure, and we will therefore use the simpler time-driven algorithm that updates the entire population in discrete time steps. The structure of the model's implementation is outlined in Algorithms 3 and 4. At each time step, every agent in every host is updated (that is we use a simultaneous update algorithm).

Algorithm 3 The main loop of the agent-based model for the evolution of fimbriation.

Time = 1
loop
  for All hosts **do**
    Determine the number, n_f, of fimbriate agents in host.
    if $n_f \geq D^h$ **then**
      Delete the entire population of the host and skip to next host.
    end if
    {Release nutrient according to the response function (2.7).}
$$R \leftarrow G + F \frac{n_f^h}{n_f^h + D^h}$$
    {Determine the number of cells N in this host.}
    $N \leftarrow$ CountAgentsInHost(thisHost)
    $f \leftarrow R/N$
    for All agents in this host **do**
      {Update the internal energy state.}
      $e \leftarrow e + f$
      $age \leftarrow age + 1$
      if $age > thresh_{age}$ **then**
        With probability $p1$ place agent in the reaper queue and skip to the next agent.
      end if
      {Reproduction places offspring in birth queue.}
      if $agent > thresh_e$ **then**
        Reproduce cell with probability $p2$.
      end if
      if agentIsFimbriate **then**
        Switch off fimbriation with probability p_{af}.
      else
        {This uses (2.5).}
        Switch on fimbriation with probability $p_f = 1 - C \frac{f^h}{f^h + K^h}$.
      end if
    end for
    {Movement between hosts.}
    With probability p_m move a randomly chosen agent to a randomly chosen host.
    Delete agents from the reaper queue.
    Place agents from birth queue into same host as parents.
    Clear reaper queue and birth queue.
  end for
  Time = *Time* + 1
end loop

When designing an update algorithm for an ABM, there are a number of choices regarding the parameters. Intuitively, one would think that, in synchronous models, the order of update would not be important (because it is synchronous). Indeed, it *should not* matter. In synchronous algorithms a potential pitfall arises in connection with the order in which rules are applied. Algorithm 3 specifies that agents first collect energy and then check their age and die if their age is above a certain threshold; reproduction happens only after checking for death. Alternatively, one could check for death only at the end of the update step. This would allow agents one additional

---

Algorithm 4 Reproduction of an agent

Agent to be reproduced is A.
{Set the energy to 0 for parent agent.}
$e \leftarrow 0$
Create new agent A'.
for All parameters p_{af}, C, K, h **do**
  Copy parameter from A to A'.
end for
{Mutate.}
With probability m change a randomly chosen parameter of A' by a small amount.
Place A' into the birth queue.

---

time step to reproduce. The difference between these two alternatives is not always great, but could sometimes make a material change to small sub-populations. The amount of energy released by the host depends on the number of agents in each host. At the same time, the switching probability of agents depends on the amount of food released by the host. During the update procedure, the number of agents will typically change, through death and birth events, which affects the amount of energy available to each agent. This means that the number of living agents at the next time step is not known before all agents have been updated. For this reason, reproduction and birth events need to be updated in a separate update loop from the assignment of energy and the updating of switching events.

One solution is to assign to agents the amount of energy that has been calculated at the previous time step. This means that the nutrients are consumed and sensed at the beginning of the time step; alternatively this could be done at the end of the time step, when agent numbers have been updated. The difference between these two possibilities is small and it does not matter which one is chosen. The important thing to remember is that the details of the update order need some careful thought in order to keep the model consistent.

A further complication, arising from the separation of agent birth from placement, is that an agent giving birth might move host in between those two steps. Should the offspring be placed in the old host or the new one? An implied assumption is that the implementation makes it possible to relate a new agent to its parent, or to its parent's new or previous host.

2.8 Confirm that the verbal description of the algorithm matches the pseudo-code in algorithm 3.

2.6.2.8 Testing the Model

Let us now analyze the behavior of the model. In the early stages of a modeling project, it is often not clear which variables will be the most revealing of the behavior of the model, or which aspects of the model one should focus on to reach an understanding of how it behaves. Once a model is programmed, the modeler will need to spend significant time exploring it, in order to understand how it behaves, to figure out which variables are most informative, and to find any surprising features that potentially suggest semantic errors. This phase can feel like time-wasting, because it does not actually contribute to a better understanding of the underlying system. However, this is not so. The detailed exploration of the model sharpens a modeler's intuition and is, in any case, an inevitable part of every modeling project.

In the present case, the total size of the population turned out to be the most important indicator. Incidentally, this is true for many models of evolutionary systems and was not entirely unexpected. One would expect adaptation to lead to a better exploitation of environmental resources, which results in a higher sustainable population number. In the present case, we would expect that evolution optimizes the number of agents that are in the fimbriate state, so as to maximize nutrient release and minimize the probability of being wiped out by a host immune response.

Indeed, Fig. 2.8 (left) shows the size of the total population (in all hosts) over time in an example run of the model. During the first 15000 or so time steps, not much seems to be happening. Then suddenly the population increases to a value between 15000 and 18000. Stochastic fluctuations mean that there are swings in the numbers of agents, but all within a well defined region. At around time step 140000, there seems to be another transition to about half a million agents in the population.

The, apparently discontinuous, jumps of the population are very much what one would expect to see in an evolutionary system. They correspond to the emergence of innovations and are commonly observed in evolutionary systems. In this sense, the results are encouraging. Yet, this is only the starting point; more work is needed to understand what is going on in the model and to confirm that the jumps are indeed due to evolution. The observed time evolution of the model could reflect some bias in the system, some delays in birth events, some stochastic fluke of a population that essentially moves randomly, or it could be simply down to a programming error. Ultimately, in large computer programs one can never be sure that the model does what it is supposed to do.

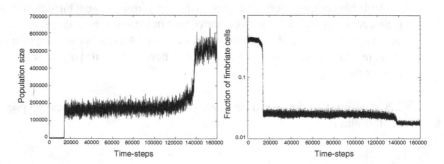

Fig. 2.8 Example of a simulation run of the fimbriation model. The *left graph* shows the population size as a function of time. The *graph on the right* shows the proportion of fimbriate agents (log scale). The following parameters were used: Size of the system 625 hosts, mutation rate 0.1, basic energy per cell 5, reproduction energy 0.2, reproduction probability 0.2, agent life time 40 time steps, death probability 0.2, maximum nutrient release before immune reaction sets in 10

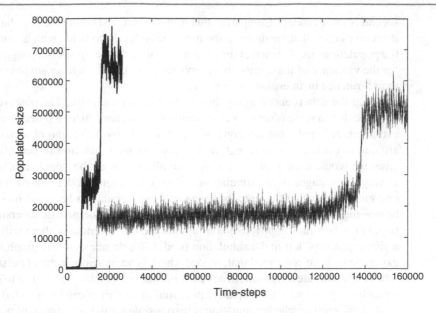

Fig. 2.9 A repeat run of the model shown in Fig. 2.8 for 30000 time steps with exactly the same parameters is shown in *bold*. The transition happens much faster which is entirely a stochastic effect. The other *curve* is identical to Fig. 2.8 (*left*) for comparison

Figure 2.9 shows a repetition of the simulation with exactly the same parameters but a different random seed. Qualitatively, the dynamics are the same but the quantitative details of the behavior are very different. Once again, there are two transitions, but these happen much earlier than in the first run. In addition, the population levels reached are higher. These differences are mostly due to the different random sequence resulting from the random number generator of the model, highlighting that a single simulation run only serves to indicate the general properties of the model. It is always necessary to repeat a simulation many times before any conclusions can be drawn from the results as a whole.

So far we have seen that there are more or less discontinuous jumps of the population in the model. The question that we need to address now is, what causes these jumps? The problem one often encounters with ABMs is that even simple systems can be rather difficult to analyze and understand. In the present case, however, we have a clear expectation of what should happen. The crucial parameter that agents need to adjust in order to be successful in their environment is the fimbriation level. Initially, all agents are just in a random state; the average fimbriation probability over the entire population should be around 0.5. The expectation we have is to see some change in that value, synchronized with increases in the population. The right-hand graph in Fig. 2.8 shows the average proportion of fimbriate agents over time corresponding to the simulation in the left-hand side. In order to make it more readable, it is presented using log scale for the fimbriation rate. The main feature of the graph is two clear drops of the fimbriation levels. The first drop is rather steep, whereas the second is more modest. Comparing this with the graph showing the population size, it becomes clear that the drops in the fimbriation levels coincide with an increase of the population size. Given that the proportion of fimbriate agents is a rough measure for the virulence of the agents, this graph suggests that the agents are able to adjust their virulence in an evolutionary process.

So far the data is encouraging, but by no means conclusive. The observed effect could be due to some other (as yet, unidentified) mechanism, or a process other than evolution. A simple, but not conclusive, way to test whether the observed effects are due to evolution is to repeat the simulations with mutation turned off. In this case, we would expect, at most, a very small growth of the population, and only at very early stages of the simulation. In such a random model without mutations one must expect that the initially heterogeneous population becomes increasingly homogeneous. This effect is simply due to the fact that the genetic diversity of the population becomes impoverished as agents die out. Eventually there will be only a single genotype left in the simulation model. We do not show the graph here, but experiments have confirmed that, indeed, there is no increase in the population size when mutations are turned off. In any evolutionary model this is a key test to generate a baseline against which the creative potential of evolution can be assessed.

A similar test of whether mutation is responsible for the adjustment of population size is to allow mutations, but to start with a completely homogeneous population (or a single agent only). In this case, the diversity is lowest at the start of the simulation, but increases over time driven by mutations. This setup tests the power of mutation to explore the space of possible behaviors. Simulation experiments show that starting with an homogeneous population restores the dynamics observed in Figs. 2.8 and 2.9, although it tends to take a bit longer before the transition from a low to a high population happens. (Again, we do not show the graphs here.)

In Sect. 2.6.2.1 we hypothesized that evolution can only work when the population of agents is partitioned into subpopulations that have limited contact with each other. Using the agent-based model we can now test this hypothesis. The hypothesis is that, if movement between populations is prevented then there should be no transition of the population size from low to high. Indeed, we would expect that the population would die out relatively quickly. Again, we tested this and, not showing the data here, we confirmed this prediction. Partitioning the population into weakly-interacting subpopulations is essential for both evolution and, indeed, survival of the population. If migration between the hosts is prevented then the total population will die out within a relatively short time.

2.6.2.9 Exploring the Behavior of the Model

All this suggests that the model truly shows an evolutionary effect, although one can never be absolutely certain, even in systems as simple as the present model. Our model seems to behave as expected and shows results that we can comfortably explain from our understanding of the system, but this is no foolproof confirmation.

Let us now take the leap of faith and accept that the model does indeed show evolution of fimbriae. Once this point is reached, the next question to be asked is, how does the model's behavior depend on the parameters? In this model of fimbriation there are so many parameters that it is essentially impossible to reasonably cover the entire parameter space with simulation. Luckily, this is not necessary either.

Many of the parameters of the model are arbitrary in that they simply scale the model. For example, the parameters F and C in (2.7) only determine how much nutrient is released and at which point the limit is reached. These parameters should be set so as to ensure that there is a sensible number of agents in each host (on average) while still maintaining acceptable run-times. Similarly, the lifetime of the agents and the amount of nutrient they require before reproducing are, to a large extent, arbitrary, as long as they do not lead to overly large or small populations. In order to be efficient in exploring the properties of the model these parameters should be kept fixed once practical values have been determined.

A parameter that recurs in most evolutionary models is the mutation rate. In general, the experience with evolutionary systems of this kind is that the precise value of the mutation rate does not matter too much, as long as it is not too small or too large. A reasonable value is best found by experimentation.[6]

While those parameters can be kept fixed once chosen, others need to be explored in more detail. As it turns out, the most crucial parameter of the model is the Hill coefficient in the response function (2.7). Figure 2.10 shows a few examples of a Hill function of the form $f(x) = x^h/(x^h + K^h)$. In the present case we chose the parameter K to be 0.5, just for illustration. From the point of view of our model, the interesting feature is that the inflection point of the function is at K and occurs for the same input value of x independently of h.

For this reason, the inflection point seems a good choice for the point at which the host response is triggered. For one, at the inflection point the slope is maximal; furthermore, this choice of trigger point of the host response also means that the optimum fimbriation level is always at the same value of K in the model and always gives the same amount of nutrient independent of the parameters C and h. The invariance of the best fimbriation levels makes it easy to compare different parameter values directly.

While the optimal fimbriation point does not change for different values of h, what changes considerably is the slope with which this optimal value is approached. The higher the Hill coefficient, the steeper the approach to the inflection point. Biologically this means that, close to the inflection point, a small change of the number of fimbriate agents could have a large effect on the host response. Such a change may well be due to a stochastic fluctuation alone.

Seen from an evolutionary point of view, the problem of E. coli/Martian mice is to adjust their switching functions such that they come close to the inflection point where the nutrient release is maximal, yet without crossing it. This task is more difficult than it seems at a first glance. One problem is the stochastic nature of the population. The number of fimbriate cells cannot be directly controlled by the cells, but only the (time-)average number of fimbriate cells. Particularly in small populations, at any given time there will possibly be quite significant deviations from the average.

---

[6]In the simulations here we used a value of 0.1 per reproduction event.

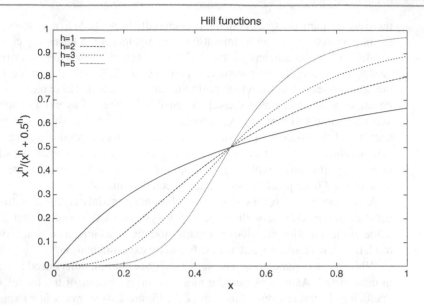

Fig. 2.10 An example illustration of a Hill function for various coefficients h. As h increases the function more and more resembles a step function

At any one time, the actual fimbriation levels can be significantly higher or lower than predicted by the mean. Therefore, the theoretical optimum point, close to the maximum nutrient release, is not feasible in practice. Stochastic fluctuations beyond the mean value would trigger a host response. Therefore, the population needs to keep some distance away from the optimum in order to avoid crossing the inflection point.

Assume now, for the sake of argument, that there is a mechanism for a sub-population to adapt towards a fitness gradient. (We have already indicated what this mechanism could be. For the moment, let us ignore the complexities and simply assume that sub-populations can evolve very much like individual organisms in a classical Darwinian individual selection scenario.) We can now ask: What does this fitness gradient look like? The main factor determining the fitness of a sub-population is the fimbriation rate. Up to a certain point, a higher fimbriation rate means that more nutrient is released, which in turn implies a higher population (we simply equate higher fitness with higher population numbers). Considering this, we see that below the optimum fimbriation point there is a positive fitness gradient for increasing fimbriation probabilities. We can assume that Darwinian actors will move along positive fitness gradients; that is, they will tend to increase their fimbriation levels. The problem starts once the population reaches the point where the host triggers a fully-fledged immune response. At this point, the fitness changes discontinuously from maximal to minimal, i.e., a small change of fimbriation takes us from the point of highest fitness to the point of worst fitness (where the sub-population becomes extinct).

This illustrates that the adaptive problem these populations face is rather difficult. The evolving populations do not "know" where the optimal fitness is, let alone that disaster strikes once they move even one step beyond that. Evolving sub-populations are thus driven up the fitness gradient, just to find themselves becoming extinct once they surpassed the point of optimal fimbriation. Luckily, for the reasons outlined above, sub-populations do not evolve that efficiently. Rather than being driven by adaptive changes within a group, the evolution of the system is mainly determined by migration between groups. This is a relatively inefficient mechanism and, once the population is globally relatively homogeneous, the rate of change will tend to be low.

The analysis of the fitness gradient shows that there is no barrier preventing the population from going "over the cliff" of optimal fimbriation. One would therefore expect that sub-populations would be prone to crossing the virulence threshold and triggering an immune response from time to time. Can we see this in the model?

Figure 2.11 shows the aggregate number of extinction events in the simulation shown in Fig. 2.9. The key aspect of this figure is the sudden decrease of the slope of the graph coinciding with sudden jumps in the population size in Fig. 2.11. This suggests that the population adapts, predominantly by decreasing the fimbriation

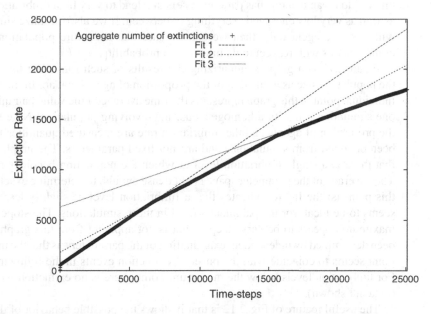

Fig. 2.11 The aggregate number of extinction events in the simulation run shown in Fig. 2.9. The *slope* indicates the rate of extinctions. The *fitted curves* in the figure are merely a guide to the eye. Clearly, the evolutionary transitions in Fig. 2.9 coincide with a reduction of the extinction rates in this model

levels. Yet, Fig. 2.11 indicates that the extinction rate never actually reaches zero.[7] There are occasional extinction events even in well adapted sub-populations. This means that becoming extinct is a part of life for our simulated agents. Evolution drives the population constantly over the cliff. Seen from the point of view of the host, this means that even commensal strains are virulent sometimes. This is simply an unavoidable result of the shape of the fitness landscape.

Next we want to know how similar individual runs are to one another, that is, what kind of variation we should expect for different random seeds? As the example simulations in Figs. 2.8 and 2.9 demonstrate, two simulation runs with the same parameters can have different outcomes. This, in itself, is an indication that the evolutionary process in this context does not necessarily lead to the optimum outcome. The question we can ask now is, how far away from the optimum will the population typically end up? Furthermore, how large is the variation between different runs of the model compared to the possible range of behaviors?

In order to address this question, we need to understand the possible behaviors of the system. One way to explore this is to take sets of hand-picked values for the parameters that are normally subject to change by evolution, and use them in simulations without evolution (i.e., with a mutation rate of 0). One should choose the parameters such that they cover the possible behaviors well. In the present case, this would mean that one has parameter sets that lead to very high fimbriation levels, as well as very low ones, and everything in between. If we also initialize simulations with a single agent only, then we can be sure that the entire population will be homogeneous with respect to its fimbriation probability.

Figure 2.12 is a graph summarizing the results of such runs. The figure shows the population size as a function of the proportion of agents that are fimbriate. Each individual point in this graph represents the time average of the value pair taken from one simulation run with a homogeneous, non-evolving population. Note that both the proportion of agents and the fimbriation rate are measured quantities that have been obtained from simulations, and are not fixed parameters. The graph suggests that there is a single fimbriation level at which the population becomes maximal. The coverage of the parameter space is not dense enough to determine exactly where this point is; the figure indicates that a fimbriation level of slightly less than 0.1 seems to be ideal, for the parameters used in these simulations. The slope near the maximum appears to be very steep. What is not apparent from this graph, but has been determined by independent examination of the parameters, is that the maximum point seems to coincide with the onset of extinction events in the following sense: For fimbriation levels below the maximum point, there is no extinction whatsoever (data not shown).

The useful feature of Fig. 2.12 is that it shows the possible behavior of the model for homogeneous populations of agents. Evolving, heterogeneous populations will do at most as well (in terms of population size) as the homogeneous ones. Due to

---

[7] Strictly, it only shows this for this particular run, but we have found this qualitative feature confirmed over all the simulation runs we performed.

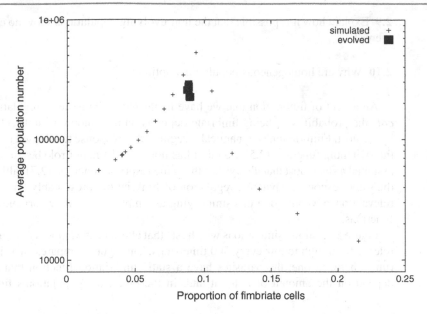

Fig. 2.12 Simulations with homogeneous populations and no evolution. This gives an indication of the achievable population sizes. Evolving populations remain well below the achievable population sizes

the peculiar dynamics of the evolving system, one would expect sub-populations to be driven over the cliff of extinction more or less continuously when the mutation rate is greater than zero. This will tend to reduce the population size of evolving populations when compared to non-evolving populations.

Figure 2.12 also shows evolving populations of agents. Two observations can be made: Firstly, compared to the possible behaviors of the model, i.e., the range of possible fimbriation levels and populations sizes, the results of the evolved populations cluster together rather tightly. Secondly, the evolving populations are sub-optimal in two senses: (i) they are below the optimality curve formed by the homogeneous non-evolving populations; and (ii) they are below the optimal fimbriation level.

From this, one is forced to draw the conclusion that evolution leads to sub-optimal outcomes, in the present case. Given the specific "fitness landscape" of the problem, this should come as no surprise. The peculiar fitness gradient that is abruptly truncated by the "cliff of extinction" leads to constant and apparently unavoidable extinction events; hence the sub-optimality (i). Stochastic fluctuations make it necessary for the average population to have a much lower average fimbriation probability. Populations that are too close to the optimum point may be pushed over the cliff; hence the sub-optimality (ii).

2.9 Explain how it is possible that the non-evolving populations show no extinction events?

2.10 Why are homogeneous populations optimal?

An aspect of fimbriation that we have neglected is the issue of regulation. In *E. coli* the probability of being fimbriate depends on the amount of nutrient released by the host. Fimbriation is dynamically regulated in response to released nutrient via the switching function, (2.5). So far, it has not played a major role because we have assumed a static host that always has the same response function, (2.7) (although not the same response). Dynamic regulation of fimbriation, presumably only becomes relevant once we abandon this simplifying assumption and allow variable response functions.

Figure 2.13 shows simulations with hosts that alternate the amount of nutrient they release from high to low every 500 time steps. The figure compares two simulation runs where, in one, the parasites have a static fimbriation function that does not depend on the amount of nutrient and, in the other, the population's fimbriation

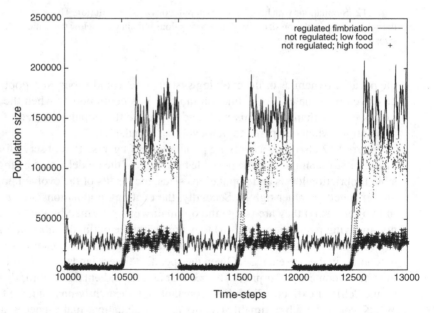

Fig. 2.13 Evolving the population under variable conditions. In these simulations, the host response changed every 500 time steps from low to high to low. The curve labeled "regulated" shows the time-course of a model that has been evolved under these conditions and is allowed a variable fimbriation rate. In contrast, the other two graphs show results from models that had a fixed fimbriation rate independent of the amount of nutrient released by the host. The labels "high" and "low" refer to the amount of nutrient to which the fixed switching rate is a response. This corresponds to the high and low nutrient rates in the "regulated" graph

function does. In practice, the three populations have been obtained from a single run with evolution enabled. We waited until this evolutionary run had reached what appeared to be a stationary behavior. We chose a time point and recorded the average parameter values for the agents that had evolved. We used this to initialize a single agent in a new run with evolution turned off. This is the curve labeled "regulated" in Fig. 2.13. The other two figures, labeled "high" and "low", were obtained as follows:

- Record the average amount of nutrient sensed by the regulated curve during the phase of high and low nutrient release respectively. From this calculate two switching rates corresponding to the two conditions.
- Start two new simulations with evolution turned off.
- Seed these simulations with a single agent only, with fixed switching rates corresponding to the high and low nutrient release conditions respectively.
- Finally, start a third simulation, also seeded with one agent corresponding to the measured parameters, but with a variable switching rate.

Note that the on-to-off switch is always unaffected by the nutrient contents, and hence is identical in all runs.

Figure 2.13 shows, somewhat surprisingly, that the regulated population is larger than the unregulated curves over the entire cycle. This is quite remarkable, given that the reaction of the adaptable population to the nutrient released at the high point is exactly the same as the one of the fixed response populations; similarly, at the low point of the cycle, one would expect the regulated population to be more or less equivalent to the population that has a fixed response to low levels of nutrients.

2.6.2.10 Summary

Even in the absence of knowledge of the values of parameters, we can still create a good understanding of the evolutionary processes of natural systems—such as Martian mice or fimbriation in *E. coli*. Of course, the computer model presented here is a crude approximation to reality. It does, however, provide some insights into the evolutionary dynamics of this kind of system. Real systems will have many additional layers of complication.

This particular example, worked through from the description of the problem to the analysis of the result, has demonstrated how to distill a computational model from the understanding of a system. We stressed that the guiding principle of this process should be simplicity. We also showed how to choose the agents, the agent-types, the interaction rules and how to represent the environment. Once the model is programmed it needs to be tested. This is a laborious process, but it can be made more efficient when a few rules are followed. We discussed how this can be done.

Even if the model developed in this section is perhaps not sufficiently accurate to reflect the "true" evolutionary history of real bacteria, at the very least it encourages us to ask further questions. An immediate question to ask is: What would happen if the host response was also subject to evolutionary change, rather then being user-determined? In particular, what types of response curves would be expected to evolve. However, it would go beyond the scope of this book to pursue this question any further.

References

1. Huse, G., Giske, J.: Ecology in Mare Pentium: an individual based spatio-temporal model for fish with adapted behaviour. Fish. Res. **37**, 163–178 (1998)
2. Bonabeau, E., Theraulaz, G., Dorigo, M.: Self-organization in social insects. Santa Fe Institute Working Paper 97-04-032 (1997)
3. Casti, J.: Would-Be Worlds. Wiley, New York (1997)
4. Ray, T.: An Approach to the Syntheses of Life. Oxford Readings in Philosophy, pp. 111–145. Oxford University Press, Oxford (1996)
5. Foundation, F.S.: GSL—GNU scientific library. https://www.gnu.org/software/gsl/. Accessed 22 June 2015
6. Kohler, T., Gumerman, G.: Dynamics of Human and Primate Societies. Oxford University Press, Oxford (1999)
7. Bak, P.: How Nature Works. Oxford University Press, Oxford (1997)
8. Venables, M., Bilge, U.: Complex Adaptive Modelling at Sainsbury. Business Processes Resource Centre (1998)
9. Tesfatsion, L.: Agent-based computational economics: growing economies from the bottom up. Artif. Life **8**(1), 55–82 (2002)
10. Wolfram, S.: Cellular Automata and Complexity. Addison-Wesley, Reading (1994)
11. Chu, D., Rowe, J.: Spread of vector borne diseases in a population with spatial structure. In: Proceedings of PPSN VIII—Eight International Conference on Parallel Problem Solving from Nature. Lecture Notes in Computer Science, vol. 3242, pp. 222–232. Springer, Birmingham (2004)
12. Dijkstra, E.W.: Chapter I: Notes on Structured Programming. Academic Press Ltd., London (1972)
13. Nowak, M.: Evolutionary Dynamics: Exploring the Equations of Life. Harvard University Press, Cambridge (2006)
14. Traulsen, A., Nowak, M.: Evolution of cooperation by multilevel selection. Proc. Natl Acad. Sci. USA **103**(29), 10952–10955 (2006). doi:10.1073/pnas.0602530103
15. Wilson, D.S.: A theory of group selection. Proc. Natl. Acad. Sci. USA **72**(1), 143–146 (1975)
16. Wilson, D.: Evolutionary biology: struggling to escape exclusively individual selection. Q. Rev. Biol. **76**(2), 199–205 (2001)
17. Sober, E., Wilson, D.: Unto Others, the Evolution and Psychology of Unselfish Behaviour. Harvard University Press, Cambridge (1998)
18. Gould, S.: The Structure of Evolutionary Theory. Belknap Press, Cambridge (2002)
19. Dawkins, R.: The Selfish Gene. University Press, Oxford (1989)

ABMs Using Repast Simphony

3

Researchers who wish to work with a computer-based model rather than a mathematical one will, sooner or later, have to build a model for themselves. They are then faced with the question of how to get started. Should they start with a blank sheet of paper and code from scratch; or take an existing similar program and adapt it; or should they use a modeling environment to help them? The answer in any particular situation will depend on a number of things: her programming ability and confidence; the availability of a suitable program to adapt; the availability of a suitable modeling environment. Building a program of any significant complexity completely from scratch is fairly rare these days—gone is the need to be able to write one's own sorting algorithms, write a linked list using pointers, or code a random number generator. Libraries of all of these programming elements are readily available for pretty-well all major programming language. Programs nowadays are usually written using extensive APIs (application programming interfaces) that considerably simplify common programming tasks. A large number of specialist APIs support particular programming areas, such as the building of graphical user interfaces (GUIs) or Internet programming. Programming ABMs is no different in this respect. Support for coding the commonly-occurring fundamental elements of a simulation—agents, their properties, their interactions, and the environments in which they exist—as well as external aspects—visual displays of the running model, data storage, visualization, and analysis—have been coded in a number of freely-available packages. Of course, it will still be the modeler's responsibility to design and code the specific details of her particular model, but the fact that much of the supporting infrastructure is already in place will allow them to spend most of her intellectual energy on the important details rather than the repetitive aspects that are common to most models. We feel, therefore, that the computer-based modeler's first port of call for creating a new model will almost inevitably be a package or environment that supports ABM programming, rather than writing a program from scratch.

© Springer-Verlag London 2015
D.J. Barnes and D. Chu, *Guide to Simulation and Modeling for Biosciences*,
Simulation Foundations, Methods and Applications,
DOI 10.1007/978-1-4471-6762-4_3

The down-side of using a modeling package is, of course, the initial hurdle of becoming familiar with the package's features and terminology. One of the purposes of this chapter, therefore, is to provide an easy entry for the reader into one particular modeling environment—Repast Simphony [1], which is part of *The Repast Suite* [2]. The Repast Simphony toolkit has extensive support for a wide range of modeling tasks. We will introduce it by exploring a subset of its most useful features through implementations of the Game of Life and Malaria scenarios that were discussed in Chap. 2. We will also include some practical exercises to help the reader gain confidence with using the toolkit.

We start, however, with a look at some of the basic elements of agent-based modeling in terms of programming terminology. This will be relevant both for researchers who wish to use a toolkit and those who want to code their model as a standalone program. Inevitably, we must assume at least some knowledge of programming and we will be using Java as the language of choice in the initial stages. The reader wishing to know more about basic, object-oriented programming in Java is directed to a general introductory text such as that by Barnes and Kölling [3].

3.1   Mapping Agent Concepts to Object-Oriented Languages

Object-oriented (OO) languages, such as Java and C++, are ideal for the implementation of agent-based models because it is possible to create a very close mapping between the ABM concepts of *agent type* and *agent instance* and the programming-language concepts of *class* and *object*. Even experienced programmers are often unsure of the exact difference between a class and an object, which is not surprising because the meanings often overlap. Their use within an ABM implementation provides us with a useful hook on which to hang an explanation. Consider the use of the word "mosquito" when discussing an ABM. Does the word mean an agent type or an agent instance? Of course it means both, so the correct answer depends upon the context in which it is used. If we say something like, "An infected mosquito transfers the infection to a human when it bites them," then we are clearly talking about the agent type because we are saying something that is true of all mosquitoes—we are saying that all infected mosquitoes pass on their infection when they bite humans. Even if the transmission were only with a certain probability, the statement would still be true of all mosquitoes so we are talking about the agent type here. Whenever we refer to the general characteristics of all agents of a particular type then this corresponds very closely to the idea of *class* in an OO programming language. We would therefore expect the class definition for a mosquito to contain source code that transfers infection to humans.

Conversely, if instead we say something like, "If the uninfected mosquito in location (6, 10) bites the human in location (7, 10) then the mosquito will become infected," we are talking about a particular instance of the mosquito type—the one at location (6, 10). We would not expect to see code in the class definition that singles out this particular location for special treatment. Whenever we refer to a particular agent we are referring to an instance of the agent type, which maps to an *object* created from the corresponding class—in fact, objects in OO programming are often also called instances and the terms "instance" and "object" are frequently used interchangeably.

However, we have to be a little careful here, because the effect being described for the particular mosquito instance at location (6, 10) is, in fact, a specific case of the more general, type-level characteristic that an uninfected mosquito biting a neighboring infected human can itself become infected. We would certainly expect to see that feature of the mosquito type represented in the mosquito's class definition, but here we happen to be talking about the behavior in the context of a specific mosquito instance.

3.1.1 Hand-Coding an Agent in Java

In Chap. 2, we noted that there are three main ingredients to any agent-based-model:

- the agents;
- the environment inhabited by the agents;
- the rules defining how agents interact with one another and with their environment.

Whether the model is programmed with the support of a toolkit or not, most of the programming effort will likely be spent on the code of the agent types, which includes their interactions. As described above, this means writing class definitions. While modeling tools might make it slightly easier to define the syntax of the code, the modeler still has to decide what is required in terms of agent features and the logic of agent interactions. Before looking at Repast in detail, we will give just a flavor of what is involved in defining an agent type from scratch, in the form of a mosquito.

Agent types consist of two main aspects:

- The properties or attributes possessed by all instances of that type. For example, whether a mosquito is infected and, if so, for how long it has been infected.
- The behaviors exhibited by all instance of that type. For instance, how a mosquito determines who to bite. Behaviors also include interactions with agents of different types. For instance, what happens to a human when bitten by a mosquito.

Sometimes, properties are held in common by all instances of the type—for instance, how long an infection period typically lasts in agents of a particular type.

Implementing an agent type means finding a way to map these all of these aspects on to program code. Fortunately, there is a simple mapping for object-oriented implementation languages:

```java
public class Mosquito {
    // The maximum infection length for all mosquitoes.
    private static final double INFECTION_LENGTH = 2.0;

    // Whether the mosquito is infected.
    private boolean infected = false;
    // The infection stage, if infected.
    private double infectionStage = 0;

    public void becomeInfected() {
        // Pick up infection from a human.
        infected = true;
        infectionStage = 0;
    }

    public void step() {
        // Update the infection stage, if infected.
        if(infected) {
            infectionStage += 1;
            if(infectionStage >= INFECTION_LENGTH) {
                infected = false;
            }
        }
        // Other actions taken on each step
        ...
    }

    public void bite(Human h) {
        // Bite a neighboring human.
        if(h.isInfected()) {
            becomeInfected();       // Pick up infection.
        }
        else if(infected) {
            h.becomeInfected();   // Transmit infection.
        }
        else {
            // Do nothing - neither is infected.
        }
    }

    public boolean isInfected() {
        return infected;
    }

    ...
}
```

Code 3.1 Partial Java implementation of the properties and behaviors of a mosquito agent

1. Each agent type will be defined as a class that encodes the characteristics of that type.
2. The properties of individual agents are implemented in the class as *instance fields*—i.e., *variables*.
3. The behaviors and interactions are implemented in the form of *methods*.
4. Where a particular property is shared between all agents of the same type it can be implemented as a single *class-level property*, shared by all instances.

Code 3.1 illustrates all of these agent-type to class mappings in the form of Java source code.

1. The class Mosquito models the mosquito agent type.
2. The agent has two individual properties: whether it is infected (a boolean true or false value) and the stage of its infection. Each agent instance will have its own copies of these properties, allowing some agents to be infected while others are not, and the infected agents to be at different stages of infection.
3. There are three behaviors and interactions in the class in the form of the methods becomeInfected, step and bite. The fourth, isInfected, is not a behavior but an enquiry method—it allows other agents to find out whether a mosquito is infected or not. Such methods are often called *accessor* methods.
4. As the time an infection lasts is a property that is the same for all mosquitoes, this has been defined in the form of a *class variable*, INFECTION_LENGTH. The static word distinguishes a shared property from the per-instance properties. The final word indicates that the value is fixed and will not change as the program is running.

The bite method illustrates interaction between a mosquito and a human agent. (The human agent's class will be implemented along similar lines to the mosquito's.) The code of the method assumes that a neighboring human agent has been identified and the behavior implements the consequences of biting the human. If the bitten human is already infected then the mosquito becomes infected. Similarly, if the bitten human is not infected before the bite but the mosquito is, then the infection is passed on to the human.

The step method is an example of a recurring theme of synchronous agent-based models: it contains the actions that will be taken on every time step of the model for every mosquito instance. In this example, any infection within a mosquito will progress and possibly expire.

Accessor methods, such as isInfected, do not alter an agent's state; they just tell us something about part of its state. In contrast, the becomeInfected method is known as a *mutator* method because it modifies (i.e., mutates) the state of an agent—a mosquito's infection properties are altered.

What our example class does not show is: (i) how the humans in the neighborhood of a mosquito are identified and passed to its bite method, and (ii) how and when the step method is called. In order to illustrate these aspects we would need to explore the simulation infrastructure as a whole, which would be too complicated for this point in the chapter. Instead, we prefer to introduce those elements when we take a detailed look at the Repast toolkit, which we will begin in the next section.

3.1 You might like to use the outline class definition in Code 3.1 to write a similar outline class for a human agent. Even if you do not consider yourself to be a Java programmer, this should still be possible if you can map the description of mosquito agents to the code elements shown. Note that the Human class will not require a bite behavior since humans pass on their infection passively and do not actively bite other mosquitoes or humans!

3.2   The Repast Suite

Repast stands for Recursive Porous Agent Simulation Toolkit. Repast Simphony [1] is part of a suite of modeling tools [2] but we shall use the term Repast loosely to refer to Repast Simphony in the rest of this chapter. Repast is a sophisticated modeling environment that supports not only the programming of models, but also their visualization and data analysis. While visualizations are often unnecessary once a model or simulation has been fully developed, their value in ensuring that a model is working as it should in the early stages of development cannot be overestimated, and Repast makes it particularly easy to create visualizations and other forms of displays. For the purposes of this chapter, we will assume that the reader has obtained the latest version of Repast Simphony and installed it according to the installation instructions.[1] Repast's developers maintain an active support mailing list that caters for both novices and experienced users.

Repast has been made available as a *plug-in* to the free open-source Eclipse Integrated Development Environment (IDE) [4]. However, it is important to appreciate that the Repast toolkit is essentially independent of Eclipse, and programs developed with it can be run without Eclipse. This becomes particularly important when we have moved beyond the development stage and want to start generating results from multiple simulation runs. As a general-purpose programming environment, Eclipse provides a convenient platform within which to create Repast programs. The downside of using Repast within a general-purpose IDE is that Eclipse contains a lot of features that don't necessarily have anything to do with agent-based modeling, and could therefore be distracting. It can also be difficult, at times, to appreciate which parts of the environment are Eclipse and which parts are Repast. The major positive, however, is that Eclipse provides a lot of sophisticated programming and debugging support that would probably be beyond the resources of the Repast developers to provide in addition to the modeling functionality. As modelers, therefore, we benefit significantly from the synergy between the two applications.

Whichever language is used to program them, agents in Repast correspond to Java objects at runtime in the way we have described above. Repast uses the same terminology as we have used up to this point: *Agents* are defined in terms of *Properties* and *Behaviors*. As we illustrated above, in object-oriented terminology, properties are the instance variables and behaviors are the methods of their class.

3.2.1   Repast Simphony and ReLogo

We will use Repast Simphony as a programming environment to support the coding of implementations of our models but instead of coding in Java we will use the related language Groovy [5]. Repast actually provides an additional layer of abstraction on Groovy, in the form of ReLogo, which is specifically designed to simplify the pro-

---

[1] At the time of writing, the current version is 2.3.

gramming of ABMs. ReLogo is partly a programming language, partly an API and
partly a set of agent-based concepts. It draws on a long history of the exploration
of decentralized systems and self-organizing phenomena that has its origins in the
influential book, *Turtles, Termites, And Traffic Jams: Explorations In Massively Parallel Microworlds* [6]. ReLogo draws a lot of its features and terminology from the
Logo programming language [7], and provides similar features to the StarLogo [8]
and NetLogo [9] programming environments.

The four key shared concepts and terminology among these related programming
environments are: *patches*, *links*, *observers*, and *turtles*:

- Patches are static (i.e., non-moving) agents arranged in a rectangular grid that are
  one way to represent the environment of a model.
- Turtles are agents that can move around within the environment. While the name
  "turtle" appears specific to a particular type of scenario, we shall see that this
  will not actually prevent us from applying our own abstractions to the properties
  and behaviors they are pre-defined to possess. A turtle has a location within the
  environment. This may be considered to be an integer coordinate pair designating
  a particular patch, or as a floating-point coordinate pair within a continuous environment. The appropriate choice will depend upon the requirements of the model.
  Turtles have pre-defined movement behaviors and a *heading* property to control
  their direction of movement.
- Observers are agents that exist outside the "physical" environment inhabited by
  the turtles. They have an overview of the model and tend to be responsible for
  setup and control of it.
- Links are agents that can be used to create network structures within a model,
  allowing both static and dynamic relationships to be maintained. We won't make
  use of them in this introductory material.

The following sections will add further detail to these concepts through the implementation of two models we discussed in Chap. 2: Conway's Game of Life and the
Malaria model.

3.3   The Game of Life Using ReLogo and Groovy

In this section we are going to walk through the creation of a full model of Conway's
Game of Life, the rules for which are described in Sect. 2.3. Modeling this will allow
us to explore the most basic features of Repast with ReLogo, as well as serving as a
gentle introduction to the Groovy programming language. This section is best read
at a computer with Repast running, so that the reader can either follow the code we
have already created, or create a version for herself. If the reader is creating their
own version then it should only be necessary to use a different name for the project
(e.g., GOL or MyGameOfLife); all other names within that project can then be the

same as we have used. The finished code of this example can also be downloaded
from this book's web site.

The Game of Life is conveniently simple to model: it consists of a collection of
identical agents arranged in the form of a grid. That means we can concentrate on
just two of the fundamental agent types in ReLogo: patches and the observer. The
final version of what we will build in this section is called GameOfLife in the
Repast workspace provided with this book.

3.3.1   Creating a New Model

A new project is started by either clicking on the *RL* ReLogo symbol in the Eclipse
toolbar (Fig. 3.1) or selecting *File → New → ReLogo Project*. Enter GameOfLife
for the new project's name and then *Finish*. The result in the *Package Explorer*
should look similar to Fig. 3.2. Repast makes it very easy to get started with a new
model because it generates a great deal of supporting code, saving us from having to
begin from scratch. For the Game of Life model, we shall be working with the two
Groovy files that have been generated in the src/gameoflife.relogo folder:
UserPatch and UserObserver. Similar files will be generated for every new
project and they will be perfectly adequate for our purposes in this one. However,
we shall also see in a later project how to make use of our own files—particularly
when working with multiple types of agent based on turtles.

Fig. 3.1 ReLogo symbols in
the Eclipse toolbar

Fig. 3.2 Folders in the
Project Explorer view of the
GameOfLife project

```
package gameoflife.relogo

import static repast.simphony.relogo.Utility.*;
import static repast.simphony.relogo.UtilityG.*;
import repast.simphony.relogo.Stop;
import repast.simphony.relogo.Utility;
import repast.simphony.relogo.UtilityG;
import repast.simphony.relogo.ast.Diffusible;
import repast.simphony.relogo.schedule.Go;
import repast.simphony.relogo.schedule.Setup;
import gameoflife.ReLogoPatch;

class UserPatch extends ReLogoPatch {

}
```

Code 3.2 Generated source of the `UserPatch` class

3.3.2   The UserPatch Class

By default, a new project is set up to support a square 33×33 2D grid of patches whose coordinates are $[-16, -16]$ to $[16, 16]$. This is a perfect starting point for the Game of Life model. Our implementation can represent one agent of the cellular automaton per patch. All we have to do is to implement the logic of a single agent in the `UserPatch` class and the initialization and time-step control in the `UserObserver` class. Code 3.1 shows the initial generated Groovy code of the `UserPatch` class to which we will be adding the properties and behaviors of the cellular agents. A key feature of the `UserPatch` class is that it `extends` an existing class of the ReLogo environment: `ReLogoPatch`. What this means, in effect, is that all the properties and behavior already defined in the `ReLogoPatch` class become directly available from the code we write inside the `UserPatch` class. We shall see the significance and value of this shortly. This is an example of object-oriented programming called *inheritance*.

A Game of Life patch-based agent maintains a simple state, indicating whether it is alive or dead (Code 3.3). Properties are defined by using the word `def` following by the name of the property, e.g.:

```
def alive
```

Notice that Groovy is very light on notation compared with languages such as Java and C++. For instance, there is no need to say what type of values the `alive` property will store. In addition, Groovy often does not require semicolons at the end of statements.

In order to visualize the states of the agents we use different colors for whether an agent is alive or dead, and set the color of the patch whenever the state of the cell is set. Since all agents will use the same two colors to visually indicate their state, we have defined `aliveColor` and `deadColor` to be class-level properties rather than per-instance properties, by including the word `static` in their definitions. The `setState` method illustrates how a particular color is associated with the cell's state.

```
class UserPatch extends ReLogoPatch {
    // Whether the cell is 'alive' or not.
    def alive
    // Colors for the two states.
    static def aliveColor = blue()
    static def deadColor = yellow()

    /**
     * Set the cell's state, and its color.
     * @param state - true (alive) / false (dead)
     */
    def setState(state) {
        alive = state
        if(alive) {
            setPcolor(aliveColor)
        }
        else {
            setPcolor(deadColor)
        }
    }
}
```

Code 3.3 Visualizing the state of a cell patch via its color

The UserPatch class effectively illustrates inheritance in action: Methods such
as blue and setPColor have been inherited from the ReLogoPatch class that
UserPatch is extending. ReLogo calls these inherited methods, *primitives*. All of
the fundamental agent types provide primitives that have been found to be generally
useful in a wide range of models, and we shall see more later throughout this chapter.
A complete list can be found in the *ReLogo Primitives Quick Reference* [10] that is
part of the documentation bundled with Repast Simphony.

Note that we did not provide a default value for alive property of a cell. Instead,
we prefer to initialize cells to a randomly-chosen true or false state. ReLogo provides
a class, called RandomHelper, that makes this easy to do so we will add the
following initialize method to UserPatch:

```
/**
 * Initialize the cell by randomly setting its state.
 */
def initialize() {
    setState(RandomHelper.nextDouble() < 0.5)
}
```

The nextDouble method returns a value in the range 0.0–1.0 inclusive, so there
is an equal chance of a cell initially being alive or dead. Note that it is also necessary
to add the following line to the beginning of the file along with the other import
lines, in order for the name RandomHelper to be recognized:

```
import repast.simphony.random.RandomHelper;
```

Rather than completing the cell agent now, it will be useful even at this very early
stage to check that what we have done so far is correct. It should already be possible
to view a grid containing a random pattern of alive and dead cells. For this we just
need to add some simple code to the UserObserver class.

```
class UserObserver extends ReLogoObserver {
    @Setup
    def setup() {
        clearAll()
        ask(patches()) {
            initialize()
        }
    }
}
```

Code 3.4 Setting up the model in the `UserObserver` class

3.3.3   The UserObserver Class

The `UserObserver` class is responsible for initialization of the model and controlling the actions that take place on each time step. The default implementation contains some useful hints for getting started but we won't make use of them here as they assume that there will be turtle agents in the model, and we are only making use of patches. Code 3.4 shows the code we have placed in `UserObserver` to initialize our model. This illustrates multiple features that will be common to models created with ReLogo:

- `UserObserver` extends `ReLogoObserver` so it inherits a lot of standard observer functionality from that class.
- We have defined a method called `setup` to initialize the model. The name of the method can be anything we want, but notice that the method has been preceded by the word `@Setup`. Words starting with the `@` character are called *annotations*. Annotations are instructions to the ReLogo system rather than code to be executed. This particular one tells the ReLogo system that the method defined immediately it after should be called to set up the model immediately before a simulation run is started—i.e., at time tick 0.
- The first action of the `setup` method is to call the method `clearAll` which is inherited from `ReLogoObserver`. The reason it calls this is that it is very common to run a model multiple times. The `clearAll` method ensures that any system state left over from a previous run is discarded. It is a good idea to adopt the habit of always calling `clearAll` as the first step of model setup.
- The next part illustrates two core features that are used throughout ReLogo code: `ask` and `patches`. Both are further examples of ReLogo primitives. `ask` is the primary means to execute a set of commands on an agent. The commands are specified between a pair of curly brackets, and it is not necessary to name the agent within the brackets. In effect, it provides a shorthand for calling methods on an agent. For instance, suppose the variable `pat` refers to one of our cell patches. Instead of writing:

```
pat.doA()
pat.doB()
pat.doC()
```

we would normally write:

```
ask(pat) {
    doA()
    doB()
    doC()
}
```

In the `setup` method, we have combined `ask` with the `patches` primitive. This idiom is a very powerful way to ask every patch in the model to perform the same set of commands, because `patches` returns a collection containing the complete set of patch agents. So the body of the call to `ask` is applied in turn to each individual member of the collection returned by the call to `patches`. We often describe this process as, "iterating over a collection of agents."

In summary, our `setup` method first discards any left over state from a previous run, and then asks every patch to initialize itself—i.e., to set itself to a random initial state. We are now in a position to visualize the initial state.

3.3.4   Visualizing the Model

To visualize the model we need to access the *Repast Runtime*, which is a separate interface used to set up and control the runs of models. Although we will access this runtime from within Eclipse, the interface is actually separate from it and can be used to run models independently. Clicking on the small black arrow immediately to the right of the green-and-white *Run* button in the Repast toolbar brings up a menu of *launchers* (Fig. 3.3). Selecting *GameOfLife Model* brings up the Runtime interface shown in Fig. 3.4. The large blank area is where the *Display* of the model will be shown once it has been initialized. For the time being, we can ignore the other two areas: the *User Panel* and the *Scenario Tree*.

A very nice feature of Repast is that we actually have to undertake no further work to create a visualization of the running model because the environment automatically provides default visualizations for all agent types. So the next step is to press the

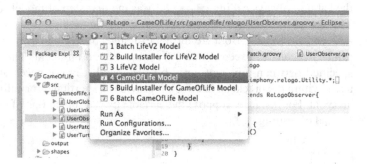

Fig. 3.3 The menu of launchers

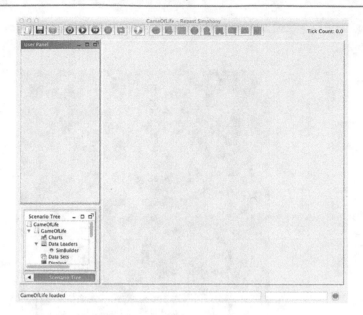

Fig. 3.4 The Repast Runtime interface

Initialize Run button. This will turn the large blank area into a display window, along with a title bar and window controls, but the initial state of the model will not yet be shown. To see that, we have to press the *Step Run* button once, which results in the method associated with the @Setup annotation being called for the first and only time—i.e., just before time tick 0 to set up the initial conditions of the model. Now the display should appear as in Fig. 3.5. The colors of the patches reflect the color settings we chose for the class-level properties aliveColor and deadColor in UserPatch. With the current state of our implementation, there is no point pressing either the *Start* button or the *Step Run* button again because we haven't yet coded what should happen on each time tick. The *Reset Run* button will return the display area to the blank state shown in Fig. 3.4 whenever we wish to re-run a model.

3.3.4.1 Inspecting Agents at Runtime

Double-clicking on a patch in the visualization brings up an inspection panel that shows the patch's properties. These include its location in the grid, pxcor and pycor, and its color pcolor, represented as a number in the range [0, 140).

3.2 Confirm that the initial state of the model is, indeed, random by resetting, initializing and single-stepping the model a few times. Is the proportion of alive and dead cells roughly equal each time, as we intended?

Fig. 3.5 An initial random state of the Game of Life cell patches

Vary the value used in the `initialize` method of `UserPatch` so that either a greater or a lesser proportion of cells will initially be alive. Confirm the effect visually via the display.

3.3 Confirm that the colors shown in the display are, indeed, the result of the colors we used in the `UserPatch`, by changing the Groovy code to use different colors and then launching the model again. The ReLogo `Utility` class defines a number of convenience names for typical colors, such as red, green, black, etc., or you can use numbers in the range [0, 140).

3.3.5 Implementing Cell State

It only remains to implement the actions of each cell agent at each time step in determining any state changes, and to coordinate these actions from the observer. However, before we do this we need to address an important issue that arises from the conflict between the assumption of concurrent updates to the cells, yet sequential processing of the cells during the update step.

3.3.5.1 The Issue of Update Ordering

In many models there is an underlying assumption that agents update their state simultaneously at each time step, so that the new state does not apply until the immediately following time step. However, in a programmatic implementation of a model, it is not practically possible for all agents to act concurrently. Rather, they update sequentially—one after the other—at notionally the same time step. Where an agent's behavior is independent of all others' behavior then this is not a problem, but this is rarely the case—agent-based models almost inevitably involve agent interaction. We pointed out in Sect. 2.3, for instance, that there will be a different outcome if one Game of Life agent updates its state during the current time step but before its neighbors have determined their own next state. What this means in practice for the model programmer is that agents often need to be written to maintain some additional properties (that are not strictly part of their state) in order to overcome the limitations of necessarily sequential execution of what should properly be simultaneous update.

The approach we will take with our Game of Life agents is to add an additional property to `UserPatch` called `aliveNextStep`. When an agent determines its new state from the state of its neighbors, it will actually store the calculated value in `aliveNextStep` rather than `alive`. That way, when an agent is asked for its current state by its neighbors when they are updating their state, it will still be able to return the "old" value from its `alive` property.

Once every agent has calculated its next state, the observer will then need to tell them to copy the value of `aliveNextStep` into `alive`, ready for a repeat of this whole process at the next time step.

3.3.5.2 Actions Repeated Each Time Step—The @Go Annotation

From the preceding discussion, we can recognise that each cell patch will need to be asked to do two things on each time step of the model: (i) determine the state it will have at the next time step without actually changing its state, and (ii) switch to the new state. We already know how to ask all the patches to carry out some actions, so Code 3.5 captures these ideas in a `step` method in `UserObserver`.

The `step` method further illustrates the `@Go` annotation. This is the way to indicate to the ReLogo system that the `step` method should be called at time step 1 and again on every subsequent time tick. In effect, it specifies the fundamental recurrent action of the model. As with the method following the `@Setup` annotation, we had a free choice over the name of the `step` method.

```
@Go
def step() {
    // Ask each cell to determine its next state,
    // based on the current state of its neighbors.
    ask(patches()) {
        determineNextState()
    }
    // Change the state of each cell to be
    // the value just determined.
    ask(patches()) {
        setNextState()
    }
}
```

Code 3.5 The @Go annotation and `step` behavior in `UserObserver`

```
count = 0
for each neighbor in the Moore neighborhood of this cell
    add 1 to count if the neighbor is alive
end for

if count is less than 2 or count is greater than 3
then aliveNextStep = false
else if agent is alive
then aliveNextStep = true
else if count is 2
then aliveNextStep = false
else aliveNextStep = true
end if
```

Code 3.6 Pseudo-code for determining a cell's next state

3.3.5.3 Determining the Next State

The process of determining the model's next state is the most complex behavior to be programmed. The process is outlined in Sect. 2.3 on p. 31. We restate it in Code 3.6 in pseudo-code that is closer to the form we want to program it in.

It is straightforward to translate this pseudo-code into Groovy, as shown in Code 3.7. Here we make use of the `neighbors` primitive which returns the set of patches in the Moore neighborhood of the current patch. The other code features introduced here are as follows:

- The variable `count` is defined as local to the `determineNextState` method. Its value is only available for the duration of the method and is lost afterwards. Hence, local variables are not the same as properties.
- Notice that the `alive` property accessed inside the `ask` call is that of the neighboring patch and not that of the patch being updated.
- The statement `count++` simply adds one to the current value of the `count` variable.

```
def determineNextState () {
    def count = 0
    ask ( neighbors ()) {
        if ( alive ) {
            count ++
        }
    }
    if ( count < 2 || count > 3) {
        aliveNextStep = false
    }
    else if ( alive ) {
        aliveNextStep = true
    }
    else if ( count == 2) {
        aliveNextStep = false
    }
    else {
        aliveNextStep = true
    }
}
```

Code 3.7 Querying neighbors to determine a cell's next state

- aliveNextStep is the additional property discussed in Sect. 3.3.5.2 and is defined along with the alive property near the beginning of the UserPatch class.

The remaining method to switch state for the next time step is simply:

```
def setNextState () {
    setState ( aliveNextStep )
}
```

which ensures that the patch's color is changed along with its alive state.

The model is now complete and can be run as before, but this time the *Start* button can be used to run the model through multiple steps. However, at least initially it is probably more useful to single step the model to observe the changes on each step more easily.

3.4 Experiment with the effect of varying the proportion of cells that are alive at the start of the model.

3.5 Change the initialization of the model so that the grid contains only a single glider pattern (see p. 34) and check that this behaves as it should.

Do this by having the initialize method of UserPatch set a cell's initial state to be *dead* (i.e., false). Then have the setup method of UserObserver use its patch primitive to access the individual patches of the glider shape and set their status to *alive*. For instance:

```
patch(0, 0).setState(true)
```

3.6 Confirm that the two-state update process described in Sect. 3.3.5.1 really is necessary by adding the following immediately before the closing curly bracket of the `determineNextState` behavior in `UserPatch`.

```
setState(aliveNextState)
```

3.7 Implement from scratch the following model of trees in a forest as a way to explore the features of ReLogo that we have covered so far. A forest is a rectangular grid of trees. A tree is a non-moveable agent that is always in one of three states: *alive*, *dead*, or *alight*. At each time step:

- A tree that is alive:

  - Has a probability, p_d, of becoming dead.
  - Has a probability, p_a, of becoming alight.
  - A tree that neither dies nor becomes alight has a probability, p_s, of changing the state of one of its neighboring dead trees to alive (equivalent to planting new seeds).

- A tree that is dead does nothing.
- A tree that is alight:

  - Changes the state of n of its neighboring alive trees to alight, where $0 \leq n \leq 8$.
  - Changes its state to dead.

Use different colors to visualize the states of the trees and use the `RandomHelper` class to generate suitable random numbers on each step.

Consider questions such as the following: Under what circumstances do all the trees in the forest die out? How high does the probability of regeneration need to be in relation to the probability of fire to prevent the forest dieing out?

Experiment with initializing the model with "fire breaks" between sections of the forest—i.e., lines of dead trees.

Consider whether there is any value in adding age-related features to the model. For instance, having trees only able to regenerate new trees if they have reached a certain age; and making the effect of very young (i.e., small) trees catching fire less significant than that of older (larger) trees. Does it make the model more "realistic"?

3.4 Malaria Model in Repast Using ReLogo

The Game of Life project introduced the basics of model creation within ReLogo but it only needed to make use of static patch agents. In this part of the chapter we will extend our exploration of ReLogo by developing a version of the Malaria model described in Sect. 2.4. In particular it will make use of turtles to implement the moving agents of the model: mosquitoes and humans.

3.4.1   The Malaria Model

Firstly, we will recap the main elements of the Malaria model from Sect. 2.4 that are relevant for this implementation.

- There are two types of agents: humans and mosquitoes.
- Agents live in a continuous rectangular environment of size $L \times L$.
- Agents can be in either of two states: infected or susceptible.
- If a mosquito is infected then all humans within a radius b of the mosquito will become infected.
- If a mosquito is susceptible, and there is an infected human within a radius b of the mosquito, then the mosquito will become infected.
- Once infected, humans remain infected for i_h time-steps, mosquitoes remain infected for i_m time steps.
- All agents move in their world by making a step no larger than s_h (humans) or s_m (mosquitoes) in a random direction.

In addition, we assume that agents do not have a limited lifetime. We will create all the model's agents at the start and the numbers remain constant throughout.

3.4.2   Creating the Malaria Model

We naturally start by creating a new ReLogo Project, which we will call *Malaria*. The Game of Life project relies heavily on default resources provided by ReLogo and source code that was automatically created as part of a new project, and the defaults saved us a lot of effort because they fitted the model well. However, for the Malaria model we will need to perform much more model-specific customization of the project because it is a much more sophisticated scenario. By default, some of the ReLogo resources are filtered out of the *Package Explorer* view within Repast and we need make them visible. This is configured via the small white down-arrow at the right of the *Package Explorer* panel (Fig. 3.6). The *Filters* option displays a panel of checkboxes and the box next to *ReLogo Resource Filter* must be *unchecked* (Fig. 3.7). When this has been done, an expanded version of the Malaria project should be visible (Fig. 3.8).

3.4.3   A Basic Human Turtle Agent

We need movable agents in the Malaria project so we will be using the features of ReLogo turtles. In addition, since we need two turtle agents—one for humans

Fig. 3.6 View filters are
accessed via the *white arrow*
shown *bottom center*

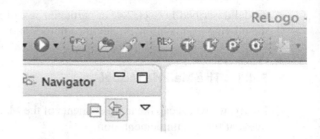

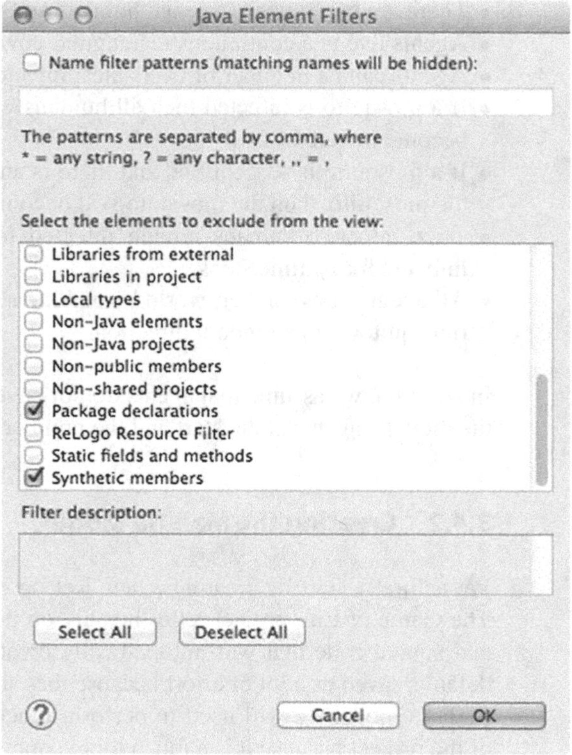

Fig. 3.7 Uncheck the *ReLogo
Resource Filter* check box

and one for mosquitoes—we cannot simply add all the agent code to the generated
UserTurtle class as we did with the UserPatch class in the Game of Life
model. Before developing the full model, we will start by creating a Human agent
with very basic functionality and illustrate some of the fundamentals of programming
ReLogo turtles.

Selecting the green *New Turtle* icon from the toolbar brings up a dialog to create a
new turtle agent type. We give it the name Human and enter malaria.relogo in
the package name before selecting the *Finish* button. A new Groovy class is created
that extends ReLogoTurtle, thus inheriting the default turtle primitives, properties
and behaviors.

Fig. 3.8 Unfiltered view of
the Malaria project

```
▼ 📂 Malaria
  ▶ 📂 src
  ▶ 📚 JRE System Library [Java SE 8 [1.8.0_4
  ▶ 📚 Groovy Libraries
  ▶ 📚 Groovy DSL Support
  ▶ 📚 Repast Simphony Development Librar
  ▶ 📂 srt-gen
  ▶ 📂 batch
  ▶ 📂 docs
  ▶ 📂 freezedried_data
  ▶ 📂 icons
  ▶ 📂 installer
  ▶ 📂 integration
  ▶ 📂 launchers
  ▶ 📂 lib
  ▶ 📂 Malaria.rs
  ▶ 📂 misc
    📂 output
  ▶ 📂 repast-licenses
  ▶ 📂 shapes
    📂 transferFiles
    📄 license.txt
    📄 MessageCenter.log4j.properties
    📄 model_description.txt
```

```
class UserObserver extends ReLogoObserver {
    @Setup
    def setup() {
        clearAll()
        createHumans(10) {
            forward(random(10))
        }
    }

    @Go
    def go() {
        ask(turtles()) {
            left(random(90))
            right(random(90))
            forward(random(10))
        }
    }
}
```

Code 3.8 Sample setup in the UserObserver class

In order to illustrate what this gives us already, we will make use of the template code that is included in the UserObserver class and remove the comments from it. We will make just one change to the setup method; instead of createTurtles(10) we will say createHumans(10). The result is shown in Code 3.8.

Whenever we define a new agent type, new primitives are automatically defined to simplify the manipulation of its instances. For instance, createHumans is provided as a primitive to create the given number of instances, and the associated code block is applied, in turn, to each newly created instance, as we saw with `ask(patches())` in the Game of Life project.[2]

By default, a new turtle will be placed at the centre of the model's space. Here, each human is told to move forward by random amount, using the `forward` and `random` primitives to provide initial separation between them. The sample `go` method uses the `turtles` primitive to ask each turtle to `turn` and `move random` amounts at each time tick of the running model.

While the actions in the sample `setup` and `go` methods in Code 3.8 do not relate directly to the Malaria model, they do provide a useful starting point that we will build on to develop the model further. If the model is launched then the visualization should look something like that shown in Fig. 3.9.[3] One thing to note is that a turtle is visualized by default using an arrow-head shape.

Fig. 3.9 Visualization of a basic turtle model

3.4.4 Model Parameterization

An important difference in complexity between the Game of Life and Malaria models is the need for *parameterization*—the ability to alter the runtime behavior of a model by varying one or more values that affect the configuration of the model. The only aspects of the Game of Life we might have wanted to vary were the size of the grid, the initial proportions of alive and dead cells and the colors used to represent cell states. All other aspects were determined by the rules of the model. For the grid size we used the default size and for the probability and colors we embedded those choices in the `UserPatch` class.

---

[2] An alternative primitive for creating a turtle is to `hatch` it, since real turtles lay eggs.
[3] Note that for the figure here we changed the color of the patches to white, which is not reflected in the code shown in Code 3.8.

In Code 3.8 it can be seen that we have already started to hard-code model parameterization values, as we used the value 10 (adopted from the template code) for the number of human agents to create. In the Malaria model there is a large number of parameters, such as numbers of agents, incubation periods and movement distances. A better approach than hard-coding parameter values across a range of classes is to group them together in one place and to give them names, so that they can be easily identified and modified as required. If we then write the source code to access these values via their names then it will be a lot easier both to read the code and to run the model multiple times using different parameter-value combinations.

There are two main places in a ReLogo project where parameters should be specified: one is the UserGlobalsAndPanelFactory class, discussed in Sect. 3.4.5, and the other is the parameters.xml file, which is discussed in Sect. 3.4.8. We will identify the following elements as candidates for parameterization in the Malaria model, based on the description in Sect. 2.4, along with indicative values, although in the initial stages of development we will tend to use smaller values for population values, for instance, for ease of testing and runtime speed:

- humanInitialInfectionProbability: 0; probability of a human initially being infected;
- humanInfectionPeriod: 40; infection period of humans (i_m);
- humanMaximumMovement: 9; maximum movement radius of humans (s_h);
- mosquitoInitialInfectionProbability: 0.5; probability of a mosquito initially being infected;
- mosquitoInfectionPeriod: 2; infection period of mosquitoes (i_m);
- mosquitoInfectionRadius: 1; infection radius of mosquitoes (b);
- mosquitoMaximumMovement: 1; maximum movement radius of mosquitoes (s_m);
- numHumans: 5000; number of humans;
- numMosquitoes: 10000; number of mosquitoes;
- size: 200; square-side size of the environment (L);
- stopTime: 5000; stop time of the simulation.

3.4.5   The UserGlobalsAndPanelFactory Class

Repast automatically creates a UserGlobalsAndPanelFactory class in a new project as one place to locate parameterization values. As its name suggests, variables defined here are intended to be globally available to all classes within the project. Code 3.9 shows the code we will add to this class to give names to the parameterization values for human agents. Given these variables we can now adjust the createHumans call in UserObserver to use the numHumans global as follows:

```
createHumans(numHumans) {
    ...
}
```

```
public class UserGlobalsAndPanelFactory
        extends AbstractReLogoGlobalsAndPanelFactory {
    public void addGlobalsAndPanelComponents() {
        addGlobal("humanInitialInfectionProbability", 0.5)
        addGlobal("humanInfectionPeriod", 40)
        addGlobal("humanMaximumMovement", 1)
        addGlobal("numHumans", 10)
    }
}
```

Code 3.9 Parameterization of Human Agents via global variables

3.4.6 The @ScheduledMethod Annotation

In Sect. 3.3.5.1 we described the importance of ensuring that state changes occurring across two time steps were not applied during the current time step, otherwise they would incorrectly affect the state changes required by the model. The same principle applies in this model, too. For instance, consider the case of two mosquitoes in the neighborhood of an uninfected human, where one of the mosquitoes is already infected but the other is not. The update ordering will significantly affect the outcome if a human's infection state is changed immediately, because if the uninfected mosquito feeds on the human first then it will not become infected, but if the infected mosquito feeds first then the other will pick up the infection from the newly infected human. In order to avoid inconsistencies such as this, we must defer making an agent newly infected until the next state of all the agents has been determined. Furthermore, when we update the infection stage of a human or mosquito we have to be careful to distinguish between the current infection status and any acquired newly at the current step. These are issues that can be managed best from the observer rather than the agents themselves. Code 3.10 shows the next stage in development of the observer to take account of this requirement.

The first point to notice is that we are no longer using the @Setup and @Go annotations for initialization and step control in the observer but @ScheduledMethod instead. This is because fine control over the passage of time is important in the Malaria model, whereas it wasn't particularly in the Game of Life model. Recall that persistence of infection in the agents is associated with time periods, so we need to synchronize updating their infection stage with the passage of time in the model and ensure that it is performed exactly once per time tick. The @ScheduledMethod annotation gives us the control we need. It is used to specify at what time a method in the observer is first called, and at what intervals.[4]

We have used the annotation with two separate methods: step and update. Both are called at regular 1-tick intervals but their starting points are offset by 0.5 so that each update action will occur in between each pair of step actions. This is one way that we can avoid the state-change problems described above—the step action determines whether new infections are transmitted or acquired and then the

---

[4]Time in Repast is measured as a double-precision floating-point number so the values in the arguments to @ScheduledMethod have to be written as 0d and 1d rather than 0 and 1.

```
class UserObserver extends ReLogoObserver {
    @ScheduledMethod(start = 0d)
    def setup() {
        clearAll()
        createHumans(numHumans) {
            setxy(randomXcor(), randomYcor())
            setShape("person")
            init()
        }
    }

    @ScheduledMethod(start = 1d, interval = 1d)
    def step() {
        ask(humans()) {
            act()
        }
    }

    @ScheduledMethod(start = 1.5d, interval = 1d)
    def update() {
        ask(humans()) {
            postStepUpdates()
        }
    }
}
```

Code 3.10 Scheduling agent activity in discrete time steps

update action changes the state of the agents ready for the next step. Separation in this way also means that it will be much easier to measure infection levels accurately if we count the numbers of infected agents only at whole-number timesteps, as we shall see later.

Notice that we have used the generic names act and postStepUpdates for the observer to interact with behaviors in the Human class. The idea is to *decouple* the design of the agent classes from the observer class—in other words, the observer does not need to encode detailed agent knowledge to control the model. Rather, the specific details are encoded distinctly in the specific agent types. For instance, the act method for human agents will cause them to move, while the one for the mosquito agents will cause them both to move and to feed on neighboring humans. We achieve a better design if responsibility for coordinating those differences is placed in the agent classes rather than the observer.

In the setup method, the actions associated with createHumans set the initial position of a human to be a random coordinate using the setxy, randomXcor and randomYcor primitives. The latter two primitives are aware of the dimensions of the model's world and will always generate valid values within its dimensions. Note that the coordinates used here are floating-point values within a continuous 2D coordinate space and not the coordinates of individual patches. In fact we don't make any explicit use of patches in this model. ReLogo allows us to interpret the default space occupied by turtles either as a rectangular grid of discrete patches, as we did with the Game of Life model, or as a continuous 2D space as we are doing here.

Finally, we have used the setShape primitive of a turtle to change the arrow-head shape used by default to visualize all turtles. The argument, "person", takes advantage of a default set of image files provided with ReLogo projects. These can be found in the `shapes` folder of the project. In fact, rather than calling `setShape` on each individual turtle as it is created, the observer offers the primitive `setDefaultShape` to set the default shape for a particular type of turtle. For instance:

```
setDefaultShape(Human, "person")
createHumans(numHumans) {
     ...
}
```

Once the position and shape of a Human agent have been set, its `init` method is called to set its initial infection stage.

3.4.7 The Human Agent

In this section we will implement the Human agent by coding details of its infection and movement. We will find that much of this effort can be re-used for the Mosquito agent.

We have already noted that careful thought is required to ensure that the infection effects described by the model are accurately implemented in the project code, and this will require further properties and methods that are not part of the model's description. The model suggests that human agents have two properties: whether they are infected and what stage of infection they are at (a simple integer counter). They also have three behaviors associated with infection: they can become infected if bitten by a mosquito; their infection stage can progress if they are infected; and they can become susceptible if they have reached the end of their infection period. They also have a movement behavior.

In order to separate the infection state at the current time step from the state at the next time step, we will add an extra property, called `newlyInfected`, and a method called `activateNewInfection` that changes the infected state between the whole-number time steps. Code 3.11 shows the completed Human agent.

The `init` method uses the `nextDouble` method of `RandomHelper` to decide whether the agent should start off with an infection and, if it does, then sets the infection stage to a random point. This is an important step to avoid introducing a bias into the model whereby all the infected agents keep step with each other, which would be unrealistic. The `random` method is used to return a value in the range [0, *humanInfectionPeriod*) and this is then adjusted by one to give a value in the range [1, *humanInfectionPeriod*].

The `act` method will be called by the observer for each whole-number time step of the model run. It simply calls the `move` method to move the human to a new location. The `move` method determines a random distance for the agent to move, and a random direction. A feature of ReLogo turtles is that they have both a position and a heading. The `left` and `right` primitive allow the heading to be adjusted by an angle expressed in degrees.

```
class Human extends ReLogoTurtle {
    def infected = false
    def infectionStage = 0
    def newlyInfected = false
    def infectedColor = orange()
    def susceptibleColor = blue()

    def init() {
        if(RandomHelper.nextDouble() <=
                    humanInitialInfectionProbability) {
            becomeInfected()
            infectionStage = 1 + random(humanInfectionPeriod)
        }
        else {
            becomeSusceptible()
        }
    }

    def act() { move() }

    def postStepUpdates() {
        updateInfectionStage()
        activateNewInfection()
    }

    def becomeNewlyInfected() { newlyInfected = true }

    private def move() {
        def distance = RandomHelper.nextDoubleFromTo(0,
                            humanMaximumMovement)
        left(random(360))
        forward(distance)
    }

    private def updateInfectionStage() {
        if(infected) {
            infectionStage++
            if(infectionStage > humanInfectionPeriod)
                becomeSusceptible()
        }
    }

    private def activateNewInfection() {
        if(newlyInfected) {
            becomeInfected()
            newlyInfected = false
        }
    }

    private def becomeInfected() {
        infected = true
        infectionStage = 0
        setColor(infectedColor)
    }

    private def becomeSusceptible() {
        if(infected) {
            infected = false
            infectionStage = 0
        }
        setColor(susceptibleColor)
    }
}
```

Code 3.11 Completed Human Class

Notice that move and other methods have the word private in their definition. The idea is to make it clear that this method can only be called by the agent itself and not by any other type of agent. While it is not strictly necessary to add this, it does make it a little clearer to the reader which methods are meant to be used by other agents in the model and which are not.

The postStepUpdates method is activated between the whole-number time ticks to update the infection stage and then apply any new infection contracted on the preceding time step. The order of these is important because the old infection state must be updated first before any new infection is recorded.

The updateInfectionStage method conditionally increments the infectionStage property and then checks its value against the global parameterization variable humanInfectionPeriod to see whether the agent has reached the end of the infection. The becomeInfected and becomeSusceptible methods are straightforward and simply set the two main infection properties. They also set the agent's color for the visualization.

At this stage it is worth launching the model to check that all the changes we have made are working as expected, before adding mosquitoes. Of course, without any mosquitoes in the model, any initial infection within the human population will completely disappear once the full human infection period has elapsed, but observing this is still useful as it does provide some evidence that the infection-stage elements of the human agents are working.

```xml
<?xml version="1.0" encoding="UTF-8" ?>
<parameters>
  <parameter name="randomSeed"
             displayName="Default Random Seed"
             type="int" defaultValue="__NULL__" />
  <parameter name="default_observer_minPxcor"
             displayName="default_observer_minPxcor"
             type="int"
             defaultValue="-16"/>
  <parameter name="default_observer_maxPxcor"
             displayName="default_observer_maxPxcor"
             type="int"
             defaultValue="16"/>
  <parameter name="default_observer_minPycor"
             displayName="default_observer_minPycor"
             type="int"
             defaultValue="-16"/>
  <parameter name="default_observer_maxPycor"
             displayName="default_observer_maxPycor"
             type="int"
             defaultValue="16"/>
</parameters>
```

Code 3.12 The default parameters.xml file, specifying the world's dimensions

3.8 Implement the model described up to this point and launch it. Use the *Step Run* button to observe the movement of the humans and the disappearance of the infection after about 40 steps. Note that the *Tick Count* shown at the top-right the display moves in increments of 0.5 as the two scheduled methods are called alternately.

3.9 Reset the model and initialize it. Double-click on one of the agents and observe the changes in its properties on each step. What happens to the agent's coordinates when it moves beyond an edge of the world? Does it disappear or reappear somewhere else? Can you explain what is happening here?

3.4.8 `parameters.xml` and the `SimBuilder` Class

Before we introduce the Mosquito class it is worth looking further at ways to capture the parameterization of models. In Sect. 3.4.5 we described how to give names to model parameterization values by locating them in the `UserGlobalsAnd-PanelFactory` class which then makes those names globally available throughout the project's classes. An alternative location for parameterization values is the file `parameters.xml` which can be found in the folder `Malaria.rs` in this project.[5] Code 3.12 shows the default contents which are mainly concerned with specifying the default coordinate space of the model's world.

The main difference between using `parameters.xml` or `UserGlobals-AndPanelFactory` to catalog parameters is that `parameters.xml` is not part of the project's source code. That makes it much easier to run the model multiple times with different parameter values because the source code does not have to be changed between each run. We will see how this is useful in practice in Exercise 3.11.

The question arises: How does the program access these values? The answer can be found in the `SimBuilder` class which is located in the `malaria.context` folder of the project (Code 3.13). The key statement in this class is the following, which reads in the information stored in `parameters.xml`:

```
Parameters p = RunEnvironment.getInstance().getParameters();
```

Then the individual values can be accessed via calls to the `getValue` method of the `Parameters` object:

```
int minPxcor = p.getValue("default_observer_minPxcor");
```

The `SimBuilder` then uses the dimensions to create a world of the specified size.

3.10 Edit the `parameters.xml` file and change the value of `default_observer_maxPxcor` from `16` to something much larger. Launch the model and confirm that world is no longer 33 × 33 as it has been previously. What do you notice about the size of the agent visualizations as a result of this change?

---

[5]The folder is named from the project name with a `.rs` suffix added.

```
public class SimBuilder implements ContextBuilder {
    public Context build(Context context) {
        if (RunEnvironment.instance.isBatch()) {
            UserGlobalsAndPanelFactory ugpf =
                    new UserGlobalsAndPanelFactory();
            ugpf.initialize(new JPanel());
            ugpf.addGlobalsAndPanelComponents();
        }

        Parameters p =
                RunEnvironment.getInstance().getParameters();

        // NOTE: minPxcor and minPycor must be <= 0
        int minPxcor = p.getValue("default_observer_minPxcor");
        int maxPxcor = p.getValue("default_observer_maxPxcor");
        int minPycor = p.getValue("default_observer_minPycor");
        int maxPycor = p.getValue("default_observer_maxPycor");
        RLWorldDimensions rLWorldDimensions =
                new RLWorldDimensions(minPxcor, maxPxcor,
                                      minPycor, maxPycor);

        LinkFactory lf = new LinkFactory(UserLink);
        TurtleFactory tf = new TurtleFactory(UserTurtle);
        PatchFactory pf = new PatchFactory(UserPatch);
        ReLogoWorldFactory wf =
                new ReLogoWorldFactory(
                    context,"default_observer_context",
                    rLWorldDimensions, tf, pf, lf);

        ObserverFactory oF =
                new ObserverFactory(
                    "default_observer",UserObserver,wf);
        Observer dO = oF.createObserver();

        context.add(dO);
        return context;
    }
}
```

Code 3.13 Default contents of the `SimBuilder` class

3.11 In the Repast runtime GUI display the *Parameters* panel by selecting *Window* → *Show View* → *Parameters*. Change one or more of the coordinate limits (but 0 must be included within the range of both dimensions), initialize the model and check that the changes have taken effect. Then stop the model, re-edit and try again. Note that the program does not have to relaunched between these changes, whereas edits to the program code would have required quitting the Runtime window and relaunching it.

In general, we will prefer to place model parameterization names and values in `parameters.xml` because it allows us to re-run a model with different values without having to change the source code. However, it is often more convenient in the early stages of development to code directly in the `UserGlobalsAndPanel-Factory` class, as we have done, and then later move the values outside the code.

```
public class UserGlobalsAndPanelFactory
        extends AbstractReLogoGlobalsAndPanelFactory {
    public void addGlobalsAndPanelComponents() {
        // Load the configuration from the parameters' file.
        Parameters p =
            RunEnvironment.getInstance().getParameters();

        addGlobal("humanInitialInfectionProbability",
                p.getValue("humanInitialInfectionProbability"))
        addGlobal("humanInfectionPeriod",
                p.getValue("humanInfectionPeriod"))
        addGlobal("humanMaximumMovement",
                p.getValue("humanMaximumMovement"))
        addGlobal("mosquitoInitialInfectionProbability",
                p.getValue("mosquitoInitialInfectionProbability"))
        addGlobal("mosquitoInfectionPeriod",
                p.getValue("mosquitoInfectionPeriod"))
        addGlobal("mosquitoInfectionRadius",
                p.getValue("mosquitoInfectionRadius"))
        addGlobal("mosquitoMaximumMovement",
                p.getValue("mosquitoMaximumMovement"))
        addGlobal("numHumans", p.getValue("numHumans"))
            addGlobal("numMosquitoes",
                        p.getValue("numMosquitoes"))
        addGlobal("stopTime", p.getValue("stopTime"))
    }
}
```

Code 3.14 Setting the values of global variables from `parameters.xml`

It remains to illustrate how values in `parameters.xml` can be made available as global variables in order to achieve the best of both worlds. Code 3.14 shows the new version of `UserGlobalsAndPanelFactory` that uses the technique seen in `SimBuilder` to obtain the parameters, from which the globals' values are then set.

3.4.9 The Mosquito Agent

We can now add the `Mosquito` agent to the model, and most of it can be added fairly easily because it has strong similarities to the `Human` agent. The distinctive elements are the mosquito-specific parameterization names and the feeding behavior—a mosquito will bite any human agents within its infection radius. Code 3.15 shows an outline of just the distinctive elements of the `Mosquito` class.

When we define a new class, ReLogo assumes that the plural of class name just adds an "s" so that it can provide primitives such as humans and `createHumans`. However, the plural of "mosquito" is "mosquitoes" so this needs to be specified via the `@Plural` annotation preceding the class header, as in Code 3.15.

The logic of the `feed` behavior requires that a mosquito identify all human agents within a defined radius. The `inRadius` primitive makes that very easy. We pass it the complete collection of human agents and it returns the set of only those that

```
@Plural("Mosquitoes")
class Mosquito extends ReLogoTurtle {
    ...

    private def feed() {
        def victims = inRadius(humans(),
                                mosquitoInfectionRadius)
        def catchInfection = false
        if(infected) {
            ask(victims) {
                if(infected) {
                    catchInfection = true
                }
                becomeNewlyInfected()
            }
        }
        else {
            // See if the mosquito is infected by a neighbor.
            def infectedHumans = victims.collect() {
                infected
            }
            catchInfection = !infectedHumans.isEmpty()
        }
        // Did this mosquito catch an infection from a neighbor?
        if(catchInfection) {
            becomeNewlyInfected()
        }
    }
}
```

Code 3.15 The @Plural annotation and feed behavior of mosquitoes

are within the specified distance. If the mosquito is infected it must then pass on the infection to the identified victims. However, there is the added requirement that if any of those neighbors is already infected, the mosquito will also (re-)acquire the infection. The logic for that is controlled via the local catchInfection variable which is set if any of the human neighbors is infected. Note the following statements:

```
def infectedHumans = victims.collect() { infected }
catchInfection = !infectedHumans.isEmpty()
```

The collect method iterates over the human agents in the victims collections and creates a new collection of those whose infected property is set to true. If the resulting infectedHumans collections is not empty then the mosquito will be infected.

Code 3.16 shows the additions required in the observer to include mosquito agents. It also makes use of the global variable stopTime to tell the ReLogo runtime at what time the model should end (via the endAt method). Notice that both the step and update methods use the general turtles primitive to call act and postStepUpdates, rather than the more specific humans and mosquitoes.

```
class UserObserver extends ReLogoObserver {
    @ScheduledMethod(start = 0d)
    def setup(){
        clearAll()
        RuntimeEnvironment.getInstance().endAt(stopTime)
        setDefaultShape(Human, "person")
        createHumans(numHumans) {
            setxy(randomXcor(), randomYcor())
            init()
        }
        createMosquitoes(numMosquitoes) {
            setxy(randomXcor(), randomYcor())
            init()
        }
    }

    @ScheduledMethod(start = 1d, interval = 1d)
    def step() {
        ask(turtles()) {
            act()
        }
    }

    @ScheduledMethod(start = 1.5d, interval = 1d)
    def update() {
        ask(turtles()) {
            postStepUpdates()
        }
    }
}
```

Code 3.16 Final version of the observer of the Malaria model

However, ordinary turtles do not have an `act` method by default and so the ReLogo editor within Repast will place an underline beneath those method calls because it cannot be sure that they are valid. This can be a useful hint that a method name has been mis-spelled, for instance. However, ReLogo double-checks at runtime for the existence of an appropriate method and, since all the turtles in the model at runtime *do* have an `act` method, this code works correctly. This version of the project now fully implements the Malaria model we have described.

3.12 Experiment with the model by running it and observing the behavior. You will obtain the most value from the experimentation if you set the values in `parameters.xml` to small values; for instance, tens of agents rather than hundreds. It is also visually interesting to remove the call to `move` from the `act` method of the humans and observe how the movement of the mosquitoes causes the infection to be carried between the static humans.

3.4.10 Runtime Memory Configuration for Large Models

Models with large numbers of agents will require a lot of memory at runtime and the default settings of the Repast environment may be insufficient to run models like this

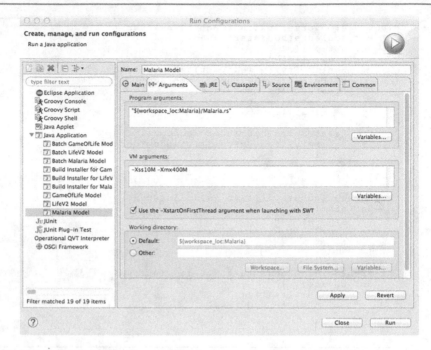

Fig. 3.10 Runtime configuration of models

one efficiently. Model configurations may be set via the *Run → Run Configurations* menu item which brings up the dialog shown in Fig. 3.10. Full details of the configuration options available from this multi-panel window are beyond the scope of this chapter, but the key one to configure runtime memory allocation may be found under the *Arguments* panel. The *VM arguments* area specifies how much memory should be made available to the *Java Virtual Machine* that runs ReLogo models. The value of 400M means 400 Megabytes. For large models, such as the Malaria one, a larger value of 2000M (i.e., 2 Gigabytes) or more may be appropriate if that amount of memory is available on the machine.

3.4.11 Recognizing the Common Elements of the Agents

When we created the Mosquito class in Sect. 3.4.9 we were able to copy most of its code from the Human class. While this process saved a lot of time, it resulted in a considerable amount of duplication of similar code. In general, duplicate code on that scale is something that is best avoided if possible, and object-oriented programming languages make it particularly easy to avoid when it is the result of strong similarities between classes such as Mosquito and Human. The duplication arises because the model consists of two very similar agents that are infectable and that move around.

The only differences between them in these respects are the length of their infection periods and the amount they can move at each time step. These relatively minor differences could easily be accommodated via additional properties in the agents, for instance, allowing the rest of the code to be shared.

The mechanism for sharing common code in object-oriented languages in cases like this is to use *inheritance*; in other words, we would define a non-specific agent class, called something like `InfectableAgent`, as a *superclass* and then the specific agent classes `Human` and `Mosquito` as *subclasses* of `Infectable-Agent`. The `Mosquito` class alone would contain the feeding behaviour that is distinctive of that particular agent. While we won't present here the code details of how this would be done, a version of the project using inheritance is provided with the book's materials.

3.4.12 Data Sets and Chart

With a large number of agents, a visualization of every agent is both very slow and not a particularly good way to obtain an appreciation of what is happening in general within the model—consider whether it is easy it is to tell if the infection levels are growing or diminishing, for instance. What we really need is an *aggregation* of the state of the different agents. This is where the *Data Sets* and *Charts* features of Repast can help us. Therefore, from now on we will abandon the spatial display and create a chart of the infected proportions of each type of agent. To support this, we will add the following method in both Human and Mosquito:

```
public int infectionCount()
{
    return infected ? 1 : 0;
}
```

Since it returns a value of 1 if the agent is infected and 0 if it is not infected, this method will allow us to generate counts of infected agents and, therefore, proportions.

Firstly, it is necessary to create a *Data Set* that can then be plotted on a chart. Data sets can also be output to file, if required. In the Scenario Tree of the Repast runtime GUI select *Data Sets*, right-click and select *Add Data Set*. A panel will be shown offering the chance to give the data set an ID and Type. We will use *Infected proportions* for the ID and select *Aggregate* from the Type dropdown menu, because we are interested in data aggregated over the collection of agents and not just from a single agent. On selecting the *Next* button we have to indicate the source of the data for this data set. We will want to plot the progress of the infection proportions over time so we select the *Tick Count* check box then select the *Method Data Sources* tab. Figure 3.11 shows the completed elements of the sources. We have entered meaningful names under *Source Name*, selected `Human` and `Turtle` from the *Agent Type* dropdown, `infectionCount` from the *Method* dropdown and `Mean` as the *Aggregate Operation*. The Data Set configuration should be saved so that it is retained with the project.

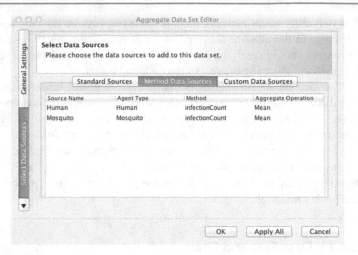

Fig. 3.11 The Data Set Editor

Once a data set has been established, a chart can be designed to make use of it via the *Charts* element of the *Scenario Tree*. Right-clicking offers *Add Time Series Chart* and *Add Histogram Chart*. The most appropriate type will clearly depend both on what data is being plotted and what information is required from the chart. We will use a time series chart to visualize how the infection proportions vary over time.

The chart should be given a name and the data set it is based on is selected from the *Data Set* drop-down. The *Chart Data* dialog offers the individual elements of the data sets for inclusion on the chart (Fig. 3.12). Once created, the scenario should be saved for future runs.

After the model has been initialized, the chart view can be accessed via *Window* → *Show View* → *Chart-name* (where *Chart-name* is the name chosen for the chart). Figure 3.13 shows the chart of infected proportions of agents for the first 1500 time steps of a single run.

3.4.13 Outputting Data with Text Sinks

The data plotted on a chart represents the results of a single run of the model and is lost at the end of the run. However, by setting up a *Text Sink* from the Scenario Tree, we can preserve the results both for future investigation and either comparative or aggregated analysis over multiple runs. We could easily perform that analysis with a tool such as *gnuplot* (Chap. 5).

Right-clicking *Text Sinks* in the Repast runtime offers *Add Console Sink* and *Add File Sink*. Selecting the latter brings up the *File Sink Editor* (Fig. 3.14). This allows a name to be given to the text sink, and the data items we wish to be recorded in it to be specified. The data items are identified from the selected *Data Set Id* drop-down

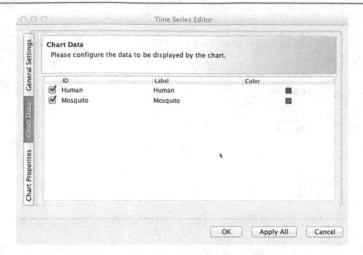

Fig. 3.12 The Time Series Editor for creating charts

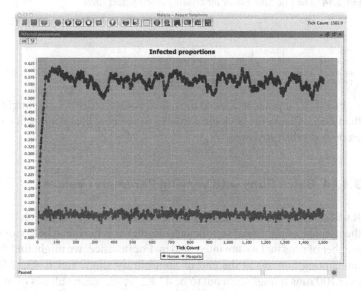

Fig. 3.13 Chart of infected proportions of human and mosquito agents (1000 mosquitoes, 500 humans, $s_h = 3$)

menu. The data items to be output should be selected in the left-hand panel and transferred to the right-hand one. The order in the right-hand panel affects the order in the output file. A following dialog window allows the file name to be chosen and formatting elements of the output file—such as the character to be used to separate the values output. A sample of the output of infection proportions might be as follows:

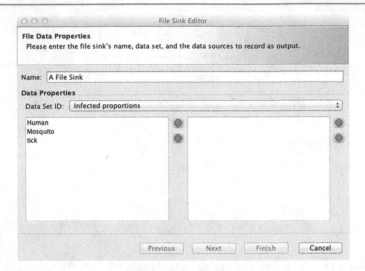

Fig. 3.14 The File Sink Editor for creating external text sinks

```
tick:  1.0,  Human:  0.0,  Mosquito:  0.465
tick:  2.0,  Human:  0.01,  Mosquito:  0.225
tick:  3.0,  Human:  0.02,  Mosquito:  0.005
tick:  10.0,  Human:  0.04,  Mosquito:  0.005
```

File sink output is written to the main project folder but you might have to refresh the folder view in the package explorer to see new files (right-click over the project name and select *Refresh*).

3.4.14 Batch Runs with Varying Parameter Values

It will often be the case that the results obtained from a single run of a model are not, themselves, of particular significance. Rather, it is the mean behavior observed over a number of runs that we are interested in. For instance, we might find that the infection rate among humans at the end of an arbitrary single run is 43.17 %, yet averaged over 100 runs it might turn out to be 47.82 %; the mean value is of more significance than any single result within those 100 runs. In addition, a full understanding of a model's behavior will often only be obtained when it is run multiple times with varying parameters. For instance, how is the mean final infection rate affected by the density of humans within the model's space? If we are to answer questions such as these in an efficient fashion with Repast then we need: (i) a way to run the model multiple times, and (ii) a way to be able to vary the parameters between runs. Running interactively is not efficient, and editing the `parameters.xml` file between runs is also not efficient. Fortunately, Repast provides a *Batch Run* feature. Batch runs can be set up and configured via the lightning bold menu item shown in Fig. 3.15. The options contain four tabbed areas, of which the main one of interest is *Batch*

Parameters (Fig. 3.16). Whoever, the *Model* pane contains the setting for the amount of runtime memory made available to the program, as described in Sect. 3.4.10 so that should also be edited appropriately.

The *Batch Parameters* pane allows the variables defined in `parameters.xml` to be set and varied from run to run. By default, each will be assumed to be *Constant* at the value given in `parameters.xml` but the drop-down allows alternatively for a *Number Range* or a *Space Separated List* of values.

If multiple parameters are varied, then there is clearly the potential for a very large number of individual runs of the program to occur on a single batch run, and this is probably best avoided. Furthermore, it is much easier to understand the effect of varying parameters if only one at a time is varied while the others are kept fixed. Only once a good understanding of the individual effects has been gained can proper insights be gained into varying parameter values in combination.

The output from batch runs will be based on the text sinks defined for a model and file output will typically be written to the `output` folder of the project. The view of this folder might need to be refreshed before the files appear there. The values

Fig. 3.15 The lightning bolt option accesses the batch run configuration options

Fig. 3.16 Parameters configuration pane for batch runs

output will need to be related to the parameter settings from that batch run and a
batch_param_map file is always generated to document the parameter values of
each batch run.

3.4.15 Summary of Concepts Relating to the Malaria Model

Working with the Malaria model offered us the opportunity to explore how to create a simulation in a way that is much closer to how we might write a program
from scratch in a modern programming language. However, we were able to take
advantage of the features that Repast Simphony offers in the form of the ReLogo
simulation framework—agent creation, spacial organization and event scheduling—
plus visualization, charting, result output and batch running. Therefore, nearly all of
our programming effort was focused on the model itself: the agents, their properties,
behaviors and interactions.

3.4.16 Going Further with Repast Simphony

With the two models we have developed in this chapter we have illustrated most of
the basic features of ReLogo within Repast that are necessary to get started writing
practical models. Nevertheless, within a book such as this, it is only really possible
to scratch the surface of what can be done with Repast Simphony. Just the API is
enormous, for instance. For the reader who wishes to develop sophisticated models
with Repast, it will be essential to make use of the official documentation available
from the Repast web site [2]. We would also strongly recommend joining the Repast
mailing list, where much help with practical issues is available.

References

1. North, M., Collier, N., Ozik, J., Tatara, E., Altaweel, M., Macal, C., Bragen, M., Sydelko, P.:
   Complex adaptive systems modeling with Repast Simphony. Complex Adapt. Syst. Model.
   1(1), 1–26 (2013). http://www.casmodeling.com/content/1/1/3
2. Argonne National Laboratory: The repast suite. http://repast.sourceforge.net/. Accessed 8 June
   2015
3. Barnes, D.J., Kölling, M.: Objects First with Java—A Practical Introduction Using BlueJ, 5th
   edn. Pearson Education, Prentice Hall (2012)
4. Foundation, E.: Eclipse integrated development environment. http://www.eclipse.org/.
   Accessed 8 June 2015
5. Groovy: Groovy—an agile dynamic language for the Java platform. http://groovy.codehaus.
   org/. Accessed 8 June 2015
6. Resnick, M.: Turtles, Termites, and Traffic Jams: Explorations in Massively Parallel
   Microworlds. MIT Press, Cambridge (1994)

7. Foundation, L.: Logo foundation. http://el.media.mit.edu/logo-foundation. Accessed 8 June 2015
8. Resnick, M.: Star logo. http://education.mit.edu/starlogo/. Accessed 8 June 2015
9. Wilensky, U.: Netlogo. https://ccl.northwestern.edu/netlogo/ (1999)
10. Argonne National Laboratory: Relogo primitives. http://repast.sourceforge.net/docs/api/repast_simphony/ReLogoPrimitives.html. Accessed 8 June 2015

Differential Equations

4

Differential equations are the bread and butter of mathematical modeling in the sciences. Many of the most fundamental equations in theoretical physics—the most basic of all sciences—are differential equations. An important activity of physicists is to formulate differential equations, solve them and interpret the solutions. Similarly, in biological modeling differential equations are of high importance. They are the right tool to describe many phenomena ranging from population dynamics and chemical kinetics through to evolution and epidemiology.

This chapter will give the reader an introduction to the use of a specific type of differential equations, namely *ordinary differential equations* (ODEs). The introduction provided here is not intended as a replacement for substantive specialist texts on this topic, such as Murray's excellent volume "Mathematical Biology"[1]. Instead, this chapter aims to introduce the reader to both the power and limitations of differential equations. The emphasis is twofold. Firstly, it will remind the reader of the basic ideas of calculus and, following from that, motivate the underlying ideas of differential equations. It will also describe some elementary methods to solve these equations. The second aim is to guide the reader through the exploration of a few examples. By the end of this tour she will have developed a basic intuition for the appropriate method to use and will be able to formulate her own models, find solutions and analyze them. At the very least, she will have the background to understand research papers on differential equation models.

We begin this tour with a brief introduction to the basic principles of calculus. This introduction is primarily aimed at motivating ideas, rather than developing a rigorous mathematical apparatus.

© Springer-Verlag London 2015
D.J. Barnes and D. Chu, *Guide to Simulation and Modeling for Biosciences*,
Simulation Foundations, Methods and Applications,
DOI 10.1007/978-1-4471-6762-4_4

4.1   Differentiation

Consider the following question:

Kurt often travels from Graz (Austria) to Hermagor (also Austria). The trip takes him exactly
2 h and 20 min. The distance between the two towns is exactly 220 km. At what speed does
he travel?

The solution to this problem is easy enough, at least at a first glance. All one needs to
do is to divide the distance traveled by the time it takes to complete the trip. Following
this we can compute the speed to be $v = 60 \times \frac{220\,\text{km}}{140\,\text{min}} \approx 94.3\,\text{km/h}$.

Straightforward as it seems at first, there is a problem with this result. The curious
reader might take a look at the actual route taken by Kurt. She will then conclude
that this value is very unlikely to reflect the actual speed at which Kurt traveled at
any one time. For example, he started in Graz, which is a small town. In continental
Europe towns have a speed limit of 40 km/h. Assuming Kurt is a law-abiding citizen
he would not drive faster than allowed by the law; we would therefore expect that
he stayed at or below the speed limit. On the other hand, a large part of the route is
a motorway (A2) with a speed limit of 130 km/h. Again, being impatient, Kurt will
normally try to keep at this speed. So, the 94.3 km/h we calculated as an average
speed, is not a very good estimate of the actual speed Kurt traveled. How can we find
a more realistic estimate?

One approach we can take is to divide the route into stages and measure the time
taken to complete these stages. As it turns, out, from Graz to Presseggen (which is a
village on the way) it took Kurt 2 h and 13 min, which corresponds to an average speed
of $60 \times \frac{213\,\text{km}}{133\,\text{min}} \approx 96.1\,\text{km/h}$. The remaining 7 km from Presseggen to Hermagor took
a mere 7 min, but at a speed of 60 km/h. The slower speed is not surprising, because
the last part of the trip is not on the motorway, but on a countryside road (B111)
with variable speed limits as it passes the villages on the way. This last estimate is
perhaps somewhat closer to the speed Kurt was traveling during the last 7 min of his
trip, before arriving; it certainly is quite different from our first estimate of his speed
as obtained above.

We can now draw a first conclusion: The speed averaged over the entire trip,
leading to an estimate of about 94.3 km/h is a poor guideline for the actual or typical
speeds Kurt traveled. As a matter of fact, he probably did not actually travel at this
speed for more than a small fraction of the entire time on the road, and was either
much faster (on the motorway) or much slower (in villages and cities). We can gain
a better idea of the actual speed traveled at various points in time by taking averages
over smaller distances. Averaging over only the last 7 km gave us a much better
estimate of the speed during the last part of the trip.

We can now take this insight further. In order to obtain a good estimate of the
actual speed traveled, we should further reduce the distance over which we average.
So, for example, Khünburg is 3.3 km from Presseggen *en route* to Hermagor. It took
Kurt exactly 3 min to cover this distance. There is no need here now to repeat the
details of how to calculate the speed given a distance and time. Instead, we will now

introduce a new notation for this very familiar calculation. Let us write the distance between the two villages as follows:

$$\frac{\text{(Dist. Khünburg-Graz)} - \text{(Dist. Presseggen-Graz)}}{\text{Time taken}} = \frac{3.3\,\text{km}}{0.05\,\text{h}} = 66\,\text{km/h} \quad (4.1)$$

By "Dist. Kürnburg-Graz" we mean the distance from Khünburg to the origin of Kurt's trip in Graz, which is 216.8 km; "Dist. Presseggen-Graz" has a similar meaning. This new notation makes it much easier to refer to a point, but is still somewhat cumbersome. If we want to measure distances very accurately, then using place names may not be sufficient precise. Hence, it might be better to refer to a position by a simpler label. Physicists often use x to refer to a specific place. We could use different indices to distinguish between different place names; so for example, x_1 could be Graz, x_5 could denote Presseggen, and so on.

This is not a complete solution, however, because we still need to worry about mapping the label to the name for the place. Instead, it is much more expressive to refer to the position in relation to the time when we were there. It is common to use a notation like $x(t)$ to indicate (in our case) the place where Kurt was at a time t since starting the trip. If we adopt this notation then clearly $x(t = 0)$, or simply $x(0)$, is the place where Kurt was at time $t = 0$ (a specific street in Graz). Similarly, $x(2\,\text{h}\,20\,\text{min})$ is Hermagor, the destination. Since it took Kurt 7 min to arrive there from Presseggen, $x(2\,\text{h}\,13\,\text{min})$ would stand for Presseggen.

This new notation is quite powerful and can make equations look much neater. We can now write (4.1) in a concise form:

$$\frac{x(2\,\text{h}\,16\,\text{min}) - x(2\,\text{h}\,13\,\text{min})}{0.05\,\text{h}} = \frac{3.3\,\text{km}}{0.05\,\text{h}} = 66\,\text{km/h} \quad (4.2)$$

Of course, with this new notation we can write the distance between any two points along the route taken by Kurt as $x(t_2) - x(t_1)$, commonly abbreviated as $\Delta x(t)$, with Δ being pronounced "delta". $\Delta x(t)$ means the distance traveled over the period of time t at some point in the journey. The delta notation is also used to indicate the difference between two moments in time. In the above case Δt would be the difference between times t_2 and t_1. With this new notation, we can rewrite (4.2) in an even more compact way.

$$\frac{\Delta x(t)}{\Delta t} = \frac{x(2\,\text{h}\,16\,\text{min}) - x(2\,\text{h}\,13\,\text{min})}{0.05\,\text{h}} = \frac{3.3\,\text{km}}{0.05\,\text{h}} = 66\,\text{km/h} \quad (4.3)$$

So far, all that has changed from (4.1) to (4.3) is the notation; the meaning has remained the same. All these equations say is how we calculate the speed given a distance and the time taken to travel the distance.

Given the power of our new notation let us now return to the question of how fast Kurt was traveling. By now, we know that our initial answer, namely that he traveled at 94.3 km/h hardly reflects his actual speed at any particular point, but rather is an average speed taken over the entire length of his trip ($\Delta x(t) = 220$ km and $\Delta t = 2$ h 20 min). Since there are varying road conditions and speed limits, in reality the speed will vary very much across the journey. In order to obtain a better estimate of the actual speed traveled at some of the time points, we had to reduce the period of time over which we averaged; in other words we reduced our Δt.

Yet, we still do not know for sure how fast Kurt was traveling at a specific point between Presseggen and Hermagor, although we know that this speed was closer to about 65 km/h than to 93.4 km/h. As we continue to decrease Δt we obtain an increasingly better idea of Kurt's actual speed at a particular point along his trip. Instead of considering the speed traveled between 2 h 13 min and 2 h 16 min into the journey (corresponding to a $\Delta t = 3$ min), we could look at the speed traveled between 2 h 13 min and 2 h 13 min 20 s since start of the journey (corresponding to $\Delta t = 20$ s). Maybe we could even consider reducing our Δt further, perhaps even to milliseconds? Where is the limit to this process?

Let us now come back to our original question from the beginning of the section: "What is the speed Kurt traveled at?" By now it should be clear that our question was not very precise. Either we should have asked about the average speed over a given time period, or we should have asked about the speed at specific time—"At what speed was Kurt traveling at time 2 h and 13 min and 0 s into the journey (and not an instant earlier or later)?"

There is an inherent paradox in this question. If we ask about Kurt's speed then, by definition, we are asking about the distance traveled. Yet asking about the speed at a particular instant of time seems meaningless because in a particular instant of time Kurt does not travel at all—he is *frozen in time* at a particular place. Therefore, it appears, the question does not make sense.

In this case, the appearance is wrong. The question *can* be asked in a meaningful way, and the discussion above has brought us half way towards being able to give an answer. As we have seen, we can generate better estimates of the actual speed traveled if we successively reduce the time over which we average, that is if we reduce Δt. So, if we were to consider a time interval of 4 min around the time we are interested then we would obtain a better estimate of the actual speed traveled than if we considered an interval of 5 min. We would obtain an even better estimate if we considered a 2 min interval, or 1 min, or even only one second around the moment we are interested in. Eventually we will reach a Δt that gives an answer that is sufficiently accurate for most practical purposes; if we keep reducing Δt then we will eventually find a time interval over which Kurt's speed is essentially constant. At least as far as car journeys are concerned, it is rarely necessary to bother with milliseconds.

In reality, hardly anyone will be interested in Kurt's travels through Austria, let alone his speed between Khünburg and Presseggen. However, the procedure by which we determine his speed in a given moment is of general importance in many areas of science. Any quantity that changes can be analyzed following the exact same procedure we used above. Instead of Kurt's speed we could consider the change of the size of a bacterial colony, or the change of temperature of a mountain lake, or the change of the concentration of a protein over time. Reducing the interval over which one averages the change of quantities to obtain the instantaneous change is a very common procedure in physical sciences, engineering, mathematics and, of course, biological modeling.

Important for the use of this idea is to develop a notation that allows us to efficiently make precise statements. The process of making the time intervals smaller and smaller is often written like:

$$\lim_{\Delta t \to 0} x(t) - x(t - \Delta t) = \lim_{\Delta t \to 0} \Delta x(t) \qquad (4.4)$$

Equation (4.4) looks very mathematical, but all it does is to formulate the idea of measuring the distance between two points. The symbol, "lim" is read as "limit" and encapsulates precisely the idea that we make Δt smaller and smaller until it *nearly* reaches 0. As we do this, we consider the distance traveled between the time t and just a small time before that, namely $t - \Delta t$.

Equation (4.4) only considers the distance traveled. As Δt becomes smaller and smaller, Kurt covers smaller and smaller distances. In the limiting case of $\Delta t = 0$, the distance traveled goes to 0, which is what we expect: In an instance of time, one is frozen in time and does not move at all. Hence, we get:

$$\lim_{\Delta t \to 0} \Delta x(t) = 0$$

At this point, the reader who is not familiar with basic calculus needs to take an intellectual leap. While it is true that reducing Δt will also reduce the distance traveled, the speed may not change. Just remember that Kurt's speed between Presseggen and Hermagor was 60 km/h traveled over 7 min; the corresponding $\Delta t = 7$ min and $\Delta x(t) = 7$ km. The much shorter distance between Presseggen and Khünburg was covered at an average speed of 66 km/h but over a much shorter $\Delta t = 3$ min and a $\Delta x(t) = 3.5$ km. So, while both Δt and $\Delta x(t)$ become smaller and smaller, their ratio—the speed—does not necessarily decrease, but could converge to a finite, non-vanishing value. So, maybe there is something wrong about the idea of being frozen in time?

We now have all the necessary notation and concepts ready to ask the question about Kurt's speed at a particular point in time t. This is simply given by the distance traveled divided by the time this has taken; averaged over a vanishingly small interval of time:

$$\lim_{\Delta t \to 0} \frac{x(t) - x(t - \Delta t)}{\Delta t} = \lim_{\Delta t \to 0} \frac{\Delta x(t)}{\Delta t}$$

Physicists normally indicate the speed at a time t by the symbol $v(t)$. And we can now define this symbol using our new notation.

$$v(t) \doteq \lim_{\Delta t \to 0} \frac{x(t) - x(t - \Delta t)}{\Delta t} = \lim_{\Delta t \to 0} \frac{\Delta x(t)}{\Delta t} \qquad (4.5)$$

Here the symbol "$\doteq$" just means "is defined by."

The problem of the notation in (4.5) is that it is somewhat tedious to write. Hence, physicists have invented a more concise way to write this equation which has the same meaning:

$$v(t) \doteq \frac{dx(t)}{dt} \tag{4.6}$$

Despite looking very different, (4.5) and (4.6) have the same meaning. The function $v(t)$ measures the *change* that $x(t)$ undergoes at a time t. In the case of a trip from Graz to Hermagor, this is just the speed at a particular time t, but we can generalize this to mean the change of any quantity. It does not even need to be the change as time passes, but it could be the change of a quantity as a different something changes. Yet, we are getting ahead of ourselves.

For the moment, it is only important to remember, that $\frac{dx(t)}{dt}$ is the instantaneous rate of change of $x(t)$ at time t Where the meaning is clear, it is quite common to abbreviate $x(t)$ as simply x.

4.1.1 A Mathematical Example

While Kurt's road trip was a good example to illustrate the core idea of rates of change, the power of (4.6) really only becomes apparent when we have a mathematical expression for the distance traveled. Let us start with the simplest case, namely linear motion, where we have $x(t)$ given by:

$$x(t) = a \cdot t \tag{4.7}$$

In this case, $x(t)$ can be represented as a straight line. The parameter a could be any number; its precise value will depend on the circumstances. Let us assume it is simply 1 (see Fig. 4.1). We could think of $x(t)$ as the distance traveled, just as in the case of Kurt's trip. Then the horizontal axis in Fig. 4.1 would represent time and the vertical axis would be the point x at the given time t. However, it could really represent anything else as well; the interpretations as "distance" and "point in time" are perhaps helpful for the moment, but not necessary as the same kind of analysis for the rate of change of any quantity can be performed no matter what x represents and what it depends on.

Let us look more closely at the concrete example of the function $x(t)$ given in (4.7). This is essentially only an instruction of how to calculate the value of x depending on t. If we set the value of $t = 3$, then $x(3) = 3a = 3$ when a has the value 1. Similarly, we can calculate values at any other time. Now, we can go through the same procedure as we did with Kurt's travels. If we want to know how much x changed from time, say, $t = 0$ to $t = 10$, then we simply calculate,

$$\frac{\Delta x}{\Delta t} = \frac{x(10) - x(0)}{10 - 0} = \frac{10a - 0a}{10} = 1a = 1$$

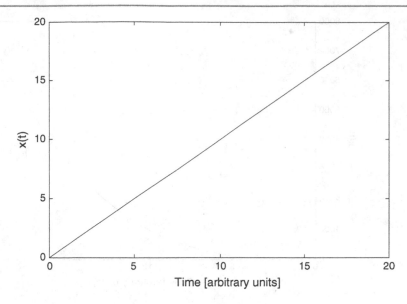

Fig. 4.1 Function $x(t)$ as in (4.7)

So, we calculated the rate of change between these two time points to be 1. Similarly, we can now calculate the rate of change between $t = 6$ and $t = 4$, and, following the same procedure, we will see that again the rate of change is a, which evaluates to 1 if we set $a = 1$. In fact, the reader can easily convince herself that the rate of change is always the same, at any point for any interval. This is not particularly surprising. The graph of the linear motion is a simple straight line, which has a constant slope throughout. The slope is just the rate of change, hence we would expect that the rate of change is constant everywhere.

Let us now look at a non-trivial example and consider a new function of $x(t)$ given by:

$$x(t) = a \cdot t^3 \tag{4.8}$$

This curve is illustrated in Fig. 4.2. Unlike the case of Fig. 4.1 this one does not have a constant slope throughout, which means that the rate of change varies; we would now expect that the rate of change depends more strongly on the interval we average over, very much as in the case of Kurt's trip. To obtain a better grasp of this, let us focus arbitrarily on the point $t = 5$. (This is exactly the same type of question as asking what speed Kurt was traveling at exactly 2 h 13 min after starting.)

At the time in question, the value of $x(t)$ can be easily calculated to $x(5) = a \cdot 5^3 = 125a$. If $a = 1$ again then $x(5) = 125$. This result is now the equivalent of the position in the above example of Kurt's trip. So, what is the rate of change ("speed") at this point? Following our now quite familiar procedure, let us start to home in on

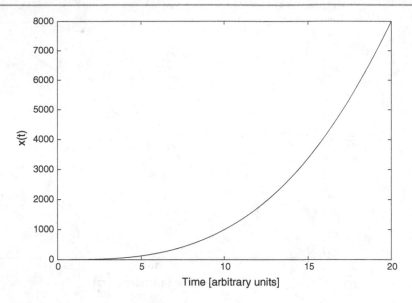

Fig. 4.2 Function $x(t)$ as in (4.8) with $a = 1$

the speed at $t = 5$. By looking at ranges centered on $t = 5$, and taking as the first interval, $t = 0$ and $t = 10$ we get:

$$\frac{x(10) - x(0)}{\Delta t} = \frac{1000a}{10} = 100a = 100$$

So, the rate of change is 100. Repeating this process, but for a $\Delta t = 4$ centered around $t = 5$ we get,

$$\frac{x(7) - x(3)}{\Delta t} = \frac{343a - 27a}{4} = 79 \cdot a = 79$$

Similarly, we get for a $\Delta t = 2$ a value of 76; for $\Delta t = 1$ a value of 75.25; for $\Delta t = 0.4$ a value of 75.04; and so on. Clearly, we move closer and closer to the rate of change at $x(5)$, but so far we have only calculated estimates.

4.1 Confirm the estimates of the slope given above.

Just as in the case of Kurt, if we want the exact value of change at the point $x(5)$, then we need to take the limit of a vanishingly small interval around that point; that is, we need to calculate $\lim\limits_{\Delta t \to 0} \dfrac{\Delta x(5)}{\Delta t}$, to obtain the expression for $\frac{dx(t)}{dt}$. As it turns out, this expression is in fact very easy to obtain by simple differentiation.[1]

$$\dot{x} \doteq \frac{dx(t)}{dt} = \frac{d(at^3)}{dt} = 3at^2 \tag{4.9}$$

We arrived at this equation simply by applying the standard rules of differentiation to the equation for the curve (4.8). Note also that we have introduced a new notation here, namely $\dot{x}$. This is "the rate of change of x"; we will use this notation extensively.

If we now want to obtain the *exact* value of change of (4.8) at the point $t = 5$, all we need to do is to calculate it from (4.9) to be

$$\dot{x}(5) = 3 \cdot a \cdot 5^2 = 75$$

when $a = 1$. So, the exact rate of change, or slope (or speed) at this particular point is 75, which is quite close to our final estimate with a $\Delta t = 0.4$, which suggested a slightly higher value of 75.04.

In differentiation, we have a very effective tool to calculate the rate of change of any function we like, and at any point. Not only in biology, but in science at large, this is perhaps the most powerful and important mathematical technique. It is used to formulate laws in physics, engineering, and crucially also in biology. Before we are able to apply this powerful tool to concrete applications, we need to generate a better understanding of what exactly it is doing and what we can do with it. It would neither be possible nor useful to go through the entire theory of calculus here, but there are a few key points that are worth highlighting and that will also be exceptionally relevant for what we will do later.

The first question we should clarify is what the "change" means. So, if at $x(5)$ the rate of change is 75, how should we interpret this number? In the case of Kurt's trip, the meaning was intuitively clear; the rate of change was a speed. In the more abstract case of a simple curve it is not so evident how we should interpret the number. The exact meaning of the derivative of a variable will, to some extent, depend on context, which can only be clarified with reference to the semantics of a particular model. Yet, there is a general sense in which we can interpret derivatives quite independently of the meaning of the variable. We will develop an understanding of this by means of a small digression.

4.1.2 A Small Digression

In pre-scientific times many people believed that the earth was flat. While today this belief seems ridiculous, it is perhaps not that absurd if considered in the light of everyday empirical evidence. If we look around us, at least on a plain, then our

---

[1]The reader who is not familiar with the rules of differentiation, is encouraged to consult Appendix A.1 for a refreshment of the rules.

planet certainly looks very flat. Similarly, when we look out on the ocean, then we cannot detect any curvature, but rather behold a level body of water. Even when we travel long distances we see no obvious evidence contradicting the assumption that we live on a flat surface (except for a few mountains and valleys of course). Even from high up on the summits of mountains (or even aeroplanes), we look down on a flat surface not a round body. So, really, pre-scientific man and woman must be forgiven for their natural (and empirically quite justified) assumption that their home planet is nothing but a big platter.

Perspectives only start to change when we take a large step backwards, in a way that was impossible until relatively recently. When seen from space, the hypothesis that the Earth is round seems much more natural given direct observation. (Presumably, the reader has not had this first hand experience, but hopefully we can take available footage on trust and believe that the Earth is actually round.[2])

The failure of pre-scientific man and woman to recognize that they dwelled on a curved object is due to an effect that is relevant in the context of differentiation as well: Even very curved bodies look straight if the observer is small and close to the curved surface. The same principle also applies in a two dimensions: if considered from close enough, most curves appear as straight lines. If we imagine the curve in Fig. 4.2 to be inhabited by some imaginary creatures, maybe of the size of bacteria, then they would undoubtedly also assume that they live on a straight line.

Mathematicians and physicists would summarize this observation in an efficient, if somewhat unexciting way: *Locally, a curve can be approximated by a straight line.* Figure 4.3 illustrates this using the example of the curve $x(t) = t^3$. As we zoom in on the point $x(5)$ (although the same thing would work with any point) the curve starts to look more and more like a straight line. This is exactly the same effect that makes the Earth look flat to us. Yet, from a technical point of view, the important question now is, what is the straight line that approximates the curve locally?

We established above that straight lines are very simple things in that their rate of change is constant and given by the parameter a in (4.7). We found this by direct calculation of the slope. Another way to determine this is to use the rules of differentiation (see Appendix A.1). If a straight line is given by $x(t) = at + b$, then $\dot{x}(t) = a$. This tells us that the slope is constant and does not depend on the time at all. (We know that already, but it is worthwhile seeing this again from a different point of view.)

We have nearly arrived at an interpretation of the *first derivative* of a function. All we need to do is to put together the two ideas, namely that: (i) a curve locally looks like a straight line, and (ii) the slope of a straight line is given by its parameter a. Locally, any (differentiable) curve $x(t)$ looks like a straight line with parameter $a = \dot{x}(t)$. This gives us the interpretation of the derivative of a function: The value of the derivative at a particular point on a curve is the slope of the straight line (*tangent*) at that point.

---

[2]Of course there are indirect ways to measure the curvature of the Earth. Based on triangulation the ancient Greeks already knew that the Earth is (at least approximately) round.

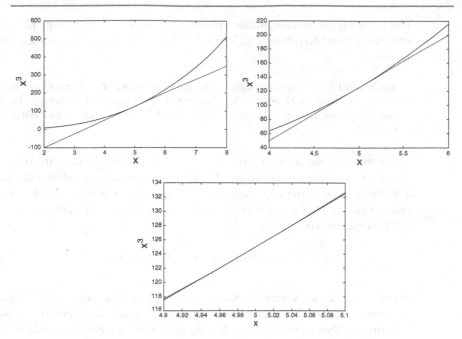

Fig. 4.3 Changing the point of view. As we zoom into the point $x(5)$, the curve in Fig. 4.2 starts to look more and more like a straight line, and we fail to detect the curvature that is so obvious to us in the wider view. The *dashed line* is a straight line with slope 75. In the bottom panel the line and t^3 are barely distinguishable

From a mathematical point of view, there is of course much more to differentiation and differentials than we are able to discuss here. The reader who really wants to obtain a deep understanding of the theory is encouraged to consult a calculus textbook. While in the long run it is necessary to develop a deeper understanding, it is still fully possible to take the first steps into a new modeling adventure without being versed in the sacred texts of mathematics, as long as one has developed some intuition for the basic concepts.

4.2   Integration

Here is another simple exercise, similar to the one we have seen before, but subtly different:

> Kurt traveled from Graz to Hermagor. The trip took him exactly 2 h and 20 min and his average speed was 94.3 km/h. What distance did he travel?

This is clearly the inverse problem to our previous exploration on p. 122, and we know the answer: Kurt traveled 220 km. So, that is easy. Now we can ask a different question:

> Kurt traveled from Graz to Presseggen. The trip took him exactly 2 h and 13 min and his average speed was 96.1 km/h. He then continued from Presseggen to Hermagor. The trip took him exactly 7 min and his average speed was 55.7 km/h. What is the total distance he traveled?

Again, we know the answer (220 km); yet if we wanted to calculate the solution we would have to figure out the distance from Graz to Presseggen, and then the distance from Presseggen to Hermagor using the average speeds and travel times we were given to get the total travel time. If we denote the distance traveled since Graz by $x(t)$ then we can write this as:

$$x(2\,\text{h}\,20\,\text{min}) = v_1 \Delta t_1 + v_2 \Delta t_2 \tag{4.10}$$

Here v stands for the average speed of a segment, and the subscript "1" stands for the segment "Graz to Presssegen" and subscript "2" stands for segment "Presseggen to Hermagor." So, all we do is to calculate the distances traveled by multiplying the average speeds with the times it took to complete the segments. No magic there.

Imagine now that, instead of being given Kurt's travel in 2 segments, we know the average speeds and travel times in much finer grained form; say we know 1000 segments. Then, again, to obtain the total distance traveled, we calculate the distances of every individual segment and sum them up. Mathematically, we write this as follows:

$$x(2\,\text{h}\,20\,\text{min}) = \sum_{i=1}^{1000} v_i \Delta t_i \tag{4.11}$$

The principle of solving this is the same as in the case of (4.10), even though the actual practice of calculating the distances for 1000 records is much more tedious than for a mere two. It is even more tedious if, instead of 1000 records, we are given 100000. Then our formula becomes:

$$x(2\,\text{h}\,20\,\text{min}) = \sum_{i=1}^{100000} v_i \Delta t_i \tag{4.12}$$

and takes 100 times as long to compute. What we did here, at every step of refinement, was to split up Kurt's trip into smaller and smaller segments, so that we obtain the required number of records. The question we want to ask now is, what will happen when we are given Kurt's trip split up into an *infinite* number of records, each giving his speed at a particular moment in time?

To see how we could deal with this situation it will be helpful to remember what v actually stands for. When we discussed the original problem (i.e., "What is Kurt's speed?") we calculated v as follows:

$$v = \frac{\Delta x(t)}{\Delta t}$$

Equation (4.12) looks almost trivial:

$$x(2\,\mathrm{h}\,20\,\mathrm{min}) = \sum_{i=1}^{100000} \left(\frac{\Delta x_i}{\Delta t_i}\right) \Delta t_i \tag{4.13}$$

More generally, for any number N of intervals, we can write this as:

$$x(2\,\mathrm{h}\,20\,\mathrm{min}) = \sum_{i=1}^{N} \left(\frac{\Delta x_i}{\Delta t_i}\right) \Delta t_i$$

It is easy to see that it really just reduces to adding the length of the individual segments of the trip, given by $\Delta x_i(t)$ (though remember that we are not directly given the length of the segments, only the speeds and times.) As we now make the individual segments shorter and shorter, we obtain more and more of them; in other words, our index i goes towards infinity, but the length and time of the individual segments Δt_i and $\Delta x_i(t)$ goes towards zero. We have already encountered this concept and introduced a notation for the speed (or more general any rate of change) in an instant. Mathematicians have introduced a new symbol for a sum that runs over an infinite number of infinitely small summands in this way, namely the integral symbol $\int_A^B$, where A and B are the limits for the summation. In the present case they would correspond to "Graz" and "Hermagor" respectively:

$$x(2\,\mathrm{h}\,20\,\mathrm{min}) = \lim_{N\to\infty} \sum_{i=1}^{N} \left(\frac{\Delta x_i}{\Delta t_i}\right) \Delta t_i = \int_{\mathrm{Graz}}^{\mathrm{Hermagor}} \dot{x}(t)\,dt$$

Using this new notation, we then arrive at our final way to write Kurt's trip as the sum of an infinite number of infinitely small segments as:

$$x(t) = \int_{\mathrm{Graz}}^{\mathrm{Hermagor}} \dot{x}(t)\,dt \tag{4.14}$$

The problem with this example, and (4.14), is that we cannot actually calculate it because we are not really given $\dot{x}(t)$ as a function. To remedy this, let us move away from the example of Kurt's trip and work with a function. For simplicity, let us choose the function given above in (4.8):

$$x(t) = at^3 \tag{4.8}$$

We also calculated the first derivative with respect to t

$$\dot{x} \doteq \frac{dx(t)}{dt} = \frac{d(a \cdot x^3)}{dt} = 3at^2 \tag{4.9}$$

Equation (4.9) gives the rate of change of the function given in (4.8) for every point in time t. As discussed above, it is the equivalent of the speed with which Kurt was traveling. In order to test the ideas about integration that we have just developed, let us assume that we "start" at time $t = 2$ and "arrive" at $t = 5$. Let us further assume that our distances are given by the function in (4.8). Direct calculation now shows us that our "distance" traveled is $(5^3 - 2^3)a = 117a$. If we were only given the function of speeds (4.9), then we can calculate the distance traveled given only the rates of change:

$$\text{Distance}(5, 2) = \int_2^5 3at^2 dt = a5^3 - a2^3 = 117a \tag{4.15}$$

This little exercise establishes two things about the nature of integration, namely: (i) that it is essentially the inverse of differentiation, and (ii) that it is essentially the process of taking an infinite number of sums over infinitely small intervals. In practice, the first observation manifests itself in the rules of integration that are exactly the opposite of the rules of differentiation (see for the details Appendix A.1). The second observation is helpful in practice when it comes to formulating problems mathematically.

4.3   Differential Equations

There are many possible uses of integration and differentiation in mathematics and physics. As far as modeling is concerned, one of the most important ones is differential equations. Much is known about this subject and many clever techniques have been developed that can aid practical modeling tasks. Getting to grips with this field requires intense studying over months. It is clearly beyond the scope of this book to present a comprehensive summary of the available knowledge about the theory and practice of differential equations, and we will limit the contents here to a vanishingly small part of this available body of knowledge. The good news is, that this vanishingly small part requires only a few key principles yet is, in many cases, sufficient to build credible and useful models of biological systems. Luckily, we live in an age where computer programs can take away much of the technical burden from us, enabling the mathematically less versed to take significant steps into a territory formerly reserved for a tiny class of mathematical high-priests.

Before addressing the topic head-on, let us take a step back. Consider the following equation:

$$4x - 5a = 0$$

This equation implicitly formulates x in terms of some numbers and the variable a. If we want to know the value of x, then we perform some simple manipulations to arrive at the solution, $x = 5a/4$.

Let us now imagine that access to x is only available through this equation:

$$4\dot{x} - 5a = 0 \tag{4.16}$$

Although we can easily calculate $\dot{x} = 5a/4$, we are really interested in x. How can we calculate x from (4.16)? The answer is by using integration. However, unlike in the case of Kurt's trip between two particular points, we want to take the *indefinite integral*; that is, we do not really want to specify the start and end points, but rather leave those open. In that we will recover the functional form of the solution instead of the length of any particular path:

$$x(t) = \int \dot{x}(t)dt = \int \frac{5}{4}adt = \frac{5}{4}at + b$$

In the indefinite integral, we always need to allow for an additional constant, here represented by b. This is a consequence of the rules of differentiation stipulating that constant factors in a function become 0 when the function is differentiated. In modeling practice, we often need to determine this additional constant. We will say more about this later and, but the moment, we will just accept that this is the case.

Equation (4.16) was our first example of a *differential equation*, in that we had the rate of change $\dot{x}(t)$ and the task was to find the function $x(t)$, of which it is the rate of change. In this particular case, the rate of change was constant; hence not surprisingly, the solution $x(t)$ turned out to be the equation for a straight line. Practical examples of differential equations tend to be more complicated, of course.

Let us now look at our next differential equation:

$$\dot{x} = ax \tag{4.17}$$

This equation is somewhat more difficult in that the simple method of rewriting used above no longer works (the reader is encouraged to convince herself of that). There are now two possibilities for solving this equation. Firstly, one might use a computer algebra system to solve it. This would then give a solution similar to:

$$x(t) = Ce^{at} \qquad (4.18)$$

Here C is some constant that still needs to be determined. It essentially corresponds to the integration constant mentioned above. We can easily convince ourselves that this solution is correct, by differentiating $x(t)$ with respect to t.

$$\dot{x}(t) = \frac{dCe^{at}}{dt} = aCe^{at} = ax(t)$$

It seems that our computer algebra system has indeed returned the correct solution. In this particular case, however, there is a simple technique available to solve this equation by hand, which is worth learning. It is called *separation of variables* and works in some of the simpler cases of differential equations. To see how, it is useful to re-write (4.17) explicitly as a differentiation:

$$\frac{dx}{dt} = ax$$

We now separate the two variables x and t by transforming the equation so that each side contains only either x or t (hence the name, "separation of variables").

$$\frac{dx}{x} = adt$$

Having done this, we can now simply integrate both sides independently to obtain the solution.

$$\int \frac{1}{x}dx = \int adt$$
$$\ln(x) = at + \text{const}$$
$$x = C \cdot e^{at}$$

To obtain the second line remember that $d\ln(x)/dt = 1/x$, and in the third line we used the fact that $e^{\ln(x)} = x$. The symbol C represents an arbitrary constant that still needs to be determined. Again, it is essentially the integration constant that always appears in indefinite integrals. We will return to how to determine it shortly.

4.2 Fill in the detailed steps from this derivation. Specifically, pay attention to trace the origin of the integration constant.

Equation (4.17) is extremely simple in mathematical terms, but luckily also extremely useful in the context of biological modeling. It can be used to describe

growth phenomena, for example of bacterial colonies. In this case, we would interpret $x(t)$ not as a "distance" but as the size of the colony. The differential equation
then states that a change in the size of the colony (its growth) is proportional to
its size. The parameter a is the *proportionality constant*. It describes how fast the
bacteria are dividing—the higher the value of a, the faster they replicate.

The assumption that growth is proportional to colony size makes perfect sense
in bacteria, because each cell is dividing independently. The more cells there are,
the more cells will be produced per time unit. This is exactly what the differential
equation states. The reader is encouraged to dwell on this for some time.

4.3 Formulate and solve a differential equation describing a (hypothetical) Martian
species of bacteria that grows proportional to the square of its own colony size and
decays proportionally to the colony size.

A key prediction of the differential (4.17) is that $x(t)$ increases exponentially.
Qualitatively this prediction is correct. Bacterial colonies grown in laboratory cul-
tures do indeed display such growth patterns—at least for a period of time until
the nutrients are exhausted, at which point growth flattens out. Our model does not
show this limit to growth or the influence of nutrients, though we will see later how
this effect could be included. For the moment we still have to clarify the role of the
constant C.

In order to see how to determine it, it is useful to consider again what exactly our
differential equation states: (4.17) tells us what the size of the population is at time
t as long as we set the value of a. So, for example, if we set $a = 1$ (arbitrarily) then
at time $t = 10$, the size of the population is,

$$x(10) = Ce^{10a} \approx C \cdot 22026.4$$

Unfortunately, this is not yet completely satisfactory, because the solution still contains the unknown parameter C, of which we neither know the value nor its meaning.
One way to interpret C is to consider it as an initial condition, that is, the size of the
colony at $t = 0$. In an actual lab-experiment this initial condition would correspond
to the size of the seed colony. Since speed of growth depends on the size of the
colony, it is crucially important to know the size of the colony at time $t = 0$ in order
to determine the size of the colony at any subsequent time point. A general law of
growth cannot contain this information (because it is general) and we have to "tell"
the model what the size of our initial colony was. How can we "tell" the model the
size of the initial colony? It all comes down to determining C.

Let us assume that we start with no bacteria at our time $t = 0$. Using our exponential growth law, this means that:

$$x(0) = 0 = Ce^{a \cdot 0} = C \cdot 1 = C$$

In this case, we get a $C = 0$, reflecting a size of the initial colony of 0. Since there are no bacteria, there is nothing that can grow, hence, we would expect that if $x(0) = 0$ then $x(t)$ remains 0 for all times t. Looking at the solution for $x(t)$, we see that this is exactly what is happening. So, no growth is a solution for our differential equation as well, although a rather uninteresting one.

If we now choose a more reasonable value for our initial condition, say, $x(0) = 1$, then we obtain a value for $C = 1$, which now fully specifies our model of bacterial growth. We could determine the size of the colony for all times ahead into the future. The astute reader could, at this point remark that if we chose a negative number for the initial conditions, e.g. $C = -1$, then we would get negative colony sizes for all times. In a mathematical sense, this is of course correct. However, biologically it makes no sense to assume that the initial colony size is negative and we must therefore disregard this possibility as irrelevant.

4.4 Using the model in Exercise 4.3, determine the integration constant for an initial colony size of $x(0) = 10$.

4.3.1 Limits to Growth

A key observation of the solution we found for the exponential growth is that it predicts a very rapid and, most of all, never-ending growth of the bacterial colony. Figure 4.4 illustrates this, but simply plugging numbers into the solution also suggests a rather scary scenario. If we start with an initial population of $x(0) = 1$, then at time $t = 10$ the population has grown to 22026.6 and at time $t = 20$ to $\approx$485165195, which is a very big number. The colony will just continue growing like this with ever increasing rates. If that reflected reality then the whole universe would be packed with bacteria by now.

Luckily this scenario is unrealistic. Exponential growth is observed in bacteria, but only for a short period of time. Sooner or later nutrient depletion, space restrictions, and possibly other factors put an end to growth.[3]

Before developing more sophisticated models of bacterial growth let us show another example of the use of differential equations in biology, namely gene expression. The particular modeling problem is to understand the time course of the concentration $x(t)$ of a protein X that is expressed by a gene G_x. Expression of a single gene can be a very complicated process requiring many different events coming together. For instance, there is the transcription step producing mRNA, followed by translation which, in itself, is an involved dynamical process. Much is known about these processes and one could try to capture all this knowledge in a model.

---

[3]Incidentally, while we have an intuitive understanding for the fact that bacterial colonies cannot sustain exponential growth for a very long time, economic growth follows a similar law. Suppose a country experienced a trend growth rate of about 2 % per year. Year after year the 2 % rate means a higher absolute growth, because the economy of which we take the 2 % has grown from the year before. Hence, (at least to a first approximation) the economy grows according to the differential (4.17), a law that we have just seen to be unsustainable for bacteria because of resource depletion.

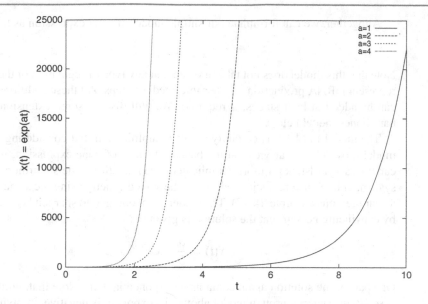

Fig. 4.4 The solution to the differential (4.17) for various values of a

As we pointed out in Chap. 1, in modeling it is nearly always better to resist the urge to try to represent everything one knows—at least in the first instance. Every good modeling enterprise should start with a very simple model that captures the basics only. If required, complexity can by built in later.

The simplest representation of gene expression we can imagine is to have a gene G_x that directly expresses a protein X at a constant rate a. We do not actually have to represent the gene itself in the model. Instead we assume that the product of the gene is created from the "void" at a rate a.

Note that in the case of bacterial growth, the rate of growth was proportional to the number of bacteria. This assumption was motivated by the observation that each cell in the colony divided. In the present case, the protein-producing entity—the gene—is constant in number. The growth rate is therefore constant as well and does not depend on the concentration of the product. Altogether, we assume that growth of the protein population is simply proportional to a.

To avoid observing unbounded growth again, we also assume that, on average, it takes a protein $1/b$ s to decay, or in other words it decays at a rate b. To give a concrete example, if $b = 0.5$ then, on average, a protein lives for two units of time. Now assume that we observe a system consisting of 1000 proteins for one unit of time. On average we would then expect to see 500 decay events per time unit. If, on the other hand, we observe a system of 2000 proteins for the same length of time, then we would expect to see, on average, 1000 decay events. This tells us that we should assume protein decay to be proportional to the number (or concentration) of protein in the system, that is decay $\propto bx$.

Altogether, we can formulate our simple model of gene expression as follows:

$$\dot{x} = a - bx \qquad (4.19)$$

Note that this model does not take into account any type of regulation of the gene or represent mRNA production or other intermediate steps. All these additional effects can be added at later stages, if required. We will discuss some extensions to this bare-bones model below.

The model in (4.19) is certainly an oversimplification, but considering a simple model provides a clear view on the basic dynamics of gene expression, which we can use as a yardstick for more complicated (and realistic) models. Furthermore, this system, unlike most realistic models, can be solved exactly using the same methods as our previous example, (4.17). The reader is encouraged to sharpen her techniques by convincing herself that the solution is given by:

$$x(t) = \frac{a}{b} + Ce^{-bt} \qquad (4.20)$$

One part of the solution again contains an exponential part. Note that, unlike in the case of our bacterial growth model above, the exponent is negative, implying decay over time rather than growth. This is reassuring. Exponential growth (or decay) is usually expected when the growth/decay rate depends on the population size. In this case, only the decay rate depends on the population size, but not the growth rate.

Furthermore, the integration constant C appears in the exponential term. We have come to associate it with initial conditions. Based on this, it seems that the exponential part of the solution relates somehow to the decay of any initial conditions that prevailed at time $t = 0$. Let us consider a concrete case and start with no protein in the system; we then obtain:

$$x(0) = 0 = \frac{a}{b} + C \quad \Longrightarrow \quad C = \frac{-a}{b}$$

Using this we obtain the fully specified solution for an initial condition of $x(0) = 0$:

$$x(t) = \frac{a}{b} - \frac{a}{b}e^{-bt} \qquad (4.21)$$

Interestingly, despite choosing a trivial initial condition, the model does not lead to trivial behavior. This is unlike the model of bacterial growth that predicted no growth if we start with no bacteria. Intuitively, this does make sense. In (4.19) the growth does not depend on $x(t)$. Hence, even if there are no proteins around, the gene still expresses them at a constant rate a.

4.5 Derive the solution (4.20) from (4.19).

4.6 In (4.20), determine C for the initial condition $x(0) = 2000$. Solve the system and discuss how the exponential term in the solution leads to the system forgetting its initial state.

4.3.2 Steady State

Once we have specified our initial conditions, we can start to consider how the protein concentration changes over time (Fig. 4.5). One thing that becomes obvious from looking at a few examples is that, after a certain transient period, the solution always tends to the same value and remains apparently unchanged from then on. This value defines the *steady state* behavior of the system. How long it takes before this long-term behavior is reached depends of course on the parameters of the system and, as Fig. 4.5 suggests, also on the initial conditions themselves.

We can easily find the steady state from our solution (4.20) by considering the behavior at $t = \infty$.

$$x(\infty) = \frac{a}{b} + \underbrace{C \exp(-b\infty)}_{=0} = \frac{a}{b}$$

A comparison with the graphical solution in Fig. 4.5 (left) shows that this seems to be correct. For $a = 1, b = 3$, the steady state is $1/3 \approx 0.333$, for $b = 4$ the graph approaches $1/4 = 0.25$ and so on.

Most differential equation models are not explicitly solvable in the sense that one can obtain a closed form expression for $x(t)$. Luckily, it is still possible to find the steady state behavior directly from the differential equation model, even in absence of an analytical solution. To see how, remember that a differential equation formulates the rate of change of a quantity $x(t)$. On the other hand a steady state is reached when the system variables cease to change. In the case of our current model of gene expression, this does not mean that the system is inert; it only means that there is a point where the expression rate a exactly balances the decay rate bx, and hence the overall number of particles no longer changes, even though proteins continue to be expressed and decay.

The fact that the system variables cease to change is expressed mathematically by setting the derivatives with respect to time to zero:

$$\dot{x}(t) = 0$$

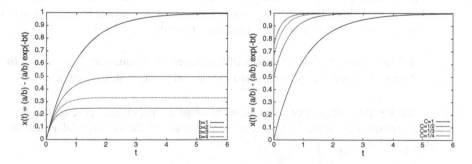

Fig. 4.5 *Left* The function $x(t)$ from (4.20) for $a = 1$ and various values of b. We chose as an initial condition $x(0) = 0$. *Right* We keep a fixed value of $a = b = 1$ but vary the initial conditions

This gives us an algebraic equation that can be solved for the system variables to find the steady state.

$$\dot{x} = 0 = a - bx \quad \implies \quad x = \frac{a}{b} \tag{4.22}$$

The reader can convince herself that the steady state in (4.22) is identical to the one obtained from the solution of the differential equation in (4.21). Steady state analysis is a powerful method for extracting information from differential equations. It often provides analytic expressions for the steady state even when it is not possible to obtain an analytic solution for the differential equation itself.

Note that, in our example, the steady state does not depend on the initial conditions. In the specific case of (4.20), the independence of the steady state solution from the initial condition can be clearly seen from the form of the solution. We have previously mentioned that the integration constant C is closely associated with the initial condition. In (4.20), C appears only in connection with an exponential factor that tends to zero as time increases. This is a clear indication that *in this particular* system, the steady state is completely independent of the initial conditions.

The independence of the steady state from the initial conditions does not hold in general. Indeed, not all systems have a finite steady state for all initial conditions. Our model of bacterial growth above, (4.17), is a case in point:

$$\dot{x} = ax \tag{4.17}$$

This differential equation only allows a steady state solution for $x = 0$. As discussed above, this is the trivial case of no bacteria, and hence no growth. For all other initial conditions, we would generate unbounded growth and the model does not approach a steady state.

Even if they exist, steady states are only reached after an infinite amount of time. Again, this can be easily seen from the solution of (4.19). The second term of the solution (4.20) becomes very small very fast (that is as time increases), but it only strictly vanishes for $t = \infty$. In practice, the state of a system can be indistinguishable from steady state after relatively short times. In this sense, "infinite amounts of time" can be very short. Hence steady state behaviors are of practical importance in biology even before infinite amounts of time have passed.

4.7 Determine the steady states of the differential equation: $\dot{x} = ax^2 - bx + c$. Assume that we are only interested in real and positive steady states.

4.8 By pure inspection, that is without using paper and pen or the help of a computer, determine at least one steady state of the differential equation: $\dot{x} = x\left(e^{\sqrt{x^5 \sin x}} - ax^b\right)$

4.3.3 Bacterial Growth Revisited

Let us now come back to bacterial growth, but drop the unrealistic assumption of infinite and inexhaustible nutrient supply. In growth experiments in laboratories it is common to supply the colony with a certain amount of nutrient and to let it grow. As the nutrient becomes exhausted, bacterial growth by necessity has to stop. We can take this into account by explicitly introducing a variable, $y(t)$, that describes the amount of nutrient available.

This means that we now need to take care of two variables: $x(t)$, describing the size of the bacterial colony; and $y(t)$, describing the remaining amount of nutrient. Clearly, we would expect that the behavior of these variables over time is coupled. The more bacteria there are, the faster the nutrient will be exhausted. We assume that the change of nutrient in time is proportional to the number of bacteria. This is an assumption, but a reasonable one and corroborated by practical experience. It is also proportional to the amount of nutrient available; clearly, if there is no nutrient left, none can be reduced. Hence, altogether, we obtain a differential equation for nutrient depletion:

$$\dot{y} = -xy \tag{4.23}$$

Note the negative sign here that indicates that nutrient is used up by the bacteria.

Having described how the nutrient changes, we still need to write an equation that links growth to nutrient availability. The idea of formulating the differential equation is similar: As discussed above, growth of bacteria is proportional to the size of the population, i.e., $x(t)$. Furthermore, we can assume that more nutrient implies faster growth, i.e., there is proportional dependence of the growth rate on the availability of nutrients. Using these assumptions we can then formulate a growth rate for the bacteria that mirrors (4.23):

$$\dot{y} = -xy \quad \text{and} \quad \dot{x} = axy \tag{4.24}$$

Note here the parameter a that describes how fast the growth of the colony is compared to the depletion of nutrients. This is perhaps the simplest example of a system of coupled differential equations. The differential equations of x and y are coupled in the sense that both $x(t)$ and $y(t)$ feature in both equations, which makes it impossible to solve them independently.

In the present case, there is a simple procedure to decouple them. The decoupling procedure is based on the observation that a variable can be found in the system that is *conserved*. Intuitively, this is simple to understand. The only way a unit of nutrient can "disappear" is by being converted into bacterium.

We can exploit this formally by solving (4.23) for y. The result $y = -\dot{y}/x$ can be inserted into the second equation of the system (4.24). We then obtain the relation

$$\dot{x} + a\dot{y} = 0$$

Another, slightly clearer, way of writing this is:

$$\frac{d}{dt}(x + ay) = 0 \tag{4.25}$$

This means that the expression $x + ay$ is constant in time, because the derivative is constant. Since (4.25) is true at any time during the evolution of the system, it is certainly true at time $t = 0$. Denoting $x(0)$ by x_0 we get:

$$x + ay = x_0 + ay_0$$

(To interpret this equation, it is useful to remember that x is an abbreviation for $x(t)$.) We can now use this equation to express y as a function of x.

$$y = \frac{x_0 + ay_0}{a} - \frac{x}{a} \tag{4.26}$$

We have now nearly succeeded in decoupling the differential equations in (4.24). All that remains to be done is to substitute the newly found expression for y into the differential equation for $\dot{x}$, abbreviating $x_0 + ay_0$ by b.

$$\dot{x} = bx - x^2 \tag{4.27}$$

We have now genuinely reduced the coupled system (4.24) to a single differential equation. We only need to solve this differential equation because we can then obtain the solution for $y(t)$ by using (4.26).

The solution, as follows, can be obtained by using the separation of variables technique again.

$$x(t) = \frac{bC \exp(bt)}{C \exp(bt) - 1} \tag{4.28}$$

4.9 Use the separation of variables technique to show that the solution for $x(t)$ is (4.28).

We still need to determine the integration constant C. As usual, we do this by specifying the initial conditions.

$$x(0) = x_0 = \frac{(x_0 + y_0)C}{C - 1}$$

$$\implies C = \frac{x_0}{-x_0 - y_0 + x_0} = \frac{x_0}{x_0 - b}$$

4.10 What happens when $x_0 = b$?

The steady state behavior can again be obtained by considering the behavior at infinite times. In this case, the exponential terms become very large and the fraction will approach 1. Therefore, at steady state, the size of the bacterial colony will be given by,

$$x(\infty) = b \tag{4.29}$$

We can obtain the same result directly from the differential (4.27) by analyzing the case of no change of the state variables, in this case $\dot{x} = 0$.

$$\dot{x} = 0 = bx - x^2$$
$$\implies x = b$$

Steady states are commonly indicated by the variable name followed by a star. In addition to $x^* = b$, the system also allows a trivial steady state given by $x^* = 0$.

Now that we have the solution for the time behavior of the bacterial colony, we can also solve for the concentration of nutrients. For this we use our solution for $x(t)$ and (4.26).

$$y = \frac{b}{a} - \frac{x}{a}$$

Just to convince ourselves of the soundness of the solution, let us check that the initial and steady state values are correct.

$$y^* = y(\infty) = \frac{b}{a} - \frac{x^*}{a} = 0$$

This can be obtained by substituting the steady state solution of $x(t)$ for x^*. For the initial value we get,

$$y_0 = y(0) = \frac{b}{a} - \frac{x_0}{a} = \frac{x_0 + ay_0 - x_0}{a} = y_0$$

For illustration, we plot the solution of our differential equation in Fig. 4.6 for various initial conditions. Note that, in this particular model, the steady state depends on the initial conditions. This is essentially due to the fact that there is a conserved property in the model, as discussed above. This is not the typical case for differential equation models in biology. Normally, cells die or molecules are broken down. This introduces dissipative elements into the model which prevent conservation relations from remaining valid.

Even when there are no conservation properties to be obeyed, it can still be the case that the steady state depends on the initial conditions. This is the case when the system supports more than a single steady state. This situation partitions the

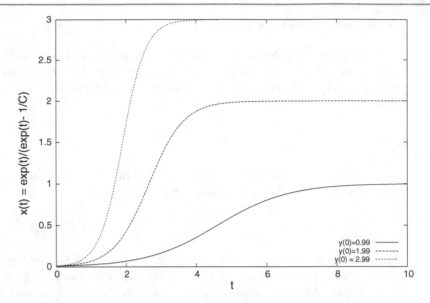

Fig. 4.6 The dynamic behavior of (4.28) for $x_0 = 0.01$, $a = 1$ and various initial amounts of nutrient

phase space of the system into so-called *basins of attraction* for the relevant steady states. We have already encountered a trivial example of this above, in the case of the bacterial growth model. Any initial condition corresponding to no bacteria being in the system causes the system to remain in this state forever. In this instance, the basin of attraction corresponds to all states $x(0) = 0$. We will see more interesting examples of this below.

4.4   Case Study: Malaria

We are now in a position to look in detail at somewhat more advanced models. As a specific system, let us consider a model of Malaria (or, in fact, any other vector-borne disease) and its spread. Extensive details of a similar model were given in Chap. 2 in the context of agent-based modeling. We have vectors of infection (the mosquitoes) that infect humans by biting them. Mosquitoes are only infectious if they have been infected themselves. We assume that both mosquitoes and humans remain infected for a certain amount of time before they return to health.

This rather qualitative information does not completely specify the differential equation model we want to formulate, but it helps us to develop a first model that will then enable us to obtain an idea of the behavior of the system. We approach the modeling process as in the examples above; that is, we start by thinking about how changes in the variables affect the spread of an infection.

Considering the infection dynamics for humans first. Qualitatively we can state that, the more infected mosquitoes there are, the more humans will become infected. To be more precise: The greater the number of infected mosquitoes there are, the higher the chance per time unit that an uninfected human becomes infected. It is unclear, at first, exactly what the dependence of the infection rate on the prevalence of infection among mosquitoes is. As a first model, and in the absence of better data, we can assume linear proportionality. Since only uninfected humans can become infected, we need to assume that the infection rates among humans also depend on the number of uninfected individuals (rather than on the total number of humans).

Next we need to decide on how humans lose their infection. In reality, there are two possible routes. Either one is cured or one dies. Once one has suffered a bout of Malaria one tends to acquire some immunity against the disease, making future infections both less likely and less severe. Including this in a differential equation model would require introducing a third category of humans beside the infected and the uninfected ones. This is fully feasible, yet we are not primarily interested in Malaria here, but in the general process of how to construct differential equation models. Therefore, we will ignore this additional complicating factor. Similarly, we will not explicitly model the fact that infected sufferers can die of the disease. Our model population will be constant and we will assume that individuals lose their infection after some time and return to the pool of uninfected individuals. Differential equations cannot, of course, keep track of individuals but always consider large amorphous populations. The way to model the "curing" process is to assume that infected humans lose their infection at a certain constant rate. Formally, the problem is equivalent to the case of protein decay above. In any time period, the actual number of curing events will depend on the number of infected individuals as well as the curing rate. Hence, altogether the total rate of curing is proportional to the fixed curing rate constant and the number of infected humans.

As far as the mosquitoes are concerned, we can make exactly the same arguments regarding the infection and curing rates, although with different proportionality constants. This reflects that the infection dynamics of mosquitoes should be assumed to be distinct, if for no other reason than that they live for a much shorter time.

Before we can formulate our model mathematically, we need to pay attention to one detail. As far as the infection rate is concerned, what counts is not the absolute number of infected individuals, but the proportion of infected individuals in the entire population. To reflect this aspect in our model let us denote the proportion of infected humans by $s(t)$; the proportion of healthy individuals is then given by $1 - s(t)$, assuming that there are no states between healthy and sick. Similarly, let us denote the proportion of infected mosquitoes by $m(t)$; in close analogy with the human population, the proportion of uninfected individuals is given by $1 - m(t)$. By adopting this form, we do not need to write separate formulas for infected and uninfected populations, since the proportion of infected and uninfected individuals can be immediately obtained from one another.

Taking all this into account, we arrive at a differential equation model for Malaria infection:

$$\dot{s} = a\,(1 - s)\,m - bs$$
$$\dot{m} = c\,(1 - m)\,s - dm \tag{4.30}$$

This model formulates our assumptions about the factors affecting the growth of infections in Malaria. The first term on the right hand side of the first line reflects the fact that the growth of the infections among humans depends on the size of the uninfected population (represented by $1 - s$) and the size of the infected mosquito population m. Note that other terms also have parameters, denoted by a, b, c and d. These system parameters assign weights to the various growth and shrinkage terms and normally need to be measured (that is, they cannot be inferred by an armchair modeler).

In this particular model, the parameter c can be interpreted physically. It summaries two effects, namely the probability of transmitting a disease should a human be bitten by a mosquito, and the probability of a human being bitten in the first place should human and mosquito "meet." The parameter a is interpreted in an analogous fashion. The parameters b and d determine the rate with which humans and mosquitoes, respectively, lose their infection.

Unlike the simpler models we have encountered so far in this section, this one cannot be solved analytically with elementary methods, and we will therefore not attempt to do so. This is a pity, but it is the norm for the typical model one tends to encounter in biology. Luckily, nowadays it is easy to obtain numerical solutions to differential equation models using computer algebra systems, such as Maxima (see Chap. 5). The disadvantage of this approach is that, in order to generate a solution, we need to specify values for all the parameters of the model (a, b, c, d). Any solution we generate is only valid for this particular set of parameter values, and the behavior can be quite different for other sets of values (Fig. 4.7). Unless the parameters are well specified, for example experimentally measured, one has to spend quite some time exploring the possible behaviors of the model by generating numerical solutions for many combinations of parameters.

Even when an analytic solution is not possible using elementary techniques alone, one can still often extract some useful general information about the behavior of the system by analyzing its steady state behavior. We can easily calculate the steady state behavior by setting $\dot{s} = 0$ and $\dot{m} = 0$ and solve the resulting system of equations in (4.30). From this we obtain a solution for the steady state behavior of the model in terms of the parameters.

$$s^* = \frac{ca - db}{c\,(a + b)} \qquad m^* = \frac{ca - db}{a\,(c + d)} \tag{4.31}$$

Figure 4.8 illustrates how the steady state varies with b when we keep the other parameters fixed at $a = 1, c = 1$ and $d = 2$. From the figure it is clear that positive solutions can only be obtained when $b < 0.5$. This can also be seen from (4.31). If we substitute the chosen values of a, c and d into the first equation, then we obtain

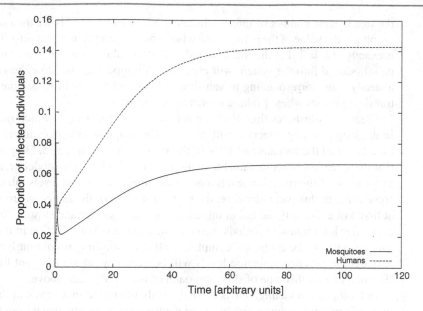

Fig. 4.7 A possible solution for the differential equation system (4.30) describing the spread of Malaria. Here we used the parameters: $a = 1, b = 0.4, c = 1, d = 2$

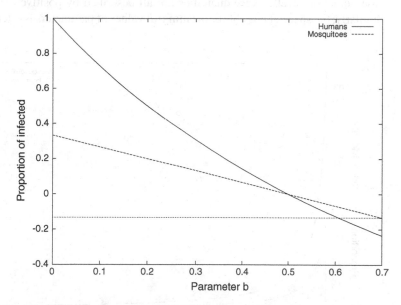

Fig. 4.8 The steady state infection levels for mosquitoes and humans of the model (4.30); these two curves illustrate (4.31). Here we kept the parameters, $a = 1, c = 1, d = 2$ fixed and varied b

the steady state number of infected humans, s^*, in terms of b. If we then set $s^* = 0$ we obtain the value of the parameter b where the steady state reaches zero. This point is exactly at $b = \frac{1}{2}$. For this value, or indeed larger values of b any infection that may be introduced into the system will eventually disappear again. Note, however, that a steady state corresponding to vanishing infection levels is fully compatible with transient periods where Malaria is present in the system.

Figure 4.9 illustrates this. Here we set the parameter $b = \frac{1}{2}$, which means that, in the long run, any infection will die out. The graph shows an example where at time $t = 0$ all the humans and 10 % of the mosquitoes are infected. Despite the fact that the infection is not sustainable in the long run, initially the model even leads to an increase of the infection levels among the mosquitoes. The levels fall only after some time. In this particular case, the rates of decline of the infection are very high at first, but eventually the fall in infections becomes slower and slower. As a result, even after long transient periods there is still a low level of infection in the system. Malaria will only be eradicated completely after infinite amounts of time have passed although, in practice, infection levels will be very low after a long but finite time. We encountered this type of slow approach of the steady state above.

In biological modeling, one is normally only interested in a subset of the mathematically possible solutions of differential equations. Typically one wishes to predict quantities such as concentrations of molecules, particle numbers or, as in the present case, infection levels. These quantities are all described by positive real numbers. Similarly, the space of possible or meaningful values of parameters is often restricted

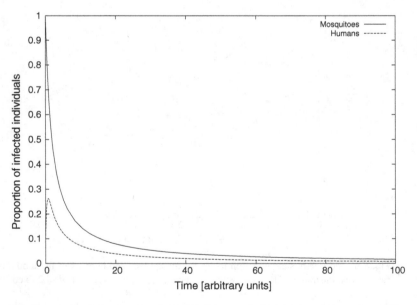

Fig. 4.9 The transient behavior of model (4.30). We started with initial conditions where all humans and 10 % of mosquitoes are infected. We used the parameters, $a = 1, b = 0.5, c = 1, d = 2$. Initially, the infection level of mosquitoes even rises before it eventually falls

by the physical meaning. For example, as far as the Malaria model is concerned, we are only interested in positive values for the four parameters a, b, c and d of the system. These all represent proportionality constants and it would not makes sense to set any of these to negative values, even though mathematically there is no reason not to do this. A negative value for any of these parameters would completely pervert the meaning of the model.

We should, however, spend some time thinking about the areas where our model might predict negative infection rates, which would make no physical sense. A glance at Fig. 4.8 reveals, rather troublingly, that for $b > \frac{1}{2}$ there will be a negative steady state. The reader is encouraged to convince herself of this using (4.31).

This is potentially a problem for the interpretation of the model with respect to the real situation. There is nothing inherently unphysical or unreasonable about assuming that $b > \frac{1}{2}$; the parameter b simply formulates how fast people recover from Malaria. It might well be that in real systems $b < \frac{1}{2}$, but this does not change the fact that $b > \frac{1}{2}$ still is a meaningful choice of parameter. Still it seemingly leads to the unphysical prediction of negative infection levels in the long run. If this were indeed so, then this would indicate a problem of the model: There is no use having a model that sometimes produces unrealistic behavior.

Let us see what happens when we increase the parameter b beyond 0.5. Figure 4.10 illustrates the dynamics predicted by the model for $b = 0.6$ for various initial conditions ranging from $s(0) = -1$ to $s(0) = 1$; (all graphs shown assume that $m(0) = 1$). All the example curves we plotted predict that, in the long run, infections among

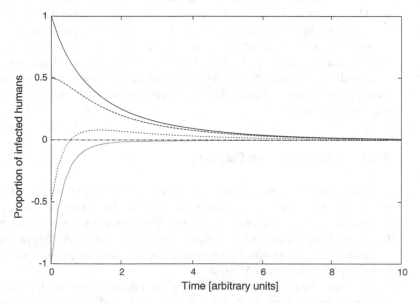

Fig. 4.10 The transient behavior of model (4.30) for various initial conditions. We used the parameters, $a = 1, b = 0.6, c = 1, d = 2$, and an initial condition of $m(0) = 1$

humans will vanish; that is, the negative steady state is not approached after all. If we
start with negative infection rates then, at least for a transient time $s(t) < 0$ but again
in the long run, infection levels tend to zero. The negative transient levels are neither
surprising nor disconcerting. After all, by choosing initial conditions corresponding
to physically unrealistic values, we provided garbage as input, and are accordingly
rewarded with garbage as output. On the other hand, if we start with the physically
realistic initial conditions, that is $0 \leq s(0) \leq 1$ then the model predicts positive
infection rates throughout, only tending to zero in the long run. This is reassuring.
It is not watertight evidence for the correctness of the model (whatever we mean by
"correct"), but at least it means that, if fed on reasonable input, the model does not
produce meaningless output.

While we can feel a little reassured about the predictions of the model in this
particular case, we should start to have some doubts about the predictions of the
steady state behavior of the model. Following our steady state analysis in (4.31) and
Fig. 4.8 we would expect the model to show a negative steady state. This is clearly
not what we observe, at least in the example depicted in Fig. 4.10.

One immediate resolution for this apparent discrepancy is that the system allows
more than two steady states, as discussed above. One of them corresponds to vanishing infection levels in both mosquitoes and humans, or simply $s* = m* = 0$. While
trivial, this is a physically realistic steady state. The second, for $b > 0.5$, corresponds
to physically infeasible solutions $s^* < 0$ and $m^* < 0$.

In order to understand the situation in more depth, it is necessary to introduce a
new concept, namely that of *stability* of a steady state point. Mathematically, much is
known about the stability of steady state points. A thorough treatment of the topic that
would do justice to the wealth of available knowledge is far beyond the scope of this
book. The reader who wishes to deepen her knowledge on this issue is encouraged
to consult a more advanced text such as Casti's excellent and very accessible two
volume "Reality Rules" [2,3]. Yet, even without entering the depths of dynamical
systems theory, it is possible to convey the main ideas. Despite being elementary this
basic understanding is often sufficient to master the demands of real world modeling
enterprises in biology.

4.4.1   A Brief Note on Stability

Let us assume a population of uninfected humans and mosquitoes, i.e., $s(t) = m(t) =$
0. Clearly, unless Malaria is somehow actively introduced into the system, the infection levels will remain at zero for all time. The question we want to ask now is the
following: What happens to the system if we introduce the infection into the system
at a small level, for example by releasing a few infected mosquitoes, or by the arrival
of a few people who carry Malaria?

There are two answers to this question: (i) It could be that an infection will gain
a foothold and establish itself. Alternatively, (ii) it might not establish itself and the
system reverts, after a short spell of Malaria, to the state where $s(t) = m(t) = 0$.
In essence, this is the intuition behind what mathematicians call the stability of a

steady state point of a system of differential equations: How does a system at steady state respond to a small disturbance? The first case, where a small disturbance to the steady state spreads and leads to the infection establishing itself in the long run, represents the case where the steady state corresponding to $m^* = s^* = 0$ is unstable. The second case, where the disturbance does not spread, the steady state is stable. In general, when modeling biological systems, we are interested in stable steady states, because these are the ones we will observe in reality. Real systems are permanently subject to more or less violent shocks that push them away from any unstable steady states they might inhabit. Unstable steady states persist just as long as a ping-pong ball would rest at the top of a convex surface, or a pool cue would remain upright balanced on its tip when exposed to a slight draught or vibration.

As it turns out, there is a fairly simple method to determine the stability of a steady state. It is based on determining the *eigenvalues* (see Appendix A.4.1) of the *Jacobian* matrix of the system at the steady state. Again, very much is known and can be said about the stability of dynamical systems (this is what systems of differential equations are sometimes called). However, much can be done with elementary methods that can be learned with relatively little effort. In the following paragraphs we will therefore introduce some methods that can be applied without going into the theory of the topic. Nevertheless, the reader is urged to eventually consult more specialized volumes to acquire more sophisticated skills backed up by a thorough understanding of the theoretical background.

Before describing how to determine the stability of steady states, we need to briefly define what the Jacobian matrix is. Assume a system of differential equations of the general form:

$$\dot{x}_1 = f_1(x_1, \ldots, x_n) \tag{4.32}$$
$$\dot{x}_2 = f_2(x_1, \ldots, x_n) \tag{4.33}$$
$$\ldots \tag{4.34}$$
$$\dot{x}_n = f_n(x_1, \ldots, x_n) \tag{4.35}$$

This system of equations is nothing but a generalized form of the differential equations we have encountered before. In the Malaria model (4.30) we have to take $x_1 \equiv s$ and $x_2 \equiv m, f_1(s, m) \equiv a(1 - s)m - bs$ and so on.

Using this form the Jacobian is defined as the $n \times n$ matrix of derivatives of f_i with respect to x_j.

$$J_{i,j}(\mathbf{x}) = \frac{\partial f_i(x_1, \ldots, x_n)}{\partial x_j} \tag{4.36}$$

In the case of $n = 3$, for example, the Jacobian would be a 3×3 matrix.

$$\mathbf{J}(x_1, x_2, x_3) = \begin{bmatrix} \frac{\partial}{\partial x_1}f_1(x_1, x_2, x_3) & \frac{\partial}{\partial x_2}f_1(x_1, x_2, x_3) & \frac{\partial}{\partial x_3}f_1(x_1, x_2, x_3) \\ \frac{\partial}{\partial x_1}f_2(x_1, x_2, x_3) & \frac{\partial}{\partial x_2}f_2(x_1, x_2, x_3) & \frac{\partial}{\partial x_3}f_2(x_1, x_2, x_3) \\ \frac{\partial}{\partial x_1}f_3(x_1, x_2, x_3) & \frac{\partial}{\partial x_2}f_3(x_1, x_2, x_3) & \frac{\partial}{\partial x_3}f_3(x_1, x_2, x_3) \end{bmatrix} \tag{4.37}$$

The basic structure of a Jacobian matrix is very simple. It is a list of derivatives of the right hand sides of the differential equations with respect to all the system variables.

It is not the Jacobian directly that provides insight into the stability properties of a steady state, $\mathbf{x}^* = (x_1^*, \ldots, x_n^*)$, but rather the set of eigenvalues $\mathscr{E}$ of the Jacobian matrix at the relevant steady state, i.e., the matrix $\mathbf{J}(\mathbf{x}^*)$. The members in the set $\mathscr{E}$ are numbers, possibly complex numbers, although we are primarily interested in the real parts of the eigenvalues. These can be directly used to classify the stability of any steady state according to the following two rules.

- If the real parts of all eigenvalues in $\mathbf{J}(\mathbf{x}^*)$ are negative, then $\mathbf{x}^*$ is stable.
- If at least one eigenvalue in $\mathbf{J}(\mathbf{x}^*)$ is positive, then $\mathbf{x}^*$ is not stable.

With this new tool in hand, let us analyze the stability of the Malaria model. Remember that the model was given by:

$$\frac{d}{dt}s = a(1-s)m - bs, \qquad \frac{d}{dt}m = c(1-m)s - dm \qquad (4.30)$$

In the Jacobian matrix, the top left entry of the matrix will be:

$$J_{1,1} = \frac{\partial}{\partial s}(a(1-s)m - bs)$$

Following (4.36), the full Jacobian of the system will be:

$$\mathbf{J}(s, m) = \begin{bmatrix} -am - b & a(1-s) \\ c(1-m) & -cs - d \end{bmatrix} \qquad (4.38)$$

We wish to evaluate the stability of the steady state $m^* = s^* = 0$; so we need to determine the Jacobian at this point.

$$\mathbf{J}(s^*, m^*) = \begin{bmatrix} -b & a \\ c & -d \end{bmatrix} \qquad (4.39)$$

Using a computer algebra system, we can now determine the eigenvalues of this matrix and obtain:

$$E_1 = -\frac{d}{2} - \frac{b}{2} + \frac{1}{2}\sqrt{d^2 - 2db + b^2 + 4ca}$$

$$E_2 = -\frac{d}{2} - \frac{b}{2} - \frac{1}{2}\sqrt{d^2 - 2db + b^2 + 4ca}$$

To illustrate the stability of the eigenvalues we must choose specific parameter values. Figure 4.11 plots the eigenvalues for the various values of b while keeping $a = 1$, $c = 1$ and $d = 2$. This is the parameter choice corresponding to Fig. 4.10. The steady state analysis predicts the existence of a negative steady state if $b > \frac{1}{2}$, which, however, the system does not seem to approach when we choose positive initial conditions

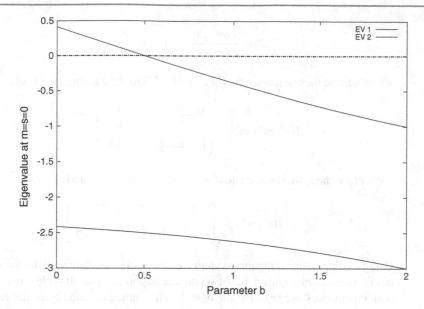

Fig. 4.11 The eigenvalues of model (4.30) at the steady state $s^* = m^* = 0$. We used the parameters, $a = 1, c = 1, d = 2$ and we varied b. For $b < 0.5$ the steady state is unstable, but it is stable for $b > 0.5$

(luckily). After performing our elementary stability analysis, we can now explain why not.

Figure 4.11 shows that one of the eigenvalues is positive which means that the steady state $m^* = s^* = 0$ is unstable for $b < \frac{1}{2}$. Introducing a small disturbance will lead to the system moving away from this steady state; hence, for this range of parameters, the system will approach the positive steady state. On the other hand, for $b > \frac{1}{2}$, Fig. 4.11 shows that both eigenvalues are negative, and hence the steady state is stable. This means that, even after "small disturbances", the system will revert to the steady state of $m^* = s^* = 0$. Incidentally, such small disturbances could, in fact, be quite large. As we see in Fig. 4.10, if we start from $s_0 = -1$ or $s_0 = 1$ the system reverts to a state where there is no infection.

The stability of the steady state is, in essence, the reason why we need not worry about our system approaching a negative steady state, as long as we start from positive initial conditions. A trivial stable steady state is an attractor for all trajectories starting from positive initial conditions; once at the stable steady state, there is no more escape from it.

We have now established that for the parameters $a = c = 1$ and $d = 2$ there is a stable steady state corresponding to $s = m = 0$ when $b > \frac{1}{2}$. This steady state is unstable when $0 < b < \frac{1}{2}$. This suggests that, in this case, the other non-trivial steady state is the stable one. We can confirm this using our Jacobian. What we are interested in is the Jacobian at the steady state that we calculated earlier for the system.

$$ s^* = \frac{ca - db}{c\,(a+b)} \qquad m^* = \frac{ca - db}{a\,(c+d)} \qquad (4.31) $$

We substitute these expressions for s^* and m^* into the Jacobian in (4.38):

$$ \mathbf{J}(s^*, m^*) = \begin{bmatrix} -\frac{ca-db}{c+d} - b & a\left(1 - \frac{ca-db}{c(a+b)}\right) \\ c\left(1 - \frac{ca-db}{(c+d)a}\right) & -\frac{ca-db}{a+b} - d \end{bmatrix} $$

We replace the symbols a, c and d with the chosen values and obtain the Jacobian:

$$ \mathbf{J}(s^*, m^*) = \begin{bmatrix} -1/3 - b/3 & 1 - \frac{1-2b}{1+b} \\ 2/3 + 2b/3 & -\frac{1-2b}{1+b} - 2 \end{bmatrix} $$

Using a computational algebra system we can easily calculate and plot the eigenvalues for the system. Figure 4.12 shows the corresponding plot. It is clear from this that both eigenvalues are negative for $b < \frac{1}{2}$. This indicates stability of the non-trivial steady states in this area. So at $b = \frac{1}{2}$ the non-trivial and the trivial steady states not only coincide, but also swap stability.

In summary, for this particular choice of parameters, the following picture is now emerging:

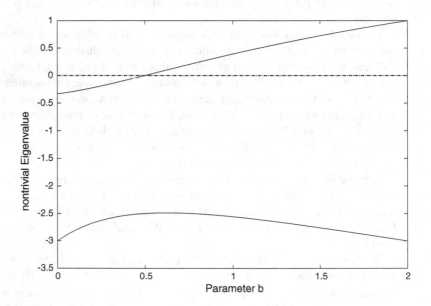

Fig. 4.12 The eigenvalues of model (4.30) at the steady states defined by (4.31). We used the parameters, $a = 1, c = 1, d = 2$ and we varied b. For $b < 0.5$ the steady states are stable, but unstable for $b > 0, 5$

- The system has two steady states, namely the trivial steady state ($s^* = m^* = 0$) and the steady state given by (4.31).
- If we choose $a = c = 1$ and $d = 2$ then the non-trivial steady state is positive and stable for $b < \frac{1}{2}$.
- The trivial steady state is stable for $b > \frac{1}{2}$.

With renewed trust in the sanity of differential equation models we can now extend our analysis to other parts of the parameter space. Without giving the details, we will find that both eigenvalues are negative if $b < \frac{ca}{d}$. Similarly, the steady state defined by (4.31) coincides with the trivial steady state when $b = \frac{ca}{d}$. This can be shown by setting the right hand sides of (4.31) to zero and solving for b. Since the steady state equations are strictly decreasing functions of b, this means that the non-trivial steady state is always positive and stable for $b < \frac{ca}{d}$ and negative and unstable if $b > \frac{ca}{d}$. We encourage the reader to work out the details of this argument to sharpen her grasp of the use of these elementary, but very useful techniques, even if she has no immediate interest in the spread of Malaria.

4.5   Chemical Reactions

Important targets of biological process modeling are chemical equations. Mathematical models can be used to analyze signal transduction cascades, gene expression, reaction diffusion systems and many more classes of biologically relevant systems. Given the importance of the field, we will provide some examples of how chemical systems can be modelled. However, again the main purpose of the case studies is to demonstrate how to translate a qualitative understanding of a system into a formal and quantitative differential equation model. We therefore recommend this section to the reader even is she harbors no particular interest in chemical process modeling.

Let us start with a simple example. Assume we have two chemical species X and Y that react to become Z with a rate of k_1. Let us further assume that the reaction is reversible; Z decays into its constituent parts with a rate of k_2:

$$X + Y \underset{k_2}{\overset{k_1}{\rightleftharpoons}} Z$$

In order to describe this set of reactions using differential equations we use the law of *mass action*. This simply states that the rate of a reaction is proportional to the concentration of the reactants. In this particular case, the rate of conversion to Z is proportional to the concentrations of X and Y. If we denote the concentration X by x, and similarly for Y and Z, then using the law of mass-action, we obtain a differential equation for the system.

$$\dot{z} = k_1 xy - k_2 z$$
$$\dot{y} = -k_1 xy + k_2 z$$
$$\dot{x} = -k_1 xy + k_2 z \qquad (4.40)$$

Taking a second glance at this system of equations with our (now) trained eye, we see immediately that the equations for $\dot{x}$ and $\dot{y}$ are the same. Furthermore, the equation for $\dot{z}$ is just the negative of that for $\dot{x}$ and $\dot{y}$. This suggests that we can perhaps describe the system with just a single differential equation.

Indeed, there is an intrinsic symmetry in the system. Whenever we lose a molecule of X, then we gain one of Z, and vice versa. Similarly, if we lose a molecule of Y, then we gain one of Z. In other words, the sum of the concentrations of X and Z and Y and Z respectively, must be constant.

$$x = X_0 - z \quad \text{and} \quad y = Y_0 - z$$

Here X_0 and Y_0 correspond to the total concentration of species X and Y in the system, and comprise both the bound and the free form of the molecule. Note that the variables $x(t)$ and $y(t)$ only measure the free form. Using these relations we can re-write the first line in (4.40):

$$\dot{z} = k_1 (X_0 - z)(Y_0 - z) - k_2 z$$

We can solve this differential equation using the separation of variables technique:

$$z(t) = \frac{k_1 X_0 + k_1 Y_0 + k_2 \theta \tan\left(\frac{1}{2} t\theta + \frac{1}{2} C\theta\right)}{2k_1} \qquad (4.41)$$

Here θ is a place-holder for a longer expression:

$$\theta \doteq \sqrt{2 k_1{}^2 X_0 Y_0 - k_1{}^2 X_0{}^2 - 2 k_1 X_0 k_2 - k_1{}^2 Y_0{}^2 - 2 k_1 Y_0 k_2 - k_2{}^2}$$

The solution in (4.41) contains an integration constant, C, that still needs to be determined. We choose the concentration of Z to be vanishing at $t = 0$, thus obtaining:

$$C = -2\theta^{-1} \arctan\left(\frac{k_1 X_0 + k_1 Y_0 + k_2}{\theta}\right)$$

For a set of parameters we have plotted $z(t)$ in Fig. 4.13. As expected, if we start from $z(0) = 0$ then the concentration of Z increases relatively rapidly, but then settles at a steady state. We leave a further exploration of the model as a set of exercises.

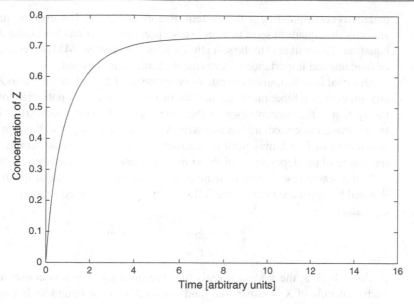

Fig. 4.13 An example solution for $z(t)$ in (4.41). We used the parameters, $k_1 = k_2 = X_0 = Y_0 = 1$

4.11 Calculate the steady state of (4.40), both from the differential equation directly and from the solution in (4.41).

4.12 Plot the solutions for $x(t)$ and $y(t)$ using the parameters in Fig. 4.13.

4.13 Without explicitly plotting the solution, what would happen if we choose a vanishing forward reaction rate, i.e., $k_2 = 0$. What would be the steady state of Z? What would be the steady states of X and Y?

4.14 Confirm your conjectures by explicitly checking this in the model.

4.5.1   Michaelis–Menten and Hill Kinetics

Let us now modify the chemical system in (4.40) and give X a catalytic property:

$$X + Y \underset{k_2}{\overset{k_1}{\rightleftharpoons}} W \overset{k_3}{\rightarrow} X + Z$$

In this reaction scheme we have a molecular species X (for example an enzyme) that catalyses the reaction $Y \rightarrow Z$. The process proceeds in two steps. The first step corresponds to the formation of an intermediate compound W of the catalyst X with the substrate Y. After some time, W decays into the product Z and the original catalyst X. We could now attempt to formulate and solve the differential equations corresponding to this system. Before doing this, however, we are going to simplify the equations so as to be able to express the reactions in a single differential equation. This will lead to the so-called *Michaelis-Menten* (MM) kinetics, which are of fundamental importance for enzyme dynamics and beyond.

Our goal is to formulate the rate of conversion of the substrate Y into Z as a one-step process as a function of the amount of the enzyme X. A potential bottleneck in the system is the concentration of the enzyme X. The conversion from Y to Z has to go through intermediate binding with X; if there are only few X around, then the availability of free X may limit the conversion rate. MM-kinetics exactly describes the nature of the dependence of the conversion rate on the basic system parameters.

Using lower-case letters to denote the concentration of the respective species denoted by upper case letters, the full system can be described using two differential equations:

$$\dot{w} = k_1 xy - k_2 w - k_3 w$$
$$\dot{z} = k_3 w \tag{4.42}$$

A closer look at the system reveals that the quantity $w + x$ is conserved; that is, each molecule of X is either free (and counted by x) or bound with a molecule of W in which case it is counted by w. Adopting a similar notation as above, we have $x = X_0 - w$.

$$\dot{w} = k_1 (X_0 - w)y - k_2 w - k_3 w$$
$$\dot{z} = k_3 w \tag{4.43}$$

Next we will make a so-called *quasi steady state* approximation. This means that we assume that the first reaction in (4.43) happens much faster than the second. Quasi-steady-state approximations of this sort can be very useful in systems where parts operate on different time-scales. They are commonly encountered in the modeling literature. In this particular case, the quasi steady state approximation is possible because we assume that the reaction rate k_3 is very small compared to k_1 and k_2. If this condition is fulfilled, then we can replace the first differential equation by a normal algebraic equation in that we set its right-hand side to zero. This effectively decouples the two differential equations. This general procedure can also be used when the system of differential equations is very large, and is by no means limited to systems with two equations only. The only condition for its successful applications is that the quasi-steady state assumption—that parts of the system operate on different time-scales—is fulfilled.

$$k_1 (X_0 - w)y - k_2 w - k_3 w = 0 \quad \Longrightarrow \quad w = X_0 \frac{y}{y + \underbrace{\frac{k_2 + k_3}{k_1}}_{\doteq K}}$$

The benefit of taking the quasi-steady state approximation becomes immediately clear now. We have managed to express w, the concentration of the intermediate product that is of no interest to us, as a function of y and the *total* amount X_0 of catalyst in the system, rather than the free amount $x(t)$. We can now substitute this into

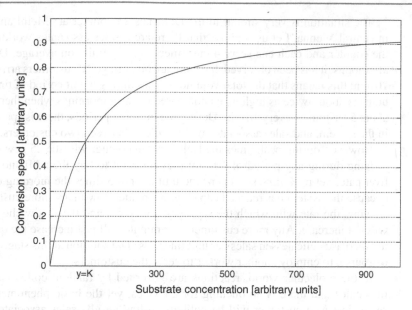

Fig. 4.14 The MM-function for $K = 100$. It describes the speed with which the substrate is converted into the product

the second line of (4.43) to obtain the equation for the Michaelis Menten equation.

$$\dot{z} = v_{max}X_0 \frac{y}{y + K} \qquad \text{(Michaelis Menten)}$$

Here we have defined $v_{max} \doteq k_3$.

Consider the two components of the rate equation. The Michaelis–Menten function, $h(y) = y/(y + K)$ changes from 0 to 1 as y goes from 0 to infinity. It describes how increased amounts of y slowly saturate the enzyme. If we assume that the concentration of the catalyst, X_0, is fixed, then we will see initially an increasing conversion speed as we add more substrate. However, as we add even more substrate, the marginal increase in the conversion speed becomes smaller the higher the concentration of substrate. At some point, the enzymes are saturated, that is a further increase of the substrate concentration leads to no increase in the conversion speed (Fig. 4.14).

The MM function $h(y)$ is actually a family of functions parameterized by the value of K. Physically, K has a very specific meaning. The MM-function takes a value of $\frac{1}{2}$ when $y = K$; hence K defines the point where the conversion speed is at half its maximum. If the concentration y increases much beyond K then the MM function saturates.

The basic intuition behind the MM kinetics is easy to understand. The concentration of X is a bottleneck in the system. Conversion of Y to Z takes some time. As the concentration of y grows, there are fewer and fewer unoccupied copies of X around and the conversion rate soon reaches capacity.

The situation is very similar to the Delicatessen counter at "Meinl am Graben" in central Vienna. Let us suppose that there are 5 sales associates working behind the counter and each can serve 4 customers every 10 min, on average. During late afternoons, it has been observed that there are on average 10 customers arriving every 10 min; this means that the total volume of sales during a unit period of time (say an hour) is about twice as high when compared to early mornings when there are only about 5 customers every 10 min. During the morning periods, there is spare capacity in the system, and sales associates will be idle in between two customers.

However, during peak times the Delicatessen counter will see 25 new customers arriving during any period of 10 min; they all want to buy their Milano salami or liver paté; yet peak time sales per time unit are not 5 times the morning equivalent because the system has reached capacity. No matter how many more customers are arriving, the rate with which the sausages can be cut and the sandwiches prepared will not increase. Any more customers per minute will just increase the queues, but not increase delicatessen sales per unit intervals. The solution at this stage would be, of course, to employ another worker to serve the customers.

In the molecular world, reactions are governed by random collisions between molecules and there is no queuing for enzymes, yet the basic phenomenon is the same. Just as a customer will be unlikely to find an idle sales associate in Meinl during peak times, so will a free molecule of Y find that the number of collisions it has with free X molecules becomes rarer as the concentration of Y increases.

The second component of the MM-kinetics is the term v_{max} defining the maximal speed with which the substrate can be converted into the product. The bare-bones MM-function $h(x)$ always varies from zero to one. The term v_{max} simply scales it to its desired range.

Empirically, it has been found that enzyme kinetics do not obey the MM-function, but rather a slightly modified kinetics. The so-called Hill function is a direct extension of the MM-function (see also Sect. 2.6.2.4, p. 59):

$$h(y) = \frac{y^h}{y^h + K^h} \qquad \text{(Hill)}$$

(Note that the MM-function is a Hill function with $h = 1$, hence we denote both as $h(x)$.) Just like the MM-function, Hill functions make the transition from 0 to 1 as the substrate concentration goes to infinity. However, a Hill function is a sigmoid function and makes this transition faster. As the coefficient h is increased, Hill functions increasingly resemble step functions. Just like the MM-function, Hill functions take a value of $\frac{1}{2}$ at $y = K$ no matter what the value of the Hill coefficient. Figure 4.15 compares some Hill functions with the MM.

In practical modeling applications one cannot normally derive the Hill coefficient from first principles and empirical data has to be used. In the context of enzyme kinetics, the Hill coefficient carries some biochemical information about the nature of the interactions at a molecular scale. A value of $h > 1$ is normally interpreted as indicating cooperativity. This means that the catalyst has more than two binding sites for the substrate and the binding of one site increases the affinity of other sites

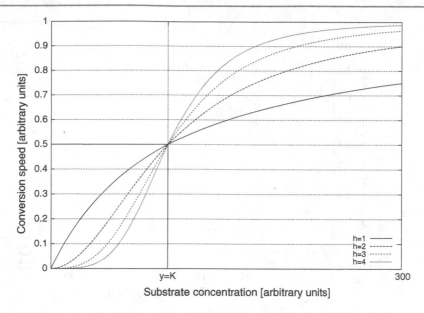

Fig. 4.15 Comparing the Hill functions with the MM-function for $K = 100$. The higher the Hill coefficient. the more step-like the function becomes

for an additional substrate. Similarly, the inverse effect—negative cooperativity—is indicated by a Hill coefficient of $h < 1$. It essentially means that the binding of one site negatively affects the affinity of further receptor sites. We will have to say more on this later on in this book when we explicitly derive a Hill-function for the binding of transcription factors to their operator sites (Sect. 6.3.3 on p. 235). For the time being, let us now show how to use the MM/Hill function in differential equations.

In biological modeling, these functions are almost ubiquitous. Unfortunately, normally differential equations involving MM or Hill kinetics are difficult to solve analytically and we have to be content with numerical solutions to our differential equations. To consider a concrete example, let us work out the model for the above system of an enzymatic reaction:

$$X + Y \underset{k_2}{\overset{k_1}{\rightleftharpoons}} W \overset{k_3}{\rightarrow} X + Z$$

Before we can do anything in terms of modeling, we have to specify how the substrate concentration $y(t)$ and the concentration of the catalyst $x(t)$ change over time. This is not clear from the above reaction scheme. For simplicity we will assume that the latter has a fixed concentration throughout, i.e., $\dot{x} = 0$ and that the former is not replenished—the initial concentration is used up over time. At this point this is an arbitrary decision. In practice, the differential equations for the substrate and enzyme are normally clear from the modeling context.

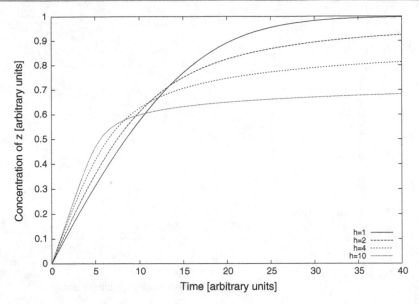

Fig. 4.16 A numerical solution for the differential (4.44) for $v_{max} = 1$, $K = 0.1$ and $x = 0.1$. Note that for higher Hill coefficients it takes longer for the substrate to be used up

Next we have to decide on the kinetic law that we wish to use to describe our catalysed reaction. For the moment we decide just on a general Hill-kinetic while remaining agnostic about the specific value of h. This leads us to a system of two differential equations.

$$\dot{y} = -v_{max}X_0\frac{y^h}{y^h + K^h}$$

$$\dot{z} = v_{max}X_0\frac{y^h}{y^h + K^h} \qquad (4.44)$$

Figure 4.16 plots some solutions for this system assuming initial conditions of $y(0) = 1$ and $z(0) = 0$. A salient feature of the graph is that the MM-kinetics lead to initially slower growth. Yet, within the time period shown in the plot, it clearly leads to the highest total conversion. Initial growth is highest for the highest Hill coefficient shown in the plot (i.e., $h = 10$), but this one also shows the poorest total amount of converted substrate after the considered time period.

This qualitative behavior can be easily explained from the properties of the MM/Hill kinetics. Hill functions become increasingly step-like as h increases (see Fig. 4.15). For our system, this means that initially, when the substrate concentration is at its maximum, Hill functions with a high h are much closer to their saturation value of 1; the MM function, on the other hand, will be comparatively low, resulting in a lower initial conversion rate. However, as the substrate is depleted and the concentration of y falls below $y = K$, a system with a higher Hill coefficient will see the conversion rate drop to zero much more quickly. The MM-function, however, is much less step like and relatively efficient conversion continues even if y is well below K.

4.15 Formulate the system of differential equations corresponding to (4.44) but under the assumption that $y(t)$ is replenished at a constant rate.

4.16 What happens qualitatively to the solution of the system (4.44) for very small and very large values of K. Confirm your conjectures by explicit numerical solutions?

4.17 What happens to the solution of the system (4.44) when $t \to \infty$?

4.18 For the system (4.44), plot the conversion rate as a function of time.

4.5.2 Modeling Gene Expression

Enzyme kinetics is an important aspect of biological process modeling. Beyond that, Hill dynamics is also commonly used to model transcription-factor (TF) regulated gene expression. As in the case of enzyme kinetics, the use of the MM/Hill function to describe the regulation of genes relies on the assumption of a separation of time-scales. We need to assume that binding of TFs to and from the operator site happens much faster than expression of gene products. Generally, this assumption is true. We will later explicitly derive the Hill equation for gene expression (see Sect. 6.2.2). For the moment, we have to ask the reader to invest faith in the correctness of our assertions.

Assume that we have a single gene that is expressed into a product x at a rate a. To make the model somewhat meaningful, we have to assume that the product is broken down with a rate b to avoid trivial and unrealistic unbounded growth rates of the concentration of the product. We have already formulated this differential equation above.

$$\dot{x} = a - bx \tag{4.19}$$

We also have found the general solution to this system.

$$x(t) = \frac{a}{b} + Ce^{-bt} \tag{4.20}$$

This is satisfactory as a very simple model of an unregulated gene. It does not capture the more important case of a regulated gene. In order to model regulated genes, we need to extend the model to include the action of a TF Y (that is present with a concentration y) on the rate of gene expression.

To begin with, let us assume that the promoter of G_x, the gene encoding protein X, is controlled by a single binding site for the TF Y which binds to its binding site with a rate constant of k_{on} and dissociates with a constant k_{off}. For the moment, let us also assume that Y is an activator. The gene G_x will then operate in two possible regimes, depending on whether Y is bound or not. If it is bound, then it will express X at a high rate a_H; otherwise it will only express Y at a "leak" rate a_L, where $a_L < a_H$. Both cases can be described separately by a differential equation of the form of (4.19). If we know the probability $P_b(y)$ that the TF is bound to its binding site, then we can combine the two differential equations into one.

$$\dot{x} = (1 - P_b(y)) \, a_L + P_b(y) \, a_H - bx$$
$$= a_L + P_b(y) \, (a_H - a_L) - bx$$

The first line just expresses the intuition that if the TF is bound to the operator site with a probability of $P_b(y)$ then the gene will express at a high rate for exactly this proportion of time. The second line is only a rearrangement of the terms on the first line. The question we need to answer now is how we can express $P_b(y)$ mathematically. As it turns out, the MM function can be used to approximate this probability:

$$P_b(y) = \frac{y}{y + K} \quad \text{if} \quad K = \frac{k_{off}}{k_{on}}$$

In this case, the MM-constant K is defined by the ratio of the unbinding rate and binding rate of the TF to its operator site. Altogether we thus obtain a differential equation describing the expression of a protein from a gene.

$$\dot{x} = \alpha + \beta \frac{y}{y + K} - bx \tag{4.45}$$

Here we renamed a_L and $a_H - a_L$ to α and β respectively, for reasons of notational simplicity. While we have not derived this expression from first principles, at least for extreme parameters it seems to make sense. For example, if we assume that the TF exists in an infinite concentration y, then Y is continuously bound to the operator site. In this case the expression rate will be $\alpha + \beta$. However, particularly for smaller K, the actual expression rate will be close to this value even for $y < \infty$. In reality, the fraction of time the TF is bound to its operator site will always be slightly below one, but probably by not very much. Clearly, the higher the binding rate k_{on} of Y to its operator site in relation to the unbinding rate, the smaller is K and (near) continuous occupation of the binding site can be achieved by lower TF concentrations.

4.19 Write down and solve numerically the full "unpacked" system of differential equations corresponding to (4.45), i.e., write out the system of differential equations for which the MM/Hill function is an approximation. Explore its dynamics and compare it to the solutions to (4.45).

In many cases a particular operator site has more than one binding site for a particular TF. Modeling these cases can be somewhat more intricate when each of the possible occupation configurations of the binding sites is associated with a different expression rate. To consider a simple case, assume that we have two binding sites and that there are only two possible expression rates, namely, the leak rate when zero or one of the binding sites are occupied, and a high rate if both are occupied. Remembering that the MM-function formulates the probability of a single binding site being occupied, we can now write down the rate equation for our system:

$$\dot{x} = \alpha + \beta \frac{y^N}{(y + K)^N} - bx \tag{4.46}$$

The high expression rate is only relevant when all N binding sites are occupied, which happens with a probability of $\frac{y^N}{(y+K)^N}$; otherwise the gene is expressed at the leak rate. While mathematically correct, this form of the kinetic equation is rather unusual and instead the gene activation function in the presence of several binding sites of a TF is more often assumed to be of the Hill form:

$$\dot{x} = \alpha + \beta \frac{y^h}{y^h + K^h} - bx$$

Similar to the case of enzyme kinetics, the Hill coefficient is normally measured rather than derived from first principles, and the Hill coefficient carries some information about the cooperativity between the binding dynamics of the underlying TF binding sites. If the TF is a monomer and it occupies N different binding sites on the operator, then $h \leq N$. The Hill coefficient h will reach its maximum $h = N$ in the case when there is perfect cooperativity between the sites. This means that the DNA-TF compound is unstable unless all binding sites are occupied. In this case, binding can only happen if N TF bind to the N binding sites simultaneously, or at least within a very short window of time. Perfect cooperativity of this kind is primarily a mathematical limiting case and in practice there will be imperfect cooperativity. This means that there will be a finite (but probably small) stability of the DNA-nucleotide compound even if only some of the N binding sites are occupied. The maximum binding times will only be achieved when all binding sites are occupied. In these cases the Hill coefficient will be smaller than N.

Often, TFs are themselves not monomers but dimers or higher-order compounds. In this case, if the Hill function is formulated in terms of the concentration of the monomer form of the TF, then h could be higher than the number of binding sites.

So far we have assumed that the TF is an activator. There is also a commonly used gene activation function for repressors, that is if the gene G_x is repressed by the TF species Y:

$$\dot{x} = \alpha + \beta \frac{K^h}{y^h + K^h} - bx$$

The Hill repressor function is simply $1 - h(x)$ and thus shares the general qualitative features of the original function. Gene expression is highest when the concentration of the repressor $y = 0$; for very high concentrations of the repressor there is only leak expression activity α.

Gene activation functions can be extended to describe a gene being regulated by multiple TFs. This is generally not difficult, but care must be taken to properly describe how the interaction between the TFs works. For example, if we have two repressor species Y_1 and Y_2 and both can independently down-regulate gene expression, then the following differential equation would be a possible way to capture the behavior.

$$\dot{x} = \alpha + \beta \frac{K^h}{y_1^h + y_2^h + K^h} - bx$$

Alternatively, if Y_1 and Y_2 only partially repress the expression of the gene, then it would be more appropriate to describe the gene activation function as a sum of two Hill repressors.

$$\dot{x} = \alpha + \beta_1 \frac{K_1^h \beta}{y_1^h + K_1^h} + \beta_2 \frac{K_2^h \beta}{y_2^h + K_2^h} - bx$$

The specific function used will depend on the particular biological context of the system that is to be modelled.

4.20 Write down an activation function for a gene that is activated by Y_1 but repressed by Y_2. Formulate the various possibilities for how the repressor and the activator can interact.

4.21 Formulate the gene activation function for the above case when the activator and the repressor share the same binding site (competitive binding).

4.6   Case Study: Cherry and Adler's Bistable Switch

Let us now look at an example to illustrate some of the ideas we have developed. A relatively simple, yet dynamically very interesting example, is the bistable genetic switch that was first described by Cherry and Adler [4]. The system consists of two genes that repress one another via their products. Let us assume the first gene G_x

produces a protein X that represses the expression of gene G_y producing Y. Similarly, Y itself also represses the expression of G_x. We can immediately write down the differential equations describing the system:

$$\dot{x} = \frac{K^h}{y^h + K^h} - bx$$

$$\dot{y} = \frac{K^h}{x^h + K^h} - by \tag{4.47}$$

Here we make the special assumption that the system is symmetric in the sense that the gene activation function of both genes has the same parameters K and h and that both products are degraded at the same rate b.

The system is too complicated to be solved analytically and it is necessary to find the solutions numerically. Given the inherent symmetry of the system—G_x and G_y are dynamically identical—we would expect that the solutions for x and y are always the same. As it will turn out though, for some parameters of the system, x and y can have different steady-state solutions. Figure 4.17 plots the steady state solutions of the system for a particular choice of parameters. Normally, we would have to specify which variables we consider but, in the present case, this is not necessary; the symmetry of the model means that x and y are always interchangeable, which does not mean that their values are always equal. In Fig. 4.17 the vertical axis records the possible steady state values as the parameter b is varied (always assuming a fixed $K = 2$).

At the high end of b, the system apparently allows a single solution only. Since we have two variables, but only one possible solution, this must mean that in this area $x^* = y^*$. At $b < 0.25$ the single steady state splits into three solutions. The dashed line are steady state solutions where $x^* = y^* = s_u$. As it will turn out, these solutions are unstable solutions and we would not expect them to be observed in a real system. Any tiny fluctuation away from this steady state would lead to the system approaching one of the other solutions.

The stable solutions s_L and s_H of the system correspond to the top and bottom (thick) lines. In this area, if one of the variables, say x, takes the steady state $x^* = s_L$ then y will take the other stable state $y^* = s_H$. Yet both steady states are possible for both variables. Which one is taken depends exclusively on the initial conditions of the system.

Biologically this can be easily understood. If, say, G_x is expressed at a high rate, then this means that the concentration of X is high in the cell. Since X is repressing Y, this also implies that the gene G_y will not be expressed at a high rate, but is repressed. Intuitively, one can thus easily understand how the symmetry of the system is compatible with the products having two different steady states. However, note that whether or not X or Y is expressed at a high rate will depend on the initial conditions of the system. If we start with a high concentration of X then we would expect that Y is suppressed and that the system remains in this steady state. On the other hand, if we start with Y initially being higher, the system will remember that and remain in a steady state defined by a low concentration of X. This is also how we need to read the results in Fig. 4.17.

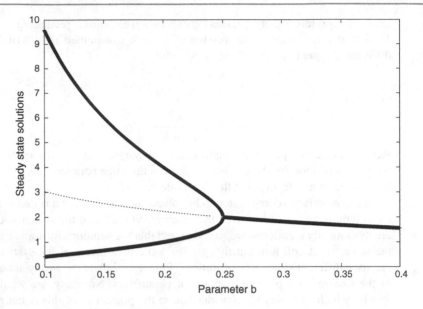

Fig. 4.17 The solutions for the system of two mutually repressing genes as defined in (4.47). We keep $K = 2$ and vary the parameter b

The dependence of the steady state on the initial conditions is illustrated in Figs. 4.18 and 4.19. Figure 4.18 shows solutions for the differential equations in phase space. The phase space shows x versus y; each point on each of the lines in this diagram corresponds to a particular time. The various curves in this figure represent trajectories of the system that start at some point and then move to one of the steady states. The sample curves in this phase-space diagram show a clear partition of the space into two halves. The curves that originate above the diagonal defined by $x = y$ end up in the steady state in the upper right corner of the figure. This steady state corresponds to G_y suppressing G_x or a high concentration of Y and a low concentration of X. Trajectories that start below the diagonal end up in a steady state where G_x represses G_y.

It is instructive to compare this phase-space representation with the normal time solutions of $x(t)$ shown in Fig. 4.19. In all curves in this figure we kept $y(0) = 2$. Depending on the initial conditions, the variable $x(t)$ tends to the high steady state, when $x(0) > y(0)$ and to the low state otherwise. The graph does not show $y(t)$, which would take the steady state x does not take.

Let us now obtain another perspective on the system and consider the stability properties of the steady state solutions. Based on our numerical examples, we would expect that the steady state $x^* = y^*$ is only stable when it is the only solution. If the

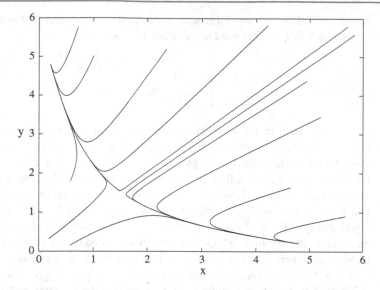

Fig. 4.18 The phase-space representation of solutions for (4.47). The parameters are $K = 2$ and $b = 0.2$. The horizontal axis represents x and the vertical axis represents y. Each line corresponds to a trajectory of the system starting from an initial condition. Depending on the initial condition, the system ends up in one of the two steady states. Compare this with the solutions plotted in Fig. 4.17

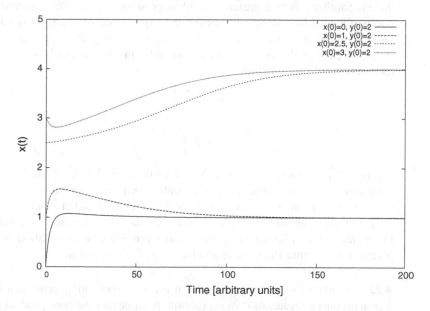

Fig. 4.19 Some sample solutions of (4.47) for different initial conditions. The parameters are $K = 2$ and $b = 0.2$ and we keep $y(0) = 2$. If $x(0) > y(0)$ then it assumes the higher steady state, otherwise the lower. Compare this with Figs. 4.17 and 4.18

parameters allow another stable state $x^* \neq y^*$ then we expect that one to be stable. To ascertain this we first need to calculate the Jacobian:

$$
\mathbf{J} = \begin{bmatrix} -b & -\dfrac{K^h y^h h}{(y^h + K^h)^2 y} \\[3ex] -\dfrac{K^h x^h h}{(x^h + K^h)^2 x} & -b \end{bmatrix}
\tag{4.48}
$$

Let us now analyze the stability of the system using the same parameters as in Fig. 4.19. We have calculated that this particular choice allows two stable steady states which we (arbitrarily) assign to $x^* = 4$ and $y^* = 1$ (we could have done this the other way around as well); Fig. 4.19 shows how these steady states are approached. To obtain the stability of the first steady state, we need to set in the values of $x^* = 4$ and $y^* = 1$ along with the parameter values $K = 2$ and $b = 0.2$. Calculating the eigenvalues of the resulting matrix then gives: $e_1 = -0.04$ and $e_2 = -0.36$. Both eigenvalues are negative, which confirms that the steady state is stable. The second steady state where X is repressed gives the same result, as we would expect.

We still need to check the stability of the steady state given by $x^* = y^* = 2.229494219$. Again, substitution of this value into the Jacobian and calculating the eigenvalues yields $e_1 \approx 0.02164$ and $e_2 \approx -0.42164$. Since one of the eigenvalues is positive, this confirms that the steady state is unstable, as suspected.

In this particular case of a system consisting of two differential equations only, there is another way to represent the behavior of the system. The *nullclines* are the curves obtained by individually setting the expressions on the right hand sides of the differential equation (4.47) to zero. They essentially show the possible steady states for each of the two equations separately; the actual steady states are at the intersections of the nullclines:

$$
x = \frac{K^h}{b(y^h + K^h)}
$$

$$
x = K \left(\frac{1 - by}{by} \right)^{\frac{1}{h}}
$$

Figure 4.20 plots these curves for the particular case of $K = 2$ and $b = 0.2$. The three possible positive and real steady states correspond to intersection points of the nullclines. From a computational point of view, it is often much easier to plot the nullclines rather than to explicitly solve the equations to obtain the steady state. Furthermore, the nullcline equations can also provide a better intuition as to which parameters influence the number and location of steady states.

4.22 Formulate the differential equation for a self-activating gene, that is a gene whose product activates its own expression. Assume that the gene product is broken down at a rate of b. Write down the differential equation and analyze the number of steady states.

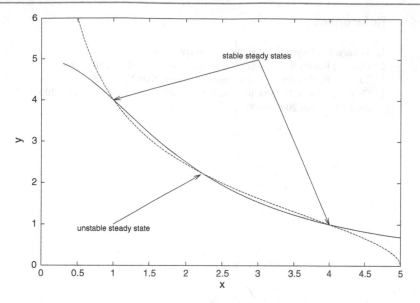

Fig. 4.20 The nullclines for the system in (4.47). The parameters are $K = 2$ and $b = 0.2$. The horizontal axis represents x and the vertical axis represents y. The intersections between the lines represent steady states. Compare this with the solutions plotted in Fig. 4.17

4.23 Formulate the system of differential equations for the import of a nutrient S that exists in the environment at a concentration s by a gram-positive bacterium. Assume that it is taken up through porins into the periplasm first and then into the cytoplasm by different porins. Assume that uptake by porins obeys Hill-dynamics. Explore different model assumptions, including that $\dot{s} = 0$.

4.24 Using the previous model, vary the volume of the periplasm and the cytoplasm. What are the effects?

4.7 Summary

Differential equations are powerful tools in biological modeling. In this chapter we have refreshed the reader's understanding of the basic ideas of differential calculus. We have also introduced some basic methods of solving differential equations. The material presented here necessarily barely scratches the surface of the body of available knowledge, yet we believe that it should be sufficient to help in formulating a number of biologically interesting models. Eventually, however, the reader may wish to consider deepening her knowledge of the subject by consulting more specialist volumes.

References

1. Murray, J.: Mathematical Biology. Springer, Berlin (2002)
2. Casti, J.: Reality Rules: I The Fundamentals. Wiley, New York (1992)
3. Casti, J.: Reality Rules: II The Frontier. Wiley, New York (1992)
4. Cherry, J., Adler, F.: How to make a biological switch. J. Theor. Biol. **203**(2), 117–133 (2000). doi:10.1006/jtbi.2000.1068

Mathematical Tools

5

Nowadays, computer algebra systems (CAS) are an essential tool for the mathematical modeler. In a strict sense, a CAS is a computer program that can perform symbolic computation; that is, manipulate mathematical expressions. Such programs can be an important aid to modelers of all skill levels. Apart from CAS as tools for performing symbolic mathematics, we will also include a second type of computational mathematics tool into the definition of CAS: Programs that can aid the modeler in producing numerical solutions to mathematical models.

We start by giving a brief overview of the range of available software, and then the bulk of the chapter will introduce one particular system—Maxima [1]—in some detail. We include some case studies that explicitly work through models encountered previously in this book. At the end of the chapter, the reader should be able to formulate and solve simple models in Maxima.

5.1 A Word of Warning: Pitfalls of CAS

Hours and days and weeks of the life of a mathematical modeler are spent looking at formulas, deriving expressions and proving theorems. Many of us enjoy this activity and feel a sense of satisfaction once the task is accomplished. Yet, particularly when the mathematics becomes more involved and the formulas longer, it is often helpful to hand over parts of the task to a CAS. Unlike us, these systems can perform the chores of mathematical modeling in an instant, without complaint and, most significant of all, without the typical errors of negligence that humans are liable to make when manipulating strings of Greek letters and strange symbols. However, while extremely useful, one must always keep a watchful eye to avoid the potential pitfalls of handing over large portions of one's own work to a machine.

Machines lack the innate laziness we have and do not mind producing lines and lines of results when the same statement can be expressed much more concisely. While most computer algebra systems have a simplification command to reduce the

© Springer-Verlag London 2015

D.J. Barnes and D. Chu, *Guide to Simulation and Modeling for Biosciences*,
Simulation Foundations, Methods and Applications,
DOI 10.1007/978-1-4471-6762-4_5

size of expressions, in practice this sometimes fails. Blind trust in the capabilities of the machine then results in long incomprehensible equations, which only serve to obscure insight rather than providing it. For this reason, it is essential to check the results of computer-generated computations by hand, wherever possible.

In the practice of mathematical modeling, it is often the very process of wrestling with the equations that leads to insight. The final result is then only the distilled version of weeks of struggle with the model. It is the secret shortcut that can only be discovered by walking backwards from the destination, but could not have been discovered from the starting point. If we hand over the struggle to the machine then we risk losing or never gaining some of that insight and perhaps miss out on the crucial formula that will then be published by our academic arch-enemy the following year (in Nature, even). It seems that the arch enemy is still computing by hand.

Undoubtedly, computer algebra systems are useful, and even more so for non-mathematicians. Built into these systems are libraries of mathematical knowledge that few non-mathematicians possess (and indeed maybe few mathematicians). Often a quick query of a CAS will deliver an instant answer that would otherwise require weeks of pestering mathematically more versed colleagues for help; or, even more frustrating, hour upon hour spent in the library on fruitless self-study. There is significant value in computer algebra systems, but only as long as the modeler remains the master and does not succumb to a naive trust in the power of the system.

While powerful, CAS still need to be systematically doubted and their results checked—not because the results are likely to be wrong but because they often are presented in a form that is not useful or hides what we are seeking. Whenever something can be calculated by hand then it should be done by hand; the computer then serves as a useful tool in confirming the result.

These cautions mainly apply to computer algebra systems whose primary purpose is to assist the user in symbolic manipulation of equations. When it comes to using computers to generate numerical solutions the situation is different. They are more of a necessity. Sooner or later, most practical modeling projects need to perform some sort of numerical computation, if only to plot an example curve for a paper to be published. Many mathematical models are so complicated that symbolic solutions cannot be obtained. In these cases there is an intense need to solve equations numerically. Computer systems that have the available algorithms to do this are then an indispensable tool for modelers.

Designing efficient algorithms to perform numerical computations is a field of research in its own right. In rare cases a modeler may need to design such a algorithms herself but, in the vast majority of cases, it would be a waste of time not to use the ready made, off-the shelf systems that are available. CAS are a library of implementations of such algorithms and can save the user months of preliminary study. It is worth bearing in mind that a great deal of sophistication has gone into the design of these algorithms and the particular implementations that underly the computer programs are complex pieces of software engineering. Still, we should always be on guard and critically examine the sanity of the results we obtain. Like all software, some implementations will contain inefficiencies, bugs and errors that may lead to wrong results. If in doubt and, in the case of very complex models as a matter of routine, it is also worthwhile cross-checking the obtained results by using two independent pieces of software. This is time-consuming, but an essential insurance against falling prey to errors in the software, artifacts produced by a quirky algorithm, but also our own mistakes in formulating the model.

This section will introduce the reader to a freely-available computer algebra system and, in addition, provide a general introduction and a few practical examples— mostly of problems that have been encountered before in the book.

5.2 Existing Tools and Types of System

There are a number of systems available that can be useful for performing mathematics on a computer. The three dominant commercial systems are MAPLE [2], MATHEMATICA [3], and MATLAB [4]. The main strengths of MAPLE and MATHEMATICA are their powers of performing symbolic computations. In addition, both have evolved from bare-bones computer algebra systems to programs that also provide a visually appealing interface capable of producing potentially publishable documents.

The main strength of MATLAB is its numerical capability. It is designed to facilitate rapid prototyping of complex numerical modeling and programming. It comes with a number of toolboxes that allow one to build on the work of previous users. In many areas of science, including biology, MATLAB has evolved to a *de facto* industry standard.

A potential drawback common to all three systems is their high cost. Universities and companies often have site licenses that cover various types of use, and may even allow installation of these programs on private machines. Student licenses may also be available relatively cheap. At the full price, these systems are very expensive which motivates the search for cheaper alternatives ... and these exist.

A well-known system that allows much of the capabilities of MATLAB is Octave [5], which has the advantage of being free. There is a high degree of cross-compatibility between Octave and MATLAB, in the sense that most code for Matlab will work on Octave, and *vice versa*. For readers familiar with Matlab who would like to undertake primarily numerical computation, Octave is certainly a possible alternative to its commercial counterpart. Another freely available system that is closely modelled on Matlab is Scilab [6]. For heavy duty mathematical modeling the ASCEND software [7] is also worth a look. It is also free but, in our experience, not as user-friendly as Octave and Scilab.

A system that is useful for some numerical tasks is R [8]. It is primarily intended for statistical computing and has a number of statistical functions built into it. Like Octave, it is freely available, can be used completely from the command line and supports a simple scripting language. In addition, there are also freely available graphical user-interfaces available for it. Unlike Octave, it is mainly intended as a tool to manipulate and analyze data, but has limited use in computationally more intense mathematical computing, such as differential equations. In our view it also suffers from a rather idiosyncratic syntax which certainly makes sense, but not to the uninitiated.

A high quality, freely available system that allows a quick and relatively reliable exploration of systems of differential equations is XPPAUT [9]. It has no analytical capabilities and is specifically meant to explore systems of differential equations. It can do nothing that cannot be done with other more generalist systems like Maple and Octave. Its main value, in our opinion, lies in the efficient way in which it allows the user to explore systems of differential equations. XPPAUT is designed to enable the user to understand rapidly how varying initial conditions impact the result. It can quickly display trajectories in phase space and even plot bifurcation diagrams. The downside is its idiosyncratic user-interface. It may take the uninitiated some time to find her way around the system. Initially, this may be frustrating, but it is well worth persevering. Should the reader anticipate using differential equation models then we would urge her to set aside an afternoon investigating XPPAUT. We are confident that the initial investment of patience will pay off in dividends of improved efficiency.

There are also a few freely available software packages that allow symbolic computation. A well-known system is Sage [10] which can be used via a web-based interface. Another system worth mentioning is Mathomatic [11] which can perform symbolic computations. It runs on very small systems as well, including handheld devices. It also offers a browser-based web service that does not require any software installation by the user.

5.3   Maxima: Preliminaries

In this chapter we will introduce the reader to another computer algebra system, Maxima [1]. It is free and can perform both symbolic and some numerical mathematics. It can be compiled for the most common operating systems. Unlike Maple or Matlab, Maxima does not come with its own graphical user interface, although there is a choice of interfaces within which Maxima can be used. The simplest approach is to run Maxima in a command-line environment. This does not provide any fancy interface, but does the job with a minimum of hassle. A simple, yet useful graphical user interface for Maxima is wxMaxima which comes with most distributions. The reader may wish to check her installation for xMaxima. Additionally there is TeXmacs [12] as a possible interface. TeXmacs is also a document preparation system in its own right. A somewhat simpler alternative interface is wxMaxima [13], which is purely a graphical user interface for Maxima. The choice between these is left to the reader. Each has its own merits and each is certainly very spartan compared to what one has become used to from commercial interfaces. While wxMaxima has a convenient, well thought out graphical interface, producing, manipulating and saving plots is perhaps easiest in xMaxima. The best mathematical output is provided by TeXmacs, although its general-purpose interface is not as user-friendly as the other two. In the remainder of this chapter we will remain agnostic about the interface to be used.

To reap maximum benefit from the rest of the section, we strongly urge the reader to choose one installation of Maxima, to actually install it and familiarize herself with the system as we progress through the examples to follow.

Before starting the description of how to use Maxima, it is worthwhile providing a general overview of the software and how it is organized. Unlike the very polished commercial products that the reader may be acquainted with, Maxima is the product of an academic open source project. Its main focus is on substance and quality. Also, it is born out of the more spartan world of UNIX where things are, to a large degree, text-based and centered on the use of a command-line interface. This might sound like an antiquated approach to software but, in reality, it has its advantages, particularly once one is used to the concept and has overcome any fear of command-line windows. In fact, the minimalism of simple interfaces has the advantage of focusing the user on the task at hand, rather than distracting with unnecessary dialogues, bells and whistles. More importantly, programs of this sort tend to be more stable and faster than their more visually-appealing cousins.

An apparent drawback is that it appears to be more contrived to extract graphical output from these programs. Maxima itself does not have its own graphical routines, yet it is still possible to use it to produce plots and to use the plots in high-quality documents such as scientific papers. To achieve this, Maxima uses a third-party program (also free) called gnuplot [14]. This is a very powerful program in its own right (and has been used to prepare many of the graphs in this book). In practice, whenever a curve is plotted from Maxima, the program produces a set of points that are then (automatically) passed to gnuplot, which in turn produces the plot window that is shown. Gnuplot can also export images in different file formats. Depending on the installation, this can include PDF, JPEG, GIF, and many others. In what follows, we will always assume that the file output format is Postscript, which is the simplest format to use and is always available with gnuplot. On OS X and Linux, display of Postscript files is unproblematic. On other platforms this may be more difficult so, as an alternative, the reader may like to choose PDF as the output format. The Maxima documentation provides details.

Unlike many of the programs the reader may be familiar with, gnuplot does not allow the saving of graphs by direct mouse-interaction. Instead, commands must be issued within the Maxima environment to write a graph to a file that can subsequently be integrated into a document, say. This may sound complicated but it turns out to be a very effective process for producing high-quality graphs.

The xMaxima program provides a viable alternative for those readers who do not wish to embark on this adventure into command-line computing. The xMaxima interface allows direct plotting within the interface, using OpenMath [15] rather than gnuplot. What is more, every graph comes with an interactive menu that lets the user manipulate the graph and save it to file.

5.4   Maxima: Simple Sample Sessions

The best way to quickly get to work with any system is to follow through a number of examples. In this section we will present some sample sessions designed to illustrate the main features of the Maxima computer algebra system. The examples will be sufficient to reproduce the mathematical models presented in this book, and to make a start with some initial mathematical modeling. However, in the long run, the serious user will need to dig deeper herself. In our opinion, the best approach for this is a "learn-as-you-go" approach, whereby the user only familiarizes herself with new concepts as and when she actually requires them.

5.4.1   The Basics

The most efficient way to acquire the basic skills necessary to become at least a novice user of Maxima is to follow the examples given here, modify them, and thus develop an intuition for the program and how to use it. The results of any computations should be reproducible independently of the particular version of Maxima used. However, the formatting of the output may vary from interface to interface, and may not be as visually appealing as the examples given here, which were produced using TeXmacs.

A few key principles of Maxima first:

- The operators for multiplication, division, addition, subtraction and exponentiation are: $*$, $/$, $+$, $-$, $\char`^$, respectively.
- The end of every statement and expression must be indicated with either a semi-colon or with a dollar sign. If the former is used then the input will be echoed back to the user; the dollar sign suppresses any output. This can be desirable when the output is expected to be long. A common error of novice users is to forget to end the line with a semicolon or a dollar sign.
- Equations, variables and statements can be named. This is done by preceding the expression with the name and a colon sign (e.g., name : expression).
- Every input and output is assigned by the program a line number of the form %ix or %ox, for input and output respectively. In this case "x" will be a number. These numbers can be used to recall previous statements.

What follows is a full Maxima session that illustrates its basic use. A look at these example interactions demonstrates that input of expressions and equations is very easy.

```
Maxima 5.36.1 http://maxima.sourceforge.net
using Lisp SBCL 1.2.10

...
```

```
(%i1) x;
(%o1) x
(%i2) x$
(%i3) eq1: x*x*x - x^3 + y - x^4;
```
$$(\%o3)\ y - x^4$$
```
(%i4) eq1;
```
$$(\%o4)\ y - x^4$$
```
(%i5) %o3
```
$$(\%o5)\ y - x^4$$
```
(%i6) eq2: a*x - z=0;
(%o6) ax - z = 0
(%i7) a: 3$
(%i8) eq2;
(%o8) ax - z = 0
(%i9) ''eq2;
(%o9) 3x - z = 0
(%i10) eq3: a*x^a = 3;
```
$$(\%o10)\ 3x^3 = 3$$
```
(%i11) quit();
The end
```

Most of the above is self-explanatory, however a short comment is necessary regarding the line marked %i3. The user will notice that the system has performed a certain amount of processing between the input and the output. Specifically, it has recognized that $x * x * x$ is the same as x^3. Often output from Maxima needs to be simplified by issuing a specific command. We will discuss this in more detail later on.

The reader should also pay particular attention to input line %i7, where we assign the value 3 to the symbol a. Recalling in %i8 eq2—an equation that contains symbol a—we find that the symbol has not been replaced by its numerical value. We can recover eq2 with the specific value of a by preceding it with double quotes (%i9). However, if we define a new equation containing a, in this case eq3, then we find that it *has* been replaced by the numerical value. Maxima sessions are ended by issuing the quit() command.[1]

One often wants to write a formula with parameters that are initially unspecified, and only later commit to specific values for these variables—for instance, to generate examples for plotting. To be able to do this, Maxima provides a way for the user to assign values to variables.

```
(%i1) eq1: x^2 - u;
```
$$(\%o1)\ x^2 - u$$
```
(%i2) eq2: ev(eq1,u=2);
```
$$(\%o2)\ x^2 - 2$$

---

[1] In the remaining example sessions we will not explicitly write the quit() command any longer.

```
(%i3) eq3:  ev(eq2,  x= sin(y));
```
$(\%o3) \sin{(y)}^2 - 2$
```
(%i4) eq4:  ev(eq3,y=1);
```
$(\%o4) \sin{(1)}^2 - 2$
```
(%i5) eq5:  ev(eq4,numer);
```
$(\%o5) -1.291926581726429$
```
(%i6) eq6:  ev(eq1,u=2,x=sin(y));
```
$(\%o6) \sin{(y)}^2 - 2$
```
(%i7) eq7:  ev(eq1,u=2,x=sin(y),y=2,numer);
```
$(\%o7) \sin{(y)}^2 - 2$

The command ev(...) is used both to replace parameters or symbols by numbers and to replace them by other expressions. Line %i3 illustrates this latter capability, where we replace x by the expression sin(y). Since Maxima is primarily a computer algebra system (rather than a fancy calculator) it tends to prefer to keep things as general and precise as possible. As a result, in %i7, even when we have specified a value for y, the system still returns the "symbolic" form of the equation. Yet, sometimes we want to know the actual number, rather than just being told that the result is sin(1)−2. In this case the command ev is used with the keyword numer to force evaluation of the expression to a numerical value.

A particularly useful feature of ev is that it allows a sequence of replacements to be specified, simply by stringing them together in a single expression, as in %i6. Note, however, that due to the way ev is processed internally by Maxima, the value 2 is not actually assigned permanently to y in %i6, whereas the variable u *is* replaced. The difference is that y is substituted into the expression, whereas u is not. Naively, one would have expected that output lines 5 and 7 produce the same result, but they do not. In order to rectify this, one needs to tell ev to add an additional evaluation step. This can be done by adding the keyword eval after the expression to be evaluated, as so:

```
(%i1) eq1:  x^2 - u$
(%i2) ev(eq1,eval,u=2,x=sin(y),y=1,numer);
```
$(\%o2) -1.291926581726429$

We have seen that Maxima allows naming of expressions so that they can be recalled later on (for example eq1 in the last session). Often we need to define customized functions. Not only does Maxima allow us to define our own simple functions, such as $f(x) = x^2 - y$, but also more complicated compound functions, which may perform very complex evaluations. The following session illustrates how a simple function can be defined.

```
(%i1) f(x,y):=x^2 - sqrt(y);
```
$(\%o1) \ f(x,y) := x^2 - \sqrt{y}$
```
(%i2) f(d,f);
```
$(\%o2) \ d^2 - \sqrt{f}$
```
(%i3) f(1,2);
```
$(\%o3) \ 1 - \sqrt{2}$

This principle immediately generalizes. So, for example, the function $f(x) = x^2$ can be defined by writing `f(x) := x^2`; note that ":=" is used for function definitions and not the simple colon assignment operator. Indeed, entire procedures can also be defined with this notation by enclosing them into a `block` environment.

```
(%i1) f(x,y):= block(
         fv1: x^2 - 6,
         fv2: y^5 - 7,
         fv1/fv2);
```
(%o1) $f(x,y) := \text{block} \left(fv1 : x^2 - 6, fv2 : y^5 - 7, \frac{fv1}{fv2} \right)$
```
(%i2) f(a,b);
```
(%o2) $\frac{a^2-6}{b^5-7}$

Within a block environment we can specify any sequence of steps to be evaluated one after the other. The last statement is returned as output. Note that statements are separated by commas rather than by dollar signs or semi-colons. (The special formatting we used in input line 1 is not essential, but chosen for readability. Blocks can be defined on a single line as well.)

Variables and names in functions are *globally* valid. In the next example, pay attention to the sequence of assignments.

```
(%i1) x: 1$
(%i2) u: s^2 - o$
(%i3) f(x,y):= block(
         u: x^2 - 1,
         v: y,
         sqrt(u + v));
```
(%o3) $f(x,y) := \text{block} \left(u : x^2 - 1, v : y, \sqrt{u+v} \right)$
```
(%i4) u;
```
(%o4) $s^2 - o$
```
(%i5) f(1,2);
```
(%o5) $\sqrt{2}$
```
(%i6) f(2,2);
```
(%o6) $\sqrt{5}$
```
(%i7) u;
```
(%o7) 3
```
(%i8) x;
```
(%o8) 1

Note how the name u is re-assigned within the function with global scope after f is called. However, variables having the same names as a function's arguments, e.g. x and y above, would not be affected by assignments within the function—x retains its value of 1, after f is called, in (%o8).

The global scoping feature is a potential pitfall and so we recommend developing a naming practice for expressions within functions that clearly distinguishes them from names outside. In the first example, we adopted an 'fv' prefix, and would correspondingly avoid such names outside of blocks; the exact choice is, of course, arbitrary.

The particular examples we have given are trivial in the sense that the usage of blocks was not really necessary in those cases. In actual modeling practice one will often come across particular problems that require some simple programming. It can then be useful to consider blocked functions. Another use of blocked functions is simply to define a block of routine evaluations that contains a number of steps. Such user-defined functions can significantly increase the efficiency of the work flow.

There are often situations when it is necessary for the system to forget variables and their values. Maxima has a simple command for this, namely kill. Using kill one can delete either specific names and functions from the system, or restart the entire system (by giving the option all). Here are some examples of selective and complete deletion.

```
(%i1) a: 3$
(%i2) f(x):= x^2$
(%i3) a;
(%o3) 3
(%i4) kill(a);
(%o4) done
(%i5) a;
(%o5) a
(%i6) dispfun(f)$
(%t6) f(x) := x^2
(%i7) kill(f);
(%o7) done
(%i8) dispfun(f)$


fundef: no such function: f
-- an error.  To debug this try: debugmode(true);


(%i9) kill(all);
(%o0) done
```

We use the command dispfun to display the definition of the user-defined function f. After "killing" f the definition is lost and so trying to display it causes an error message by the system. The kill command can also be used to kill multiple names and functions in one go, by stringing them together in the list of arguments, e.g., kill(f,a,b).

Maxima has a very efficient help system built into it. Help on commands and their syntax, along with examples, can be accessed by typing ? followed by the command of interest. (Note that a space between the question mark and command is essential.) Here is an example:

```
(%i1)? quit;
```

```
-- Function: quit ()
   Terminates the Maxima session.  Note that the
   function must be invoked as 'quit();' or
   'quit()$', not 'quit' by itself.  To stop a
   lengthy computation, type 'control-C'.  The
   default action is to return to the Maxima prompt.
   If '*debugger-hook*' is 'nil', 'control-C'
   opens the Lisp debugger.  See also *note
   Debugging::.
```

(%o1) true

5.4.2 Saving and Recalling Sessions

When using the command line interface, it is important to be able to save a day's work. Maxima offers a number of ways to keep a record of a session or part of a session.

Perhaps the simplest way to save a session is to create a record of the input and output commands. This can be done using the writefile command:

```
(%i1) writefile("test.txt")$
NIL
(%i2) a: x^2$
(%i3) closefile()$
NIL
```

This will cause Maxima to record to file anything appearing on the screen *after* the command has been issued. The argument to writefile is the name of the file (in quotes) to be written which. It is important to issue the command closefile at the end of the session (or the desired output), as failing to do so may result in no file being written. The command closefile does not need any arguments because only one writefile session can be open at any one time.

One might of course sometimes forget to issue the writefile command. In this case, the entire input and output of a session can be recalled by issuing the playback() command.

The major disadvantage of writefile is that its output cannot be loaded back into Maxima. There are two options to overcome this. Firstly, the command save writes the session to a file in Lisp format—Lisp is the programming language in which Maxima is implemented. There are a number of options to control what exactly is saved by save. The two most useful are to save all user-defined functions, and to save everything.

In the following example session we define a name and a function, and save it to a file:

```
(%i1) a: 34$
(%i2) f(x,y):= x - y$
(%i3) save("test.lsp",functions)$
```

If you are not sure where the file is saved to, use a semicolon rather than $ at the end of the save command which will cause the full filename to be shown.

We now start a new session and load the session file saved before using the load command:

```
(%i1) load("test.lsp");
(%o1) test.lsp
(%i2) dispfun(f);
(%t2) f(x,y) := x − y
(%o2) [%t2]
(%i3) f(2,3);
(%o3) −1
(%i4) a;
(%o4) a
```

After loading the file test.lsp we have the user-defined function f(x,y) available. However, since we only specified the keyword functions above, the name a is not defined in the new session. Had we specified the keyword all instead, then the value of a would have been saved as well.

The second way to save sessions is to use the stringout command. This command writes session files in Maxima format rather than in Lisp format. The clear advantage of this is that the output files can be manipulated in a text editor by the user. Again, what exactly is saved can be controlled by various options. Typing stringout("test.msc",input) will write the entire input history to the file test.msc. If, instead of input, functions is specified, then only the function definitions are written to file. Finally, the keyword values would cause Maxima to write out names and their values only (i.e., things such as a: 23, etc.).

Files written with stringout can be loaded into the system using the batch command, for example, batch("test.msc"). In addition to "machine generated" content, the batch command can also execute human written files that are in the same format.

5.5   Maxima: Beyond Preliminaries

So far we have only described how to use Maxima, but we have not really put it to any particular use. One of the main attractions of computer algebra system is their ability to do things like solving equations, computing derivatives, find integrals and so on. All this turns out to be very easy in Maxima.

5.5.1 Solving Equations

To begin with, let us define an equation and solve it; to make the example more rewarding, we have chosen an equation that has two solutions:

```
(%i1) eq1: x^2 - a*x + b$
(%i2) sol: solve(eq1,x);
```
$$(\%o2)\ \left[x = -\frac{\sqrt{a^2-4b}-a}{2}, x = \frac{\sqrt{a^2-4b}+a}{2}\right]$$
```
(%i3) sol[1];
```
$$(\%o3)\ x = -\frac{\sqrt{a^2-4b}-a}{2}$$
```
(%i4) sol[2];
```
$$(\%o4)\ x = \frac{\sqrt{a^2-4b}+a}{2}$$
```
(%i5) x;
(%o5) x
(%i6) x: rhs(sol[2]);
```
$$(\%o6)\ \frac{\sqrt{a^2-4b}+a}{2}$$
```
(%i7) x;
```
$$(\%o7)\ \frac{\sqrt{a^2-4b}+a}{2}$$
```
(%i8) ev(x,a=1,b=0.01,numer);
(%o8) .9898979485566356
```

The command `solve` returns a list of solutions. The elements of this list can be addressed using the index notation as exemplified by input line (%i3); so `sol[1]`, refers to the first element of `sol`. Note how we use the `rhs` command to extract the right-hand side of the equation. It is important to appreciate that the solution to an equation is a list of *equations*, and not a list of assignments. For this reason, it is necessary to invoke the `rhs` command to extract the actual value of x rather than the equation x = `sol[2]`.

5.1 Predict (without actually trying it out) what the effect would be of assigning `sol[2]` directly to x rather than its right hand side.

More often than not we would like to solve *systems* of equations. That can also be achieved with the `solve` command, but we need a slightly modified syntax. In the following example session we solve a system of two equations:

```
(%i1) eq1: x^2 - b = y$
(%i2) eq2: x*y - b = c$
(%i3) sol: solve([eq1,eq2],[c,b]);
```
$$(\%o3)\ \left[\left[c = (x+1)\,y - x^2, b = x^2 - y\right]\right]$$
```
(%i4) sol[1][1];
```
$$(\%o4)\ c = (x+1)\,y - x^2$$
```
(%i5) sol[1][2];
```
$$(\%o5)\ b = x^2 - y$$

Note that we need to enclose the equations within square brackets. Also, one must never forget to tell the system which variables the system should solve for.

5.2 Solve equation eq1 in the example above for x, y and assign the solutions to the variables d1, d2, d3, d4.

One often encounters equations, or sets of equations, that are too complex to be solved analytically. In those cases it is then necessary to find an approximate numerical solution. Unfortunately, solve may not always be the best tool to use for this as it either returns quite incomprehensible solutions or does not find any solutions at all.

Maxima provides a number of other tools that can be used to solve equations numerically. The commands realroots and allroots find the roots of polynomials in one variable only, that is expressions of the form $a_0 + a_1x^1 + a_2x^2 + \cdots$. Expressions of more general form can be solved by find_root. For heavy duty solving the additional packages newton and mnewton provide solving capabilities for equations and systems of equations respectively.

Unlike the alternatives, realroots and allroots do not require an initial guess and are therefore the preferred method if they can be used. In the following example, we will transform our initial expression into a polynomial using denom which returns the denominator of an expression. We also use rat to bring our initial expression into a convenient rational shape:

```
(%i1) eq1: x^3/(342 + x^3) + 45*x - 200;
```
$$(\%o1) \ \frac{x^3}{x^3+342} + 45x - 200$$
```
(%i2) eq1: rat(eq1);
```
$$(\%o2) \ /R/ \ 45x^4 - 199x^3 + 15390x - 68400/x^3 + 342$$
```
(%i3) eq1: eq1*denom(eq1);
```
$$(\%o3) \ /R/ \ 45x^4 - 199x^3 + 15390x - 68400$$
```
(%i4) realroots(eq1);
```
$$(\%o4) \ \left[x = -\frac{234804647}{33554432}, x = \frac{148978867}{33554432} \right]$$
```
(%i5) allroots(eq1);
```
$$(\%o5) \ [x = 6.061573739019288i + 3.490014868130795, \cdots]$$

This is a very useful tool, but unfortunately very limited in its usage. For example, applying allroots directly to %o1 would result in an error message. We would also be in trouble if we attempted to find the solution to an equation involving trigonometric functions, or exponentials, logarithms, and the like. In these cases we need to use the other numerical solvers that Maxima puts at our disposal. In the following example we will solve the equation $\sin(x^2) = \exp(x)$. In the case of find_root we need to specify the interval in which the system searches for solutions. A special feature of this command is that it will only accept end-points if the solutions corresponding to them are of opposite signs. To make this clear. Our equation can be written as follows:

$$f(x) \doteq \sin(x^2) - \exp(x) = 0$$

To find solutions using find_root we need to provide an interval $[x_1, x_2]$, such that $f(x_1)$ and $f(x_2)$ are of opposite sign. In so far as it exists, find_root will then return a solution for the equation within this interval. More conveniently, newton only requires a single point to be specified near which it will look for solutions. In the following session we will look for solutions for $f(x)$ in the interval $[-1, 0]$ using find_root and near the point -1 in the case of newton. The newton command is only available if we load an extra package, via the load command. This package comes with the standard distribution of Maxima.

```
(%i1) eq1: sin(x^2) - exp(x)$
(%i2) find_root(eq1,-1,0);
(%o2) -.7149689691896562
(%i3) load(newton)$
(%i4) newton(eq1,-1);
(%o4) -7.149689725341498b-1
```

5.3 Use the command ev to check that the numerical solutions we have generated do indeed satisfy eq1.

5.4 Find other solutions for eq1. (There are many!)

5.5.2 Matrices and Eigenvalues

There are two ways in which matrices can be entered in Maxima: The command entermatrix(n,m) interactively first asks whether the matrix is diagonal, symmetric, antisymmetric, or general. Once this is decided, the system will ask for each element of the matrix to be entered. This is very much self-explanatory and we will not demonstrate this here. The reader is urged to try this command for herself.

The second way is to enter a matrix in 2d form using the matrix command. The command matrix([x,y,z],[a,b,c],[1,2,3]) creates the following matrix ϕ.

$$\phi = \begin{pmatrix} x & y & z \\ a & b & c \\ 1 & 2 & 3 \end{pmatrix}$$

A few of the most common matrix operations are pre-defined in Maxima. The '$*$' operator means an element-by-element multiplication between two matrices. To perform a proper matrix multiplication, the '.' operator must be used. We are particularly interested in matrices because we want to be able to calculate Jacobian matrices and their eigenvalues. In order to do this, the linearalgebra package must be loaded. It comes with the standard Maxima distribution and should be available without the need of a separate download. In the following example session we will first enter two matrices and multiply them. Then we will define three functions, generate the

Jacobian and finally calculate the eigenvalues using the `eigenvalues` command built into Maxima.[2]

```
(%i1) ma1: matrix([a1,a2],[a3,a4]);
```
$$(\%o1) \begin{pmatrix} a1\ a2 \\ a3\ a4 \end{pmatrix}$$
```
(%i2) ma2: matrix([b1,b2],[b2,b4])$
(%i3) ma1*ma2;
```
$$(\%o3) \begin{pmatrix} a1b1\ a2b2 \\ a3b2\ a4b4 \end{pmatrix}$$
```
(%i4) ma1.ma2;
```
$$(\%o4) \begin{pmatrix} a2b2 + a1b1\ a2b4 + a1b2 \\ a4b2 + a3b1\ a4b4 + a3b2 \end{pmatrix}$$
```
(%i5) kill(all)$
(%i1) f(x,y,z):= x^2 - exp(y) + z$
(%i2) g(x,y,z):= z^4 - x^3$
(%i3) h(x,y,z):= x + y + z$
(%i4) load("linearalgebra")$
(%i5) ma1:jacobian([f(x,y,z),g(x,y,z),\
                     h(x,y,z)],[x,y,z]);
```
$$(\%o5) \begin{pmatrix} 2x & -\%e^y & 1 \\ -3x^2 & 0 & 4z^3 \\ 1 & 1 & 1 \end{pmatrix}$$
```
(%i6) eig(va,vb,vc):=
          eigenvalues(ev(ma1,x=va,y=vb,z=vc))$
(%i7) eig(1,2,3)$
(%i8) float(%o7);
(%o8)[[-9.288231884808303 (.8660254037844386 %i - 0.5)...],
        [1.0,1.0,1.0]]
(%i9) realpart(%o8[1][1]);
(%o9) 8.015704168881044
(%i10) imagpart(%o8[1][1]);
(%o10) -3.936133465594463
(%i11) imagpart(%o8[1][3]);
(%o11) 0
```

We have suppressed the output of %i7, where we calculate the eigenvalues for specific values of x, y and z, because Maxima would return an exact but rather long list of expressions. In %i8 we ask for the result in floating point representation. This gives us a list of lists. The second list ([1.0, 1.0, 1.0]) represents the multiplicity of the eigenvalues. The eigenvalues themselves are contained in the first list. The first two are complex numbers. The example shows how to extract the real and the imaginary part using `realpart` and `imagpart`. The third eigenvalue in the list (≈ -13.03) is real and hence we would expect a vanishing imaginary part. This is confirmed by %o11.

---

[2]We have omitted much of the long output in line %o8.

There are many more features in Maxima to help the fledgling mathematician with her tasks. This additional material, while important, is peripheral to the central aim of this book. We will leave it therefore to the reader to explore it at her own perusal, should she so be inclined.

5.5.3 Graphics and Plotting

An important part of any modeling activity is the visualization of results. As mentioned in the preliminary remarks, Maxima needs to use the external plotting program gnuplot [14]. Should the reader not be able to generate any plots despite following the instructions given in what follows, then it is likely that gnuplot is not installed. Unlike most mathematics commands in Maxima, the syntax for plotting is not as intuitive as the rest of the system. This is partially due to the fact that any plotting facility must be very flexible, in order to allow users to customize plots according to individual needs and requirements.

Before discussing plotting in general, a warning is issued to users of xMaxima. As mentioned above, this interface comes with its own plotting interface which, in many ways, is more user-friendly than the standard one. However, since we suspect that most readers will use the gnuplot interface and xMaxima is compatible with both, we will only discuss plotting with gnuplot.

The first steps in plotting with Maxima are very simple. The two main commands are `plot2d` and `plot3d`. For two-dimensional plots, `plot2d` can draw multiple curves within a single plot. With plotting in three dimensions, the capabilities of `plot3d` are more restrictive; essentially, it can only plot a single function. For more advanced or demanding users there are some external plotting libraries that provide additional plotting capabilities. One of these is Mario Riorto's Maxima-Gnuplot interface [16]. This can be loaded by issuing the command `load("draw")`. In the remainder of this section we will limit ourselves to the standard plotting commands provided by Maxima.

In its basic form, plotting in two dimensions is intuitive. The command:

`plot2d(x^2,[x,1,2])`

plots x^2 for values of x ranging from 1 to 2. The function x^2 can be replaced by any user-defined function. If we want to compare two curves graphically, say x^2 and x^3, then we can plot both, by issuing the command:

`plot2d([x^2,x^3],[x,1,2])`

Note that we enclosed the two function definitions in square brackets.

In addition to plotting functions, `plot2d` can also plot discrete points or parametric functions. The former can be plotted by using the command:

`plot2d([discrete,lst])`

Here `lst` is a list of entries of the form `[i_x, i_y]`. Alternatively, the discrete points can also be plotted using `plot2d([discrete,lstx, lsty])`. Here `lstx` is a list of values for the horizontal axis and *lsty* is a list of values for the vertical axis. It is not necessary for *lstx* and *lsty* to be of same length. Loosely speaking, in Maxima lists are a number of values enclosed by square brackets. It is perhaps clearest to illustrate lists by giving examples. In the case of two different lists, each of the list should be of the overall form:

```
lstx: [1,2,3,4,5,6,7,8,9,10]$
lsty: [1.1,1.2,1.7,1.3,1.1,1.5,1.2,1.2,1.0,1.6]$
```

If, on the other hand, the syntax with a single list is chosen, then this list should be of this form:

```
lst: [[1,1.1], [2,1.2], [3,1.7], [4,1.3], [5,1.1],
      [6,1.5], [7,1.2], [8,1.2], [9,1.0], [10,1.6]]
```

(The reader who is interested in exploring discrete plots in more depth may be interested in the makelist command described below in Sect. 5.5.4 on p. 197.)

The command plot2d can also be used to produce parametric plots; that is, plotting two functions against one another. The syntax is very similar to discrete plots. For example, the following plots $\sin(t)$ and $\cos(t)$ against one another on the horizontal and vertical axes respectively:

```
plot2d([parametric,sin(t),cos(t),[t,1,2]])
```

In parametric plots it is important *not* to use x or y as the independent variable, as this will confuse the system. Parametric plots can take additional options of the form [x,2,3] to control the range of the horizontal and [y,0,3] of the vertical axis; note that x and y are here the names of the axes and not the variables. In fact, if either x or y are used as the free variables in the functions to be plotted, then this will cause Maxima to limit the range of the relevant axis. Beware!

So far we have only discussed the core capabilities of plot2d and shown how to bring up a plot window in an interactive session of Maxima. Real-world plotting requirements usually require more fine-tuning of the figures than that. To achieve this one must specify options in the plot command. Options are always given within square brackets following the variable that one wants to print. There are two broad classes of options that plot2d accepts, namely genuine Maxima plot options, and options that are gnuplot options. The latter are, of course, only relevant if the plotter of choice is actually gnuplot (which may not be the case for xMaxima users).

We will first describe the genuine plot2d options.

logscales The option for logscales is [logx] and [logy] to set the x and y axis into log scale. For example, the command

```
plot2d(x^2,[x,1,100000],[logx],[logy])
```

will plot the function x^2 with both x and y axis being log scale. Specifying only one of the options allows the production of semi-log plots. Gnuplot enables the user to set/unset log scales interactively as well. By clicking on the plot window and keying in "l" (i.e. the lower-case letter "L") it is possible to toggle between normal and log scale for the y-axis. Similarly, if the mouse is close to the x-axis, then keying in the upper-case letter "L" will toggle the x-axis between log and normal mode.[3]

Legends Particularly when several plots are plotted into the same window, it is desirable to mark each curve with a label. This can be done using the `legend` option. For example, the command:

```
plot2d([x^2,x^3],[x,1,2],[legend,"sq","cube"])
```

will label the curves appropriately. Conveniently, the program automatically selects different colors and/or linestyles to make it possible to distinguish between the curves. By default, the legend key will be displayed in the top right corner of the plot window. Often this is not convenient and clashes with the curves. In these cases it is possible to manually change the position of the legend. This requires gnuplot options and will be described below.

Axis labeling The option `[xlabel, "label"]` labels the x-axis with the text `"label"`. For example:

```
plot2d[x^2,[x,1,2],[xlabel,"x"]]
```

Analogously, `ylabel` is also available.

Range of vertical-axis The range of the vertical axis is automatically determined by `plot2d`, yet sometimes this needs overriding. This can be done by supplying an option such as `[y,2,3]`; in this case the range of the horizontal axis is the interval $[2, 3]$. This also works when the free variable to be plotted is called y. In this case the horizontal axis is determined by the second y-range. So for example, to plot y^2 in the interval $[0, 3]$ while only restricting the range of the horizontal axis to values between 1 and 7, the following needs to be entered:

```
plot2d[y^2,[y,0,3],[y,1,7]].
```

Plot styles Conveniently, `plot2d` automatically chooses colors and styles for curves, particularly when multiple curves are plotted into the same window. This choice can be overridden with the `style` option. For example,

```
plot2d([x^2,x^3],[x,1,2],[style,[lines,3,4]])
```

plots x^2 and x^3 with styles 3 and 4 respectively. When plotting on screen, these styles are different colors, but when plotted to file they may be converted to dotted or dashed lines, and the like (the line-style depends on the format in which it is saved). Replacing the `style` option above with `[points,3,4]` would plot points rather than connected curves (which is rather pointless in this case, but could be useful sometimes). Finally, `[style,[linespoints]]` would plot points connected by lines. This latter option accepts numbers to modify

---

[3]This will only work if gnuplot is used in X11 mode, which may not be the default setting on some systems.

size, style, and color of the points and the lines. When two curves are plotted into the same figure each may have its own set of options. To plot the first one using lines and the second one using points one needs to specify the option [style, [lines], [points]]. Many other combinations are possible, of course.

Plotting to file There is a very simple way to save the plot to file. The option [ps_file, "myFirstplot.ps"] within a plot2d command will save the plot in a file called myFirstPlot.ps in Postscript format. On OS X this file can be opened by double-clicking on the file. On other platforms this may be more involved and require additional programs. Users who are not comfortable handling Postscript files are advised to use the gnuplot options below to save the output of a plot in a different format.

Readers who have their own favorite plotting program may be interested in the maxout.gnuplot_pipes file. Whenever Maxima produces a plot it writes the sample dots that will be used to plot the figure into this temporary file; most likely this is located in the home directory of the user (at least on OS X and Linux systems). The points in this file can be used directly to generate the plots using third-party software.

Since Maxima uses gnuplot as the plotter, it is also possible to harness some of the power of gnuplot directly from within Maxima to produce high quality, highly customized plots. This is particularly useful when there is a need to export the plot in file formats other than the standard provided by plot2d. To do this it is necessary to pass options directly to gnuplot. The simplest option is designed for output to terminals[4] with no graphics capabilities, as follows:

```
plot2d(x^2,[x,1,2],[gnuplot_term,dumb],
                    [gnuplot_out_file,"art.txt"])
```

This will plot a pseudo-graphic using keyboard characters into the file art.txt. It can be viewed using a text editor. Besides terminal-type dumb, there are a number of other gnuplot terminals. These include JPEG, Postscript, PNG, PDF and many others. However, not all are installed by default on every system, hence some of the commands might not work. We give a few examples of options to illustrate the possibilities. To save space, we will henceforth not repeat the gnuplot_out_file option, but this must always be supplied to specify the output file.

[gnuplot_term, "png size 123 123"]
Will plot the graph in PNG format and in color and uses the size 123×123 pixels.
[gnuplot_term, "postscript enhanced color"]
Plots in Postscript format, but generally produces very high quality results. Chang-

---

[4]In its widest sense, "terminal" is best thought of as gnuplot-jargon for *file format*.

ing the keyword `color` to `monochrome` produces very good black-and-white graphs.
`[gnuplot_term,pdf]`
Produces output in PDF format but requires the relevant driver to be installed.

We have noted before that it might be necessary to adjust the position of the legend in the plot. This can be done by providing options directly to gnuplot, using the `gnuplot_preamble` option. So, for example, the legend can be set at the bottom right corner by providing the option `[gnuplot_preamble,"set key bottom right"]`, where `bottom right` could be abbreviated to `b r`; `top`, `right` and `center` are available analogously. The legend can also be completely suppressed by `[gnuplot_preamble,"set nokey"]`. Finally, a last useful option is `title`:
`[gnuplot_preamble,"set title 'Display' "]`
which causes gnuplot to place "Display" at the top of the plot window.

Figure 5.1 shows an example using the following set of options:

```
plot2d([y^2, y], [y, 1, 2],
        [legend, "Plot 1", "Plot 2"],
        [xlabel, "amount of x"],
        [ylabel, "f(x)" ],
        [style, [lines, 2], [points, 2]],
        [gnuplot_term,
         "postscript enhanced monochrome"],
```

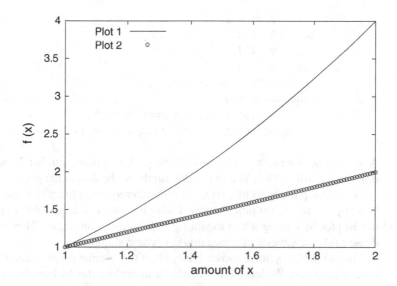

Fig. 5.1 A 2-D plot illustrating multiple style options

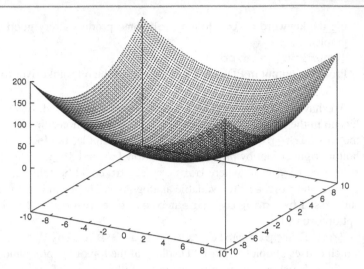

Fig. 5.2 A sample 3-D plot

```
        [gnuplot_out_file, "examplePlot.eps"],
        [gnuplot_preamble, "set key t l"]);
```

The command for 3-D plots in Maxima is `plot3d`. Its syntax is very similar to `plot2d`, so much of the understanding we have gained so far will apply. The basic plot shown in Fig. 5.2 can be created using a command such as

```
plot3d(x^2 + y^2,
        [x, -10, 10],
        [y, -10, 10],
        [legend, false],
        [grid, 100, 100],
        [gnuplot_term,
          "postscript enhanced monochrome"],
        [gnuplot_out_file, "3dexample.eps"]);
```

A relevant option to control the size of the grid, that is the number of output points, is `[grid,100,299]`. The larger the numbers, the denser the grid.

When not plotting to file, on most computer systems the `plot` commands will bring up a separate gnuplot window. Using the mouse it is possible to interact with the 3d plot by turning it and exploring it from different angles. However, for very dense grids the gnuplot interface might become sluggish.

Most of the 2d options, and certainly all of the gnuplot options, can be used in the case of `plot3d`. We leave it to the reader to explore this by herself.

5.5.4   Integrating and Differentiating

A priority for the biological modeler is to be able to differentiate and integrate expressions and functions. The principles are very simple. The command $\mathtt{diff(eq1, x)}$ returns the derivative of *eq1* with respect to x. We can extend this to $\mathtt{diff(eq1, x, 2)}$ which returns the second derivative, while $\mathtt{diff(eq1, x, y)}$ returns $\frac{\partial^2}{\partial x \partial y}$ eq 1.

The syntax for integration is equally simple. The command $\mathtt{integrate(eq1, x)}$ returns the non-specific integral, whereas, $\mathtt{integrate(eq1, x, 1, 2)}$ returns $\int_1^2$ eq1 dx. The basics barely need any more introduction, as the reader will have acquired some intuition about how to use Maxima commands. There are a few tricks, however that can be quite useful in some contexts. In the following example we differentiate and integrate an expression, then use the $\mathtt{kill(all)}$ command to clear the memory of Maxima—this removes all the stored items—and, finally, we demonstrate how we can use derivatives as unspecified expressions without evaluating them. This can be quite useful, for instance if the derivatives are very long expressions. We have already encountered the keyword $\mathtt{eval}$ and $\mathtt{numer}$ to manipulate what the $\mathtt{ev}$ command is doing. In the following example we introduce the keyword $\mathtt{diff}$ that forces derivatives to be calculated explicitly:

```
(%i1) eq1: x^a - sin(x)$
(%i2) integrate(eq1,x);
Is a equal to -1?  n;
```
$$(\%o2)\ \cos(x) + \frac{x^{a+1}}{a+1}$$
```
(%i3) diff(%,x);
```
$$(\%o3)\ x^a - \sin(x)$$
```
(%i4) kill(all)$
(%i1) eq1;
```
$$(\%o1)\ eq\ 1$$
```
(%i2) g(x):= diff(f(x),x)$
(%i3) eq1: f(x)/g(x);
```
$$(\%o3)\ \frac{f(x)}{\frac{d}{dx}f(x)}$$
```
(%i4) ev(eq1,f(x)=x^2 -(x^2/(x^2 + K)),K=1);
```
$$(\%o4)\ \frac{x^2 - \frac{x^2}{x^2+1}}{\frac{d}{dx}\left(x^2 - \frac{x^2}{x^2+1}\right)}$$
```
(%i5) ev(eq1,f(x)=x^2 -(x^2/(x^2 + K)),K=1,diff);
```
$$(\%o5)\ \frac{x^2 - \frac{x^2}{x^2+1}}{-\frac{2x}{x^2+1} + \frac{2x^3}{(x^2+1)^2} + 2x}$$

Note the difference between %o4 and %o5. Whereas in the latter the differential is evaluated, in the former it is not. The difference is due to the inclusion of the $\mathtt{diff}$ keyword.

Suppressing the evaluation of an expression is often desirable and is achieved by preceding it with an inverted comma: ′. This can be useful in many contexts, but is particularly necessary when formulating differential equations in Maxima. The evaluation of a suppressed expression can be forced by preceding it with two inverted commas.

The following example uses the command `ode2` to solve the differential equation $\dot{x} = a - bx$.

```
(%i1) eq1: 'diff(x,t) = a - b*x;
```
$(\%o1) \frac{d}{dt}x = a - bx$
```
(%i2) sol: ode2(eq1,x,t);
```
$(\%o2) \; x = \%e^{-bt}\left(\frac{a\%e^{bt}}{b} + \%c\right)$
```
(%i3) eq1: ic1(sol,x=10,t=0);
```
$(\%o3) \; x = \frac{\%e^{-bt}\left(a\%e^{bt}+10b-a\right)}{b}$
```
(%i4) plot2d(rhs(ev(eq1,a=1,b=2)),
             [t,0,6],[xlabel, "t"], [ylabel,"x(t)"],
             [gnuplot_preamble,"set nokey"])
```

Using the inverted comma operator, this equation can be simply entered into Maxima like a normal algebraic equation. The `ode2` command takes three arguments, namely the equation itself (in this case `eq1`), the function for which it should be solved (here x) and the variable with respect to which the function is differentiated (here t). The system then returns the general solution that contains an arbitrary integration constant (`%c`). A convenient feature of Maxima is the command `ic1` that allows specification of the initial conditions, and returns the specific solution of the differential equation. In the example below, we choose an initial condition corresponding to $x(0) = 10$.

Unfortunately, solving differential equations analytically is only of limited use. Biological models are nearly always too complicated to be solved by Maxima. Nevertheless, it is of course still good to know how to solve differential equations analytically in Maxima, even if in most cases we will need to resort to solving our system numerically.

The command `rk` can be invoked to solve ordinary differential equations using the 4th order Runge-Kutta method. This method has its limitations and may not always return stable results for some more difficult (stiff) systems. In these cases it may be necessary to use more powerful tools such as XPPAUT [9]. For the vast majority of models, however, `rk` will be sufficient.

As a first example, let us solve again the differential equation $\dot{x} = a - bx$, this time numerically. To obtain a solution, `rk` must be provided with (in this order) the differential equation, the dependent variable, the initial value (e.g. $x(0)$) and a specification of the independent variable and its range. An example of the latter would be: `[t,0,10,0.1]`. The first entry specifies the independent variable; the second and third are the interval over which the solution is found; the last entry determines the number of points that are generated. The smaller this number, the more will be generated. The output of `rk` is a list of points that can be plotted using `plot2d` (and the `discrete` option) or exported and plotted with an external program.

Concretely, to obtain a solution to our differential equation with $a = b = 1$, we need to first load the dynamics package, by typing load(dynamics). This package should be pre-installed on most systems. The command rk always outputs all solution points. It is therefore best to suppress output by ending a line with a dollar sign and to assign the output to a variable, say sol.

```
sol: rk(1-x,x,10,[t,0,6,0.1])$
```

The variable sol contains now a list of points corresponding to the solution of the differential equation $\dot{x} = 1 - x$ for $t = 0 \ldots 6$ and for an initial value of $x(0) = 10$. A basic plot of the solution can be generated using plot2d([discrete,sol]) (Fig. 5.3).

5.5 Solve $\dot{x} = a - \mathrm{b}x$ numerically and produce a few sample plots of the solution.

The command rk can also be used to solve systems of differential equations. The syntax generalizes straightforwardly to several equations, yet the options relating to the variables and the initial conditions must be supplied within square brackets. As an example, let us solve

$$\dot{x} = xy - x \tag{5.1}$$

$$\dot{y} = y\frac{x^2}{x^2 + 10} - y \tag{5.2}$$

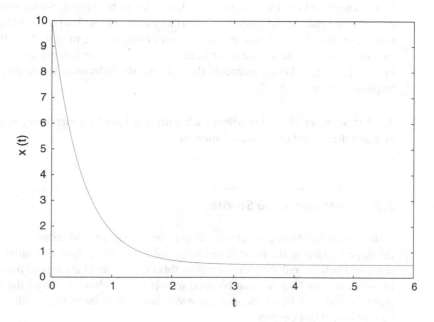

Fig. 5.3 The solution for $\dot{x} = a - \mathrm{b}x$ solved in Maxima. We used the parameters, $a = b = 1$

In Maxima this system can be numerically solved as follows:

```
(%i1) eq1: x*y - x$
(%i2) eq2: y*(x^2/(x^2 + 10)) - y$
(%i3) load(dynamics)$
(%i4) sol: rk([eq1,eq2],[x,y],[1,1],[t,0,10,0.1])$
(%i5) sol1:makelist([sol[i][1],sol[i][2]],
                              i,1,length(sol))$
```

Lines %o1 and %o2 define the right hand sides of the differential equations. When using rk for systems of equations it is essential to provide the list of dependent variables in the correct order. Since eq1 defines how x changes, x must be supplied first in the list of variables. The same applies to the list of initial conditions. The option for the independent variable t is the same as in the case of a single differential equation.

In the case of two equations, the output of rk is a list, of which each entry is itself a list of three points, corresponding to [t,x,y]. If there are more than two equations to be solved then the output is expanded accordingly. A problematic consequence of this is that plot2d cannot accept this format directly. It is therefore necessary to extract a list of pairs from the list of triples. This is done in %i5 above. We will not describe the details of list manipulation in Maxima, but the command makelist will be useful as it takes an expression and converts it into a list.

In our case, the expression is itself a list, namely the list corresponding to the first and second entry of each list in the output of rk. This corresponds to the values of t and x. The second entry in the command, i, names the index; the third selects the first element; and the fourth selects the last index to be written. So altogether, %i5 takes the first and second entry from each entry of the output list produced by rk, and puts these into a new list. It does that for every entry in sol ranging from the first to the last. To extract the solutions for y, instead, one can simply replace sol[i][2] by sol[i][3]. Having extracted the sublists, the individual solutions can again be plotted using plot2d.

5.6 Extract from sol a list where each entry is a list whose first entry is a solution of x and the second entry is a solution of y.

5.6   Maxima: Case Studies

In the remainder of this chapter we will illustrate the usage of Maxima using examples developed earlier in the book. The biological meaning of these examples has been discussed before, and we will not repeat this discussion. Our purpose here is solely to illustrate how Maxima can be used to solve the technical side of the modeling problem. This will be a welcome and useful addition to the listing mode that has so far dominated this chapter.

5.6.1 Gene Expression

We start with the simple model of gene expression given by the differential (4.19).

$$\dot{x} = a - bx \qquad (4.19)$$

We have already used Maxima to solve a similar system using `ode2` in the previous section. We will not repeat this. Instead, we will now try to reproduce this result by "manually" generating a solution using the separation of variables technique (see Sect. 4.3).

To do this we will need to integrate each side of the equation:

$$\frac{dx}{a - bx} = dt$$

This gives us an algebraic equation that we can solve for x:

```
(%i1) expr1: (a - b*x)^(-1);
```
$$(\%o1) \ \frac{1}{a-bx}$$
```
(%i2) leftS: integrate(expr1,x);
```
$$(\%o2) \ -\frac{\log(a-bx)}{b}$$
```
(%i3) eq1: leftS = C + t;
```
$$(\%o3) \ -\frac{\log(a-bx)}{b} = C + t$$
```
(%i4) solve(eq1,x);
```
$$(\%o4) \ \left[x = \frac{a-\%e^{-bC-bt}}{b}\right]$$
```
(%i5) tmpv: rhs(%[1]);
```
$$(\%o5) \ \frac{a-\%e^{-bC-bt}}{b}$$

We have now generated the general solution. Comparing this solution with (4.20) shows that the solutions are not the same. This is not a problem, however, as the difference between the solutions is merely a result of the differing definitions of C here and in (4.20). Once we have specified our initial conditions, the results should be the same.

Let us now do exactly that. In Sect. 4.3.1 we chose as our initial condition $x(0) = 0$. We make the same choice here. To determine C we need to equate the general solution ($\%o5$) to zero, solve for C and re-insert the resulting expression into the general solution. We continue our session.

```
(%i6) tmpC: solve(ev(tmpv,t=0),C);
```
$$(\%o6) \ \left[C = -\frac{\log(a)}{b}\right]$$
```
(%i7) C: rhs(tmpC[1]);
```
$$(\%o7) \ -\frac{\log(a)}{b}$$
```
(%i8) solution: ''tmpv;
```
$$(\%o8) \ \frac{a-a\%e^{-bt}}{b}$$

This is precisely the same solution we obtained in Sect. 4.3.1.

5.6.2 Malaria

Let us now progress to a more challenging case study. In Sect. 4.4 we developed a model for the spread of Malaria consisting of a system of coupled differential equations (4.30).

$$\dot{s} = a(1 - s)m - \text{b}s \tag{4.30}$$
$$\dot{m} = c(1 - m)s - \text{d}m$$

We first solve the system of equations numerically using the parameter values from Fig. 4.7, e.g. $a = 1, b = 0.4, c = 1, d = 2$. We will not actually reproduce the plot here, but we will provide the commands necessary to bring up a basic plot during a Maxima session.

```
(%i1) eq1: a*(1-s)*m - b*s$
(%i2) eq2: c*(1-m)*s - d*m$
(%i3) (a:1, b:0.4, c:1, d:2)$
(%i4) (en1: ''eq1, en2: ''eq2)$
(%i5) sol: rk([en1,en2],[s,m],[0.1,0.1],[t,0,120,0.1])$
(%i6) length(sol);
(%o6) 1201
(%i7) sol[1201];
(%o7) [120.0, 0.1428549797077344, 0.06666568841299914]
(%i8) sol[1201][2];
(%o8) 0.1428549797077344
(%i9) res1: makelist([sol[i][1],sol[i][2]],
                      i,1,length(sol))$
(%i10) res2: makelist([sol[i][1],sol[i][3]],
                       i,1,length(sol))$
(%i11) res1[1201];
(%o11) [ 120.0, 0.1428549797077344 ]
(%i12) plot2d([[discrete,res1],[discrete,res2]],
              [gnuplot_preamble,"set key b r"]);
```

Line %i5 performs the actual integration of the result. The input lines thereafter only check the format of the output of the solver and process the results to get them ready for plotting. The variables res1 and res2 are lists of lists. The first entry of each is the time and the second entries are the values of $s(t)$ (res1) and $m(t)$ (res2). A comparison of %o7 and %o11 will clarify the structure of the output of rk.

We continue our session by checking the stability of the solution:

```
(%i13) with(linearalgebra)$
(%i4) J: jacobian([eq1,eq2],[s,m]);
(%o14) ⎡ -am - b  a(1 - s) ⎤
       ⎣ c(1 - m) -cs - d  ⎦
(%i15) kill(a,b,c,d)$
(%i16) sol2: solve([eq1,eq2],[s,m]);
```

(%o16) $\left[\left[s = -\frac{bd-ac}{(b+a)c}, m = -\frac{bd-ac}{ad+ac}\right], [s = 0, m = 0]\right]$

(%i17) Jss: ev(J,s=0,m=0,simp);

(%o17) $\begin{pmatrix} -b & a \\ c & -d \end{pmatrix}$

(%i18) eigV: eigenvalues(Jss);

(%o18) $\left[\left[-\frac{\sqrt{d^2-2bd+4ac+b^2}+d+b}{2}, \frac{\sqrt{d^2-2bd+4ac+b^2}-d-b}{2}\right], [1, 1]\right]$

To do this, we first calculate the Jacobian; remember that this requires the additional package linearalgebra. Then we calculate the steady state solutions of the system directly from the set of differential equations by setting the right hand sides to zero. At this point, we would like to do this without specifying the values of the parameters. We therefore need to make Maxima forget the values by issuing the command kill(a,b,c,d). We then substitute the trivial solution into the Jacobian and calculate the eigenvalues. The first list gives the two eigenvalues of the system. This is the solution we reported earlier in Sect. 4.4.1 on p. 152. The second list is the multiplicity of the eigenvalues.

5.6.3 Cherry and Adler's Bistable Switch

In Sect. 4.6 we considered the bistable switch. The system was given by a pair of differential equations (4.47).

$$\dot{x} = \frac{K^h}{K^h + y^h} - bx \qquad (4.47)$$

$$\dot{y} = \frac{K^h}{K^h + x^h} - by$$

In the next session example we wish to determine the steady states for various parameters, and determine their stability. To get there we first calculate the steady state points of the system by setting the right-hand sides to zero. We can solve this system in several ways. One possibility is to use solve, which would return all results. A better solution is to use realroots which will return only real solutions (and we are only interested in those). The problem is that realroots can only be applied to polynomials. So, we need to bring the equations into the form of a polynomial. To get there, we define the function ssv ("steady state values"):

```
(%i1) eq1: K^h/(K^h + y^h) - b*x$
(%i2) eq2: K^h/(K^h + x^h) - b*y$
(%i3) ssv(pK,ph,pb):=
    block( fgl1: ev(eq1,K=pK,h=ph,b=pb),
        fgl2: ev(eq2,K=pK,h=ph,b=pb),
        fy: solve(fgl2,y),
        ffgl: ev(fgl1,y=rhs(fy[1])),
        ffgl2: rat(ffgl),
        ffgl: ffgl2*denom(ffgl2),
```

```
                     fres: realroots(ffgl),
                     ffres:makelist(float(fres[i]),i,1,
                                 length(fres)))$
(%i4) ssv(2,2,1/10)[1];
(%o4) x = .4174242913722992
```

The function `ssv` first assigns numerical values to the equations `eq1` and `eq2`. Secondly, it substitutes `eq2` into `eq1` obtaining a new equation called `ffgl`. It then converts `ffgl` to a canonical rational expression. Multiplying this by its denominator yields a polynomial that can be solved using `realroots`. Finally, the last step of `ssv` is to return the results in a convenient list structure.

So far we have only defined a function to calculate the steady states. Next we want to define a new function that can generate a whole list of solutions for different parameters. We decide to keep the parameters K and h fixed, and only vary b. For this we define the function `genres` that takes as inputs our choice for these parameters, and then generates results for various values of b. To do this we use a `for` ... `step` ... `thru` ... `do` construct, that allows us to define loops in Maxima.

```
(%i5) genres(pK,ph):= block( finL: [[0,0]],
      for i:1/10 step 1/100 thru 1/2 do (
           tmpres: ssv(pK,ph,i),
           for j:1 step 1 thru length(tmpres) do(
                finL:append(finL,[[i,rhs(tmpres[j])]]))),
           finL: delete([0,0],finL),  finL)$
(%i6) finalV: genres(2,3)$
(%i7) plot2d([discrete,finalV],[style,[points]]);
```

In the `genres` function, `i` is an index variable that simply counts where we are in the loop at a particular time. During each loop it will be incremented by a value specified after the `step` keyword; in this case we chose it to be $1/100$. It then executes a command (or set of commands) as specified after the `do` keyword. Notice that the commands in the loop's body must be enclosed within parentheses. The loop stops when the index variable `i` reaches the number specified after the `thru` keyword; in this case $1/2$.

The task of this loop is to generate results for various values of b. An inspection of the function definition below shows that the value of `i` is used as a value for b. The inner loop in `genres` (with index variable `j`) puts the results from `ssv` into a list called `finL`. Each member of the list is itself a list of the shape `[bvalu,ssvalu]`, where `bvalu` is the value of b (or `i`) and `ssvalu` is a steady state value for this particular parameter.

In `%i6` we call `genres`, which returns the list of steady states in convenient shape to be plotted by `plot2d` as discrete points. We do not actually show the resulting plot (but see Fig. 4.17 on p. 170); the reader who has actually typed in our code will notice that the command on `%i8` would also plot some negative steady states, which are, of course, undesired. The simplest solution is to restrict the range of the plot to positive values.

Finally, we determine the stability properties at $b = 1/3$:

```
(%i8) load(linearalgebra)$
(%i9) J: jacobian([eq1,eq2],[x,y])
```

$$(\%o9)\quad \begin{pmatrix} -b & -\dfrac{hy^{h-1}K^h}{(K^h+y^h)^2} \\ -\dfrac{hx^{h-1}K^h}{(K^h+x^h)^2} & -b \end{pmatrix}$$

```
(%i10) Jv:  ev(J,K=2,h=2,b=1/3);
```

$$(\%o10)\quad \begin{pmatrix} -\dfrac{1}{3} & -\dfrac{8y}{(y^2+4)^2} \\ -\dfrac{8x}{(x^2+4)^2} & -\dfrac{1}{3} \end{pmatrix}$$

```
(%i11) eigenvalues(ev(Jv,x=rhs(ssv(2,2,1/3)[1]),
              y=rhs(ssv(2,2,1/3)[1]) ))$
rat : replaced − 0.08059943715242143 by
  −3597169/44630200 = −0.08059943715242146
(%i12) float(%);
(%o12) [[−0.6172337333901516, −0.04943293327651504],
  [1.0, 1.0]]
```

First we generate the Jacobian, then we substitute the steady state values for the particular set of parameters into the Jacobian to be able to determine the stability. We know from the bifurcation diagram (Fig. 4.17 on p. 170) that at $b = 1/3$ only one steady state exists. Line %o12 shows the two eigenvalues and their multiplicities. Since both eigenvalues are negative, we conclude that the steady state point is stable.

5.7   Summary

Maxima is a very powerful computer algebra system. Compared to many other systems, it is intuitive to learn, easy to use, fast and portable. Best of all, it is freely available.

In this chapter we have provided an introduction to Maxima. This is enough to solve the problems discussed earlier in this book and to get started on an independent exploration of this system. The sheer breadth of Maxima means that only the most important commands and concepts have been discussed here and much has been left out. While using the system, the reader will nearly certainly encounter problems and be "stuck" for some time. In our experience, Maxima is very user friendly and contains few of the idiosyncrasies that plague other systems. Nonetheless, being stuck is part of the research process, as frustrating as it seems!

There are a number of web-based resources to help the reader. One of the best points of departure is the Maxima home page [1] which provides a number of links to tutorials and manuals of varying lengths, depths and qualities. Particularly valuable is the Maxima reference manual that lists all the Maxima commands and packages with a brief description.

An aspect that has been nearly completely left out here is the ability to program Maxima. Most users will probably not wish to write extensions to Maxima in Lisp. On the other hand, most will, sooner or later, write more complex procedures using the Maxima command language. In a rudimentary way we have already given an example above, but it is not difficult to write more advanced and powerful procedures and programs in Maxima.

References

1. Maxima, a computer algebra system: http://maxima.sourceforge.net. Accessed 24 June 2015
2. Maplesoft: Maple. http://www.maplesoft.com. Accessed 24 June 2015
3. Wolfram: Mathematica. http://www.wolfram.com/mathematica. Accessed 24 June 2015
4. Mathworks: Matlab. http://mathworks.com/products/matlab/. Accessed 24 June 2015
5. Eaton, J.W., et al.: Gnu octave. http://www.gnu.org/software/octave. Accessed 24 June 2015
6. Enterprises, S.: Scilab. http://www.scilab.org. Accessed 24 June 2015
7. Ascend modelling environment: http://ascend.cheme.cmu.edu. Accessed 24 June 2015
8. The R project for statistical computing: http://www.r-project.org. Accessed 24 June 2015
9. XPPAUT: http://www.math.pitt.edu/~bard/xpp/xpp.html. Accessed 24 June 2015
10. SageMath: http://www.sagemath.org. Accessed 24 June 2015
11. Mathomatic: http://www.mathomatic.org. Accessed 24 June 2015
12. GNU TeXmacs: http://www.texmacs.org. Accessed 24 June 2015
13. wxMaxima: http://andrejv.github.io/wxmaxima/. Accessed 24 June 2015
14. Gnuplot: http://www.gnuplot.info. Accessed 24 June 2015
15. OpenMath: http://www.openmath.org/. Accessed 24 June 2015
16. Riorto, M.: A Maxima-Gnuplot interface. http://riotorto.users.sourceforge.net/gnuplot/. Accessed 24 June 2015

If we model a system using differential equations then we make a number of implicit assumptions about the nature of the system. These assumptions will sometimes be correct, sometimes incorrect but reasonable, and sometimes incorrect and unreasonable. At the heart of the method of differential equations is the idea of *infinitesimal* changes. Remember that a statement of the form:

$$\dot{x} = f(x)$$

means that the state variable x changes during every infinitely small window of time dt like $f(x)dt$. Assume now that our differential equation describes the change in concentration of a molecular species due to chemical reactions. The assumption underlying any such differential equation is that the concentration is a continuously varying quantity; so it could take a value of 0.3012 or 0.4919 or anything in between, above or below. Fundamentally, of course, chemical systems are made up of individual particles. Concentrations are just the number of particles per volume and are, therefore, only approximately continuous. This means that there is discrepancy between the assumptions made in deterministic models and reality. This discrepancy by necessity leads to an error of the model whenever nature happens to be discrete.

It clearly makes no sense to speak of a change of the number of molecules by 0.001 or, indeed, any other non-integer change of particle numbers. Yet it is precisely such a fractional change that a differential equation model may describe. Indeed, as the system approaches the steady state, changes in the particle numbers will become successively smaller, and eventually require changes of concentrations that correspond to fractions of molecules.

To see the origin of the modeling error, consider a system where molecules of type A are converted into molecules of type B. Assume that a differential equation model predicts that, at time 10, 4.7361 molecules of A will be converted into B. Being discrete, the real system can only change by either exactly 4 or 5 molecules. Which one is the right solution? The value from the differential equation gives us a hint that the conversion of 5 molecules is more likely, but we cannot be sure. In either case, even if up to $t = 10$ the differential equation model has accurately described our system (which it most likely has not) then at least at time $t = 10 + \Delta t$ the

© Springer-Verlag London 2015

D.J. Barnes and D. Chu, *Guide to Simulation and Modeling for Biosciences*,
Simulation Foundations, Methods and Applications,
DOI 10.1007/978-1-4471-6762-4_6

prediction will necessarily be at odds with reality. Our assumed natural system can, by its very nature, not follow the differential equation. Depending on whether 4 or 5 molecules have been converted, the differential equation will predict a higher or lower concentration of *A* molecules than exist in the real system.

When we choose to model the change of discrete entities using a formalism that is inherently continuous in nature (differential calculus), then it is clear that there will be modeling errors. This tells us that the differential equation models are always inaccurate models when they describe discrete systems. The relative importance of that inaccuracy will depend on the size of the system. A relevant measure could be the number of particles in the system. The higher this number, the better the deterministic models. For very small systems, deterministic approaches are often too inaccurate to be useful.

This problem is further exacerbated by truly stochastic effects in biological systems. More often than not, real systems do not follow a deterministic law but their behavior has elements of randomness. On average, these systems may often follow a deterministic law and are accurately described by differential equations. Fundamentally though, their apparently deterministic behavior is the emergent result of a myriad of random effects. In very large systems, the stochastic nature of these systems may not be noticeable. For smaller systems, however, these fluctuations become increasingly dominant. The actual behavior of these small systems will significantly deviate from their mean behavior. This deviation of the actual behavior of a system from the mean behavior is often referred to as noise, and is the result of stochastic or statistical fluctuations. In the literature, these words are largely used interchangeably to describe one and the same phenomenon.

This means that small and discrete systems with stochastic fluctuations are typically not very well described by differential equation models. There are, of course, huge advantages to be gained from modeling systems using differential equations. Theoretically, this method is very well understood and, often, statements of considerable generality can be deduced from differential equation models. Moreover, computationally they are relatively cheap to analyze numerically, particularly when compared to explicit simulation methods. Altogether, there are significant arguments in favor of using differential equations to model systems. Hence, even if a system has significant stochastic effects and only approximately follows a deterministic path, one may still prefer differential equations to explicit stochastic models. In modeling, the question is never whether or not a model is wrong, but always, how useful it is.

If our system consists of millions of molecules that interact with one another then it will still be a system consisting of discrete entities. At the same time it will, for all practical purposes, behave very much like a continuous system. In such cases, the differential equation approaches will remain very accurate. The problem starts when the number of particles is not very high, and when the discrete nature of the particles starts to manifest itself in the phenomenology of the system. In biology, such cases are common. Transcription factors, for example, are often present in copy numbers of thousands or maybe only hundreds of molecules. In these cases, the deviation of the actual behavior from the deterministic behavior predicted by the differential equation can be quite substantial.

An example of a system where random fluctuations are often taken into account is gene expression. There are many layers of stochasticity in the system. For one, the gene may be controlled by a stochastic signal (for example, a fluctuating transcription factor concentration). Secondly, binding of any regulatory molecules to their sites is stochastic. Thirdly, even under constant external conditions, the actual expression of the protein is a stochastic process, in the sense that the number of proteins produced will fluctuate from one time interval to the next. Finally, the breakdown or loss of proteins from the cell is also a stochastic process in this sense. Taking into account that the total molecule number may be very low (hundreds of molecules rather than millions) the fluctuations of the protein concentration may dominate the system and cannot be ignored any longer.

Protein expression is but one example of an area in biology where stochastic effects are often important. There are many more. Noise and stochastic fluctuations occur at all levels of organization in biology and their importance is increasingly recognized. Consequently, over recent years significant research effort has gone into both developing new methods and tools to model noise, and to apply those methods to biological problems. Much progress has been made in understanding and describing stochastic systems in general. As a result, there are a number of mathematical techniques, as well as computational tools and algorithms, available to deal with stochastic systems.

The focus of this chapter is to give an introduction into mathematical techniques to deal with stochasticity. As (nearly) always, mathematical approaches are preferable to simulations. Unfortunately, when it comes to stochastic systems the mathematics involved is rather difficult and even relatively basic models require a level of mathematical skill that is normally the reserve of the expert. Moreover, as with most mathematical modeling methods, realistically sized problems quickly test the limits of tractability and will challenge even the skills of the specialist. Given the importance of these approaches it is nonetheless useful to understand the basics of stochastic modeling techniques, as it will provide some essential background material to understand the primary literature in the field.

Within the scope of this book it is only possible to give a taster of what can be done and what has been done mathematically. The reader who wishes to deepen her mastery of the subject will find Gardiner's "Handbook of Stochastic Methods" [1] a useful starting point.

6.1   The Master Equation

The basis of many mathematical models of stochastic systems is the so-called *master equation*. The name "master equation" is a bit misleading. In essence, it is a differential equation, or set of differential equations, that formulates how the probabilities associated with a particular system change over time. One way to think about it is to assume that a system can be in a number of states, $s = 1, 2, 3, \ldots, n$, each with a certain probability P_s. Concretely, the P_s could represent the probability that there are s proteins in the cell, or that s molecules are bound to the DNA, and so on. The numbers here are just labels and should not be understood to refer directly to some aspect of the model. We have here assumed that there is only a finite number of such states, but one could equally imagine that the index runs from $-\infty$ to $+\infty$.

The master equation is then a set of differential equations, each of which describes how the probability of being in a particular state s changes over time.

$$\dot{P}_s = f_s(P_1, \ldots, P_n) \tag{6.1}$$

As is usual, we have suppressed the time dependence of the probabilities, and we have also assumed that the states are labeled from 1 to n. In principle, the solutions to the master equation give us the probabilities of being in the various states as a function of time, which is the information one normally wishes to have in stochastic modeling applications. The problem is that, for the moment, we do not know the function f_s; but we will describe below how this can be found. Another problem is more serious: Even when f_s is known, in most cases it is not possible to solve the master equation exactly. In some cases good approximations can be found, but this is nearly always difficult. In the vast majority of cases one needs to resort to simulation to solve the master equation. Yet, despite being rarely of great utility, it is still worth being familiar with the concept of the master equation because it plays a very important role in stochastic modeling in biology.

One way to think of the master equation is to regard it as a *gain-loss equation* of probability similar to the normal differential equations encountered in Chap. 4. An example to illustrate this idea is the random walk in 1 dimension. The basic setting is that there is an infinite number of squares lined up such that each square has exactly two neighbors. The "random walker" can, at any one time, occupy exactly one of these squares. From its current position the walker can move to one of the two adjacent sites with a rate of k, or choose to remain where it is. (We assume transition rates and, hence, continuous time. The same problem could also be formulated in discrete time.) The relevant states for this problem are the occupied squares, that is we can assume the system to be in state s at time t if the random walker is on the square labeled s at time t. Since we assumed an infinite number of squares, for this particular problem it is then also most suitable to let the state labels run from $-\infty$ to $+\infty$.

To formulate the master equation for this problem it is helpful to first remember what exactly we want to know: In the absence of any information, it is impossible to make any statements about the whereabouts of the random walker, except that it could be anywhere. The situation changes when we start to take into account that, at time $t = 0$ the random walker was on square 0. Given this information/assumption, what we are interested in is the probability that the random walker is at square 45, say, at some point in time, say 10 minutes after starting. Mathematically, this is often formulated using the notation from conditional probabilities. The expression $P(s, t|0, 0)$ denotes the probability that the random walker is on square s at time t *given* that it was at square 0 at time 0.

We cannot write down the probability of this outright. However, what we can do is to write down how this probability changes. If we assume that at time t the random walker is actually at square s, then we know that, with a rate of k, it will move to the left or to the right. Altogether, it thus moves away with a rate of $2k$. If we express this in terms of probabilities, we can also say that the probability of being on s reduces with a rate of $2k$. Of course, the random walker does not just disappear, but moves on to one of the adjacent sites. One man's loss is another man's gain, so to speak. Our focal square s is the neighbor of the two other sites and can accordingly profit from their loss. From each of the adjacent sites it will experience an influx with a rate of k. Altogether we can thus formulate the change of probabilities at each site in terms of the rates as a differential equation:

$$\partial \frac{P(s, t|0, 0)}{\partial t} = kP(s - 1, t|0, 0) + kP(s + 1, t|0, 0) - 2kP(s, t|0, 0) \tag{6.2}$$

This is the master-equation for the random walk.

Often one sees a more compact way of writing the master equation using the *jump operator*, $\mathbb{E}$, which transforms expressions in terms of s into expressions in terms of $s + 1$; the inverse of this operator, $\mathbb{E}^{-1}$ transforms s into $s - 1$. The jump operator is simply a convenient abbreviation tool. We will not use it any further within this book but merely bring it to the reader's attention as she will almost certainly encounter it in the literature. Using this operator we obtain a different form of the master equation (6.2):

$$\partial \frac{P(s, t|0, 0)}{\partial t} = k(\mathbb{E}^{-1} + \mathbb{E} - 2)P(s, t|0, 0) \tag{6.3}$$

6.1 Show how to derive (6.3) from (6.2).

Let us consider another simple example. Assume that the probability of an event happening is proportional to some waiting time. For example, if we "sit" on a molecule in a well stirred vessel, then the probability $P_c^{\Delta t}$ of colliding with another molecule within the next time unit Δt is (more or less) proportional to Δt. Using m as a proportionality constant, we can express this formally as:

$$P_c^{\Delta t} = m\Delta t \tag{6.4}$$

Assume that we are interested in the number of collisions since time $t = 0$. Let us denote by $P(k, t)$ the probability that at time t there have been exactly k collisions between the molecule we are sitting on and others. We do not know this probability but, given our assumption in (6.4), we can formulate how it changes:

$$P(k, t + \Delta t) = P(k, t)(1 - P_c^{\Delta t}) + P(k - 1, t)P_c^{\Delta t}$$
$$= P(k, t)(1 - m\Delta t) + P(k - 1, t)m\Delta t \tag{6.5}$$

This equation can be interpreted very easily. It consists of the sum of: the probability that there have already been exactly k collisions by time t (given by $P(k, t)$) and that another does not follow within Δt (given by $(1 - P_c^{\Delta t})$); and the probability that there have been $k - 1$ by time t (given by $P(k - 1, t)$) and that a single collision follows within the next Δt. One could imagine that within Δt there could be more than a single collision, which would complicate the calculation considerably. A relatively easy way out of this problem is to assume that the time interval, Δt, is so short that more than a single collision is unlikely enough to be ignored as a possibility. This is precisely what we do. We can now further transform (6.5):

$$\frac{P(k, t + \Delta t) - P(k, t)}{\Delta t} = m\left(P(k - 1, t) - P(k, t)\right) \tag{6.6}$$

Taking the limit $\Delta t \to 0$ we obtain a differential equation for the probability of observing k collisions within a time period of t:

$$\frac{\partial P(k, t)}{\partial t} = m\left(P(k - 1, t) - P(k, t)\right) \tag{6.7}$$

This is a master equation. At a first glance, it seems very similar to the simplest differential equations we have encountered in Chap. 4 and we would expect it to be easily solvable using elementary methods. However, a closer examination reveals a complication. The first argument to P on the right-hand side of the equation is not the same as that on the left-hand side. The symbols P on the right hand side are therefore different functions of t. Since the left-hand side of the differential equation and the right-hand side do not contain the same functions, our approaches from Chap. 4 are not applicable, at least not immediately.

To solve this master equation we must resort to a trick, using the so-called *probability generating function*:

$$G(x, t) \doteq \sum_{k=0}^{\infty} x^k P(k, t) \tag{6.8}$$

This function is a transformation of the probability function P. The reader will notice the new variable x that is introduced by the transformation, while the original variable k (over which the sum is taken) is dispensed with. The easiest way to think of x is not to think about it at all, and just take it as a formal variable without any intrinsic meaning! As it will turn out, the master equation (6.7) will be very easy to solve for this transformed function. Once a solution is obtained we can transform back to obtain the desired solution. This approach works because the generating function sums over the offending variable k of the (as yet unknown) probability function of observing exactly k collisions. At the same time, it leaves the time variable t untouched. Since x is not dependent on time, there is a clear relationship between the time derivative of G and the probability function:

$$\frac{\partial G(x, t)}{\partial t} = \sum_{k=0}^{\infty} x^k \frac{\partial P(k, t)}{\partial t}$$

This expression is simply obtained by differentiating (6.8) with respect to time. We can now reformulate the differential (6.7) in terms of the probability generating function. To do this, we first multiply (6.7) by x^k, and then take the sum over all k:

$$\sum_{k=0}^{\infty} x^k \frac{\partial P(k,t)}{\partial t} = \sum_{k=0}^{\infty} m \left(x^k P(k-1,t) - x^k P(k,t) \right)$$

$$= m \left(\sum_{k=0}^{\infty} x^k P(k-1,t) - \sum_{k=0}^{\infty} x^k P(k,t) \right)$$

$$= m \left(x \sum_{k=0}^{\infty} x^k P(k,t) - G(x,t) \right)$$

$$\frac{\partial G(x,t)}{\partial t} = m(x-1)G(x,t) \tag{6.9}$$

We can now solve (6.9) using the methods of Chap. 4:

$$G(x,t) = G(x,0) \exp\left(m(x-1)t\right) \tag{6.10}$$

The first question to address is to determine the value of $G(x,0) = \sum x^k P(k,0)$. Given the meaning of P we must assume that at time $t = 0$ there have, as yet, been no collisions; we have, so to speak, reset the counter at $t = 0$. Hence, $G(x,0) = P(1,0) = 1$.

Now we have a solution for the probability generating function, but not yet for the probability density itself. To obtain that, we will use the Taylor series for $e^x = \sum x^n/n!$ to expand the right hand side of (6.9) around 0:

$$G(x,t) = \exp\left(mt(x-1)\right) = \sum x^k P(k,t)$$

$$= \frac{1}{\exp(mt)} + \frac{(mt)\,x}{\exp(mt)} + \frac{(mt)^2\,x^2}{2!\,\exp(mt)} + \frac{(mt)^3\,x^3}{3!\,\exp(mt)} + \frac{(mt)^4\,x^4}{4!\,\exp(mt)} + \cdots$$

$$= \exp(-mt)\frac{\sum x^k (mt)^k}{k!} \tag{6.11}$$

The solution can now be obtained by comparing the coefficients of the expansion of $G(x,t)$ with the coefficients of the sum $\sum x^k P(k,t)$. We then obtain the probability function for every k:

$$P(k,t) = \frac{(mt)^k \exp(-mt)}{k!} \tag{6.12}$$

The reader may recognize this expression as the Poisson distribution.

6.2 Fill in the missing steps to get from (6.11) to (6.12).

A similar method can be used to solve the master equation for the random walk (6.2) above. Remember that we denoted by $P(n,t|0,0)$ the probability of the random walker being at n at time t given that he started at 0 at time $t = 0$:

$$\frac{\partial P(n,t|0,0)}{\partial t} = kP(n-1,t|0,0) + kP(n+1,t|0,0) - 2kP(n,t|0,0) \tag{6.2}$$

This equation can be solved using the *characteristic function*, which is similar to the probability generating function:

$$G(x, t) = \langle \exp(ixn) \rangle = \sum_{n=-\infty}^{\infty} \exp(ixn) P(n, t) \qquad (6.13)$$

where i is the complex unit, that is the square root of -1. Note that here we have omitted the initial conditions on the probability to simplify the notation. As before, we can transform the differential (6.2) into a differential equation of the characteristic function by multiplying both sides with $\exp(ixn)$ and summing over all n. Following a similar procedure as in the derivation of (6.9), we obtain a new equation:

$$\frac{\partial G(x, t)}{\partial t} = k(\exp(ix) + \exp(-ix) - 2)G(x, t) \qquad (6.14)$$

6.3 Fill in the details of how to reach (6.14).

Again, this can be solved for $G(x, t)$ using the methods from Chap. 4:

$$G(x, t) = G(x, 0) \exp \left[(\exp(ix) + \exp(-ix) - 2) kt \right] \qquad (6.15)$$

As in the first example, we can now determine that $G(x, 0) = 1$ by choosing our initial conditions properly. Unfortunately, this leaves us with an unpleasant exponential that has an exponential as its argument. What makes the situation worse is that this argument is not even real, but complex. However, things are not as bad as they seem. To solve this we can use the fact that $\exp(\pm ix) = \cos(x) \pm i \sin(x)$:

$$G(x, t) = \exp \left[(2 \cos(x) - 2) kt \right] \qquad (6.16)$$

To further process this, we expand the cosine around $x = 0$ using a Taylor series:

$$cos(x) = 1 - \frac{x^2}{2!} + \frac{x^4}{4!} + \cdots \qquad (6.17)$$

We next make an approximation to keep the problem simple. Instead of keeping the entire Taylor expansion, we will only keep terms of order x^2 and lower. To truncate Taylor expansions in this way is a very common technique in physics to simplify expressions. It usually means making a mistake, in that the truncated expansion is only an approximation to the original function. One hopes that this mistake is very small and does not matter in practice. In any case, a small inaccuracy is mightily outweighed by the massive simplification gained.

If we keep only terms of order x^2 and lower in the expansion of (6.17), then we obtain in (6.16):

$$G(x, t) = \exp \left[-x^2 kt \right] \qquad (6.18)$$

We have now derived a solution for the characteristic function. As explained above, this does not help us very much unless we can somehow transform it back into probabilities. In this particular case, however, this is not necessary, because the specialist would immediately recognize (6.18) as the characteristic function of the

Gaussian distribution. However, we will not expect the reader to take this on trust. In the following paragraphs we will check that the specialist is correct and explicitly calculate the characteristic function of the Gaussian.

Consider a Gaussian of the form:

$$f(z) = \frac{1}{\sqrt{2\pi\sigma^2}} \exp\left(-\frac{z^2}{2\sigma^2}\right)$$

Using a computer algebra system (Chap. 5), we can explicitly calculate the characteristic function of the Gaussian (which we will call F now to avoid confusion):

$$F(x) = \int_{-\infty}^{\infty} \exp(ixz) f(z) = \exp\left(\frac{-x^2\sigma^2}{2}\right) \tag{6.19}$$

This characteristic function is of the same overall form as (6.18). The good thing about (6.19) is that it is formulated in terms of the variance σ^2. Equation (6.18) is of the same overall form, but it is not apparent what σ^2 is within it. We can find that out by directly comparing the right-hand sides of (6.18) and (6.19). We find that (6.18) is a Gaussian with $\sigma^2 = 2n^2 kt$. This seems to solve the problem. Unfortunately, there is another problem. A Gaussian distribution would suggest a continuous random walk, rather than a discrete random walk. We can therefore regard the Gaussian as a solution for our problem in the limiting case of n very small and k very high.

$$P(n, t|0, 0) = \frac{1}{\sqrt{4\pi kt}} \exp\left(-\frac{n^2}{4kt}\right) \tag{6.20}$$

6.4 Fill in the details of the comparison between (6.18) and (6.19).

6.5 (advanced) Somewhere in the derivation of (6.20) we silently converted the discrete random walk into a continuous solution. Locate where this happened.

The solution given by (6.20) is not quite satisfactory. We started off with a discrete space random walk problem and ended up with a solution for continuous space, in that we identified the (approximated) characteristic function (6.18) with the characteristic function of the Gaussian, which is a continuous distribution function. What we ideally want is a solution that gives us the probability of being at $n = 1, 2, 3, \ldots$, only, and not the probability of being at the impossible values of $n = 3.2342$ or $n = 0.012$.

To obtain our desired discrete probability we will try again. To do this we will use a similar approach as above for the Poisson distribution. For convenience, let us substitute the place-holder z for the exponentials in the exponent in (6.15):

$$G(x, t) = \exp\left[(z - 2) kt\right] \tag{6.21}$$

We can then Taylor expand $G(x, t)$ around $z = 0$:

$$G(x, t) = \exp(-2kt)\left[1 + ktz + \frac{k^2 t^2}{2!}z^2 + \frac{k^3 t^3}{3!}z^3 + \cdots\right] \tag{6.22}$$

To be able to obtain the expression for the individual $P(n, t)$ we need to compare the coefficients of this sequence with the general expression of the probability generating function (6.13). In order to find the probability $P(n, t)$ we need to collect all the terms in (6.22) that are of the same order in $\exp(ix)$. In essence this is the same procedure we used above to derive the Poisson distribution, but in this case it is slightly complicated by the fact that each z is a place-holder for a sum of exponentials, that is $z^q = (\exp(ix) + \exp(-ix))^q$. This means that each z^q is a potentially very long expression containing many different powers of $\exp(ix)$. We can write the general procedure to explicitly represent this expression using the binomial coefficients[1]:

$$z^q = \sum_{j=0}^{q} \binom{q}{j} \exp\left(ix(q - 2j)\right)$$

This formula is not a definition or approximation, but simply the rule for how to expand expressions of the form $(\exp(ix) + \exp(-ix))^q$. (Note that in this formula the letter i is not an index variable, but the complex unit.) Replacing the z terms by their definition and inserting the whole expression back into probability generating function (6.22), gives a double sum:

$$G(x, t) = \exp(-2kt) \sum_{q=-\infty}^{+\infty} \frac{k^q t^q}{q!} \sum_{j=-\infty}^{q} \frac{q!}{j!(q-j)!} \exp\left(ix(q - 2j)\right) \qquad (6.23)$$

6.6 Derive (6.23) explicitly.

Here we wrote the binomial coefficient explicitly in terms of factorials. All that is left to do now is to extract from (6.23) all the terms for which the exponential is $exp(ixn)$, that is all the pairs of indices for which $q - 2j = n$. The coefficients of those terms are then the probability we are looking for; this is simply due to the definition of the characteristic function. For every specific index q there will be at most one j such that that $q - 2j = n$, from which we can express the sought after probability using a single sum only. We can obtain it be replacing q by $n + 2j$ throughout in (6.23):

$$P(n, t) = \exp(-2kt) \sum_{j=-\infty}^{n} \frac{k^j t^j}{j!(n+j)!} \qquad (6.24)$$

Besides these two examples, there are a number of cases that can be solved exactly, and many more that can be solved approximately. Presenting these methods would go well beyond the scope of this book and the interested reader is referred to Gardiner's fantastic book on this subject [1]. The main aim of the presentation here is

---

[1] The binomial coefficient $\binom{n}{k}$ is defined as the number of ways to choose k elements from a set of n. It can be calculated as follows:

$$\binom{n}{k} \doteq \frac{n!}{k!(n-k)!}.$$

to present the basic ideas of the master equation and to enable the reader to have a passive understanding of the subject matter, that should enable her to understand the theoretical literature in the field. Unfortunately, a full mastery of the techniques presented here requires a significant degree of specialization.

6.2 Partition Functions

The master equation is of great importance in stochastic modeling of biological processes. There are also a few other approaches that can be of use and are technically and conceptually more accessible. One of them is the partition function approach which can be used to compute equilibrium distributions of some systems. The clear disadvantage of this method is that it cannot be used to understand how a stochastic system evolves in time from a specified initial state. This is in contrast to explicit solutions of the master equations that do offer this. The partition function can be mathematically rather involved, which often makes it difficult to draw conclusions from the formula directly; hence numerical analysis is often necessary.

The basic idea behind the approach is very simple. It rests on the identification of some "macro-state" that corresponds to (normally) more than one micro-state. What constitutes macro and micro is entirely case specific. Normally we are interested in the probabilities of observing the macro-states. Typically, the macro-state will be something like, "the operator site is bound," "pathway X is activated," or a similar observable phenomenon. The word "micro" normally implies "small scale," but this should not distract us here. In the present case, the micro-states are only small in the sense that they somehow realize the macro-state.

For the moment, let us describe the basic concept in abstract. We can consider the macro-states as an equivalence class of micro-states; that is, a set of micro-states that are essentially the same with respect to the particular purpose that motivates the modeling exercise. For example, a micro-state could be a particular distribution of transcription-factors over the DNA; a corresponding macro-state could then be the set of all distributions such that a specific operator site of gene X is bound.

In the simplest case, we can assume that all micro-states are equally likely. If we are interested in the probability of observing a particular macro-state, then all we need to do is to calculate the proportion of micro-states that realize the particular macro-state we are interested in. We can do this for every macro-state, if we wish. In most realistic cases, the situation is a bit more complicated in that not all micro-states will be equally likely. In this case we associate each micro-state with a weight and take this weight into account when calculating the proportions.

Let us consider an example to clarify this. Imagine a royal court of C people, consisting of a king, a queen and the lower members of the court ($C - 2$ courtiers). Let us assume the entire court to be decadent, in fact so decadent that they take no interest whatsoever in the outside world and practically never leave the palace. It is rather easy for them to stay "at home" since their palace has many rooms. Somebody might now be interested in modeling which rooms the king, a queen and the court are dwelling in throughout the day. The partition function method is a convenient tool in this case. To obtain a first, simple model, let us assume that every person in the palace moves randomly between the rooms.

The first question to ask is the mean occupancy of every room; that is, the average number of people in every room. This is, of course, simply given by C/N. A somewhat more challenging question is to ask about the probability that there is exactly one person (either the king or the queen or any of the courtiers) in a particular room, say room 1. In this case, we identify two macro-states: one corresponds to exactly one person being in room 1, and the other covers all other cases. It is clear that each of these two macro-states can be realized by many micro-states. For example, the queen could be the person in room 1, or the king, or any of the courtiers. As long as only one person is in the room in question it does not matter what goes on in all the other rooms.

Given our assumption that each and every micro-state has the same probability, we can then answer questions about various probabilities by considering all possible configurations that fulfill a particular condition, and then dividing this number by the number of all possible configurations. For example, we could ask about the probability of finding exactly one person in room 1. This is true when $C - 1$ persons are in room 2 and one in room 1. It is also true, if $C - 2$ people are in room 2, one in room 3 and exactly one in room 1; and so on.

Let us make this more precise and calculate the number of all possibilities (i.e., micro-states) first. This can be obtained by observing that each member of court (king and queen included) can occupy only one of the N rooms at any particular time. If, for the moment, we only consider the king, then there are clearly N possible configurations. Taking into account the queen, then for each of the N configurations of the king, the queen has N configurations of her own; hence altogether there are N^2 ways to arrange the king and queen. Just by extending the argument, we see that there are $Z \doteq N^C$ arrangements of the entire court over the rooms of the palace. Z is sometimes called the partition function.

The next question to ask is: How many configurations are there such that there is exactly one person in room 1? This is nearly the same as asking how many configurations there are to distribute $C - 1$ people over $N - 1$ rooms. The answer to this is of course $(N - 1)^{(C-1)}$ (the reasoning is exactly the same as in the case of C people in N rooms). To obtain the desired number of configurations, we need to take into account that there are C possibilities to choose the person occupying room 1 (the king, the queen, etc...). Hence, altogether we have $C(N - 1)^{(C-1)}$ different micro-states compatible with the macro-state of room 1 being occupied by exactly one person. Dividing this by the total number of configurations gives us the sought probability:

$$P(\text{exactly 1 person in room 1}) = \frac{C(N - 1)^{(C-1)}}{N^C} \qquad (6.25)$$

We can also calculate the probability that exactly 2 people are in room 1. The reasoning is similar, although slightly more involved. We need to choose two members of court to occupy room 1. This number is given by the binomial coefficient,

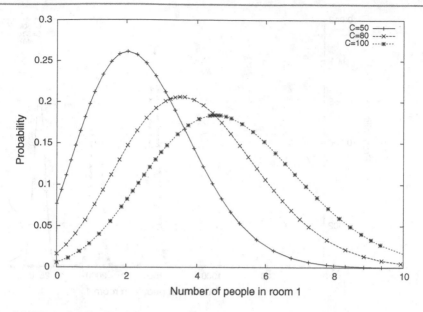

Fig. 6.1 The probability distribution of finding a given number of members of the court in room 1 according to (6.27). Results for courts of size 50, 80 and 100

i.e., $\binom{C}{2}$. Then for each of these choices there are $(N-1)^{(C-2)}$ possibilities to distribute $C-2$ members of court over the remaining $N-1$ rooms. Altogether, we get:

$$P(\text{exactly 2 persons in room 1}) = \binom{C}{2} \frac{(N-1)^{(C-2)}}{N^C} \qquad (6.26)$$

More generally, we obtain for exactly k people occupying room 1:

$$P(\text{exactly k persons in room 1}) = \binom{C}{k} \frac{(N-1)^{(C-k)}}{N^C} \qquad (6.27)$$

Here we require that $k \leq C$ for the formula to make sense. Figures 6.1 and 6.2 illustrate the probability distributions for various values of C in (6.27).

6.7 What is the probability of having either, 1 or 2 or 3 ... or P people in room 1? Check that the calculation yields the expected result.

6.2.1 Weighted Configurations

One of our key-assumptions above—that the king and queen and the courtiers are equally likely in every room—is, of course, unrealistic. In reality, there will be some rooms that they prefer to others (in the sense that they stay in them for longer). This will change the result. In order to be able to include this into our model, we must

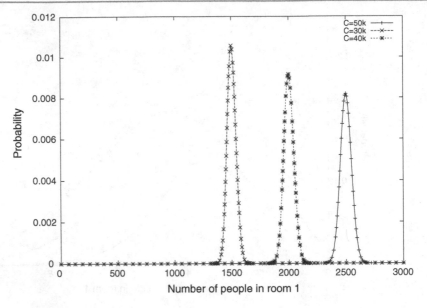

Fig. 6.2 The probability distribution of finding a given number of members of the court in room 1 according to (6.27). Results for large courts of sizes 30000, 40000 and 50000. These large courts are likely to provoke an uprising of the people, as in France in 1789 AD

assign weights to each configuration. Let us say that the members of court prefer room 1 over all the other rooms. Then we give a higher weight to every configuration where somebody is in room 1 than to configurations where they are in other rooms. The precise nature of this weight does not matter, as long as it is a clear measure of the preference. We are therefore free to choose weights. For reasons that will become clearer later, we formulate our weights as exponential functions:

$$w_i \doteq exp\,(-G_i)$$

Here the index i represents the room. We take $G_i \leq 0$ so that the lower the value of G_i the higher the weight of room i. To keep things simple we only distinguish between two cases, at least for now:

$$G_i = \begin{cases} a & \text{if } i = 1, \\ b & \text{otherwise} \end{cases} \qquad a < b \leq 0$$

The weights need to be reflected in the partition function Z, which is no longer of the simple form $Z = N^C$. Specifically we must distinguish between configurations with different numbers of court members in room 1. Since all the other rooms have equal weights we do not need to distinguish between those. This leads to the new partition function:

$$Z = \sum_{k=0}^{C} \underbrace{\binom{C}{k}(N-1)^{(C-k)}}_{T1}\underbrace{\exp\,(-ka-(C-k)b)}_{T2} \qquad (6.28)$$

To understand this equation it is helpful to compare it with (6.27), which simply formulated the total number of configurations. Equation (6.28) essentially does the same thing, but must assign different weights to different configurations. Specifically, occupation of room 1 carries a higher weight than occupation of any of the other rooms, and it is therefore necessary to consider how many possible ways there are to occupy room 1. The basic idea of (6.28) is very simple. It first counts all the possibilities where room 1 is unoccupied. This would be the first term of the sum corresponding to the index $k = 0$. It then considers the case of exactly 1 person being in room 1, and $N - 1$ people distributed over the other rooms. This corresponds to the index $k = 1$, and so on. To obtain the partition function that includes the weights, it must be summed over all possible k, that is over all possible configurations, ranging from nobody being in room 1 to every member of court dwelling there, and multiply the result with the appropriate weights. In (6.28) the term T1 counts the number of possible configurations and the term T2 gives the appropriate weight. The term T1 is in essence (6.27). In the term T2, k represents the number of persons in room 1; each of those carries a weight of a; hence the total weight is $\exp(-ka)$. Similarly, all persons that are in a room other than room 1 carry a weight of b.

To obtain the probability of room 1 being occupied by exactly k members of court, given a preference a for room 1, we obtain then, in close analogy to (6.27):

$$P(\text{exactly k persons in room 1}) = \frac{1}{Z}\binom{C}{k}(N-1)^{C-k}\exp\left(-ka - (C-k)b\right)$$

$$(6.29)$$

Figure 6.3 shows the probability of finding room 1 occupied with a given number of people; it illustrates the effect of weighting. As the preference for being in room 1 increases, the mean of the distribution moves to higher values and room 1 receives more than its fair share of dwellers. On the other hand, where the weights are equal the peak of the distribution is always at C/N, as can be see in Figs. 6.1 and 6.2.

Let us now take the next step and assume that there are altogether three preferences. Suppose that the members of court have a certain preference for room 1, a different one for room 2, and a third for the remaining rooms in the palace. In essence, this case can be treated just like the simpler case with only two different preferences. One just imagines the occupation of the first room fixed, and then sums over all possible ways to occupy the remaining rooms. This is precisely the solution in (6.28). To obtain the sought partition function it is necessary to let the occupation of the first room vary from 0 to N. Given that we have already worked out how to formulate the partition function Z for the case of only two preferences, the case of 3 or more preferences is a straightforward generalization. Assume that there are k persons in room 1, l persons in room 2 and $C - k - l$ persons distributed over the remaining rooms. Conceptually, we can think of the number of people in room 1 fixed at k and simply re-apply the same formula we have used in (6.28), with a total number of $n - k$ people. Then we sum over all possible values of k, that is all possible occupations of room 1. If we assume preferences $G_1 = a$, $G_2 = b$ and $G_j = c$ when $j \neq 1, 2$ then we obtain the new partition function:

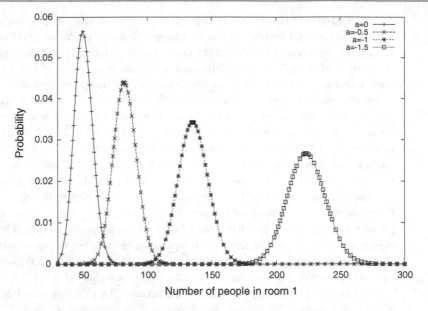

Fig. 6.3 The probability distribution of finding a given number of members of the court in room 1 with various weights. We assume here: a court of size $C = 100000$; $N = 2000$; the weight $b = 0$; and we range a from 0 to -1.5. Clearly, as we decrease a the probability distribution shifts to the right, because it becomes more likely for members of the court to be in room 1

$$Z = \sum_{k=0}^{C} \sum_{l=0}^{C-k} \binom{C}{k} \binom{C-k}{l} (N-2)^{C-k-l} \exp\left(-ka - lb - (C-k-l)c\right) \quad (6.30)$$

Using this partition function, we can again calculate the probability of observing exactly k persons in room 1, exactly l people in room 2:

$$P(k, l) = \frac{1}{Z} \binom{C}{k} \binom{C-k}{l} (N-2)^{C-k-l} \exp\left(-ka - lb - (C-k-l)c\right) \quad (6.31)$$

6.8 Extend (6.31) to include 4 or more different preferences for rooms.

6.9 Plot the distribution for the occupation of room 2 for different preferences, a, b and varying C, N.

So far we have not distinguished between the royal couple and the courtiers. We would expect that the king and queen have different preferences and access rights from the other members of the court. We would now like to reflect this in the calculation of probabilities, using preferences to formulate their different movement patterns.

The best way to think about the problem is to assume that there are two different "populations"; in our case population 1 consists of the king and queen, and population 2 represents the courtiers. Since we can consider these sub-populations as independent, there is not much work left to do. All that needs to be done is to formulate the partition function for each sub-population. This is a formula along the lines of (6.31). Multiplying the partition functions for each of the populations then gives the total partition function. So, if Z_{P_1} and Z_{P_2} are the partition functions for each sub-population, then the total partition function is given by $Z = Z_{P_1} Z_{P_2}$. We leave the details of working this out as an exercise to the reader.

6.10 Assuming that the king and queen have preferences G_1 and G_j for room 1 and all the other rooms respectively, and the courtiers have preferences G_3 and G_4 for rooms 3 and 4 and G_j ($j \neq 3, 4$) for all other rooms. Calculate the probability of finding all members of court in room 3. Obtain numerical values for different choices of the size of the court and the preferences.

6.2.2 Binding to DNA

There are a number of biological problems that can be attacked using partition function techniques. One very prominent problem concerns transcription-factor (TF) occupation of specific binding sites. The problem of binding-site localization of TFs is an interesting one, since a particular regulatory protein needs to find one (or a few) specific binding site among literally millions of non-specific sites. There are usually only a few specific sites for a TF on the DNA, but even away from the specific binding sites DNA is sticky for TFs and they will bind, albeit poorly, to every nucleotide sequence. Non-specific binding times are much shorter, on average, than specific binding. Yet, while each individual interaction may not last for a long time, there are so many non-specific sites that collectively they could be very important. One of the questions one may ask is: How do TFs "find" their specific binding sites among all the non-specific ones?

The process of TFs "searching" for their specific sites is often modelled as a random walk, whereby the individual TFs randomly move across the DNA, dissociating from one site to associate to another one. The association/dissociation rates will depend on how well the particular sequences of the DNA match the TF-binding motif. To treat such a random walk model adequately one needs to use more sophisticated models, perhaps using a master-equation approach. The partition function *ansatz* that we introduced in the last section is a more course-grained approach, but it can give some initial insights into the role of non-specific binding sites. The approach is very similar to the example above of distributing members of court over N rooms. We could take members of the court as TFs and the rooms as binding sites. A modification is necessary, namely that each binding site can take at most one TF. This affects both the kind of questions we ask, but we must also re-formulate the partition function and the number of configuration we consider.

The simplest case is again a single TF and one specific binding site. The micro-states would be the various possibilities to distribute TFs over the DNA, while the macro-states are the binding states of the specific site. In this case we can write down our partition function as:

$$Z = (N - 1)\exp(-G_n) + \exp(-G_s).$$

Here G_n and G_s are the binding free energies for the non-specific and the specific sites respectively. They take the role of the preferences of the members of the royal court.

To generalize this, we make the assumption that it is not possible to distinguish between the molecules. This assumption is not strictly correct, but one can show that for the resulting probability this does not matter, as long as one does not break the assumption of indistinguishability in the formulation of the set of configurations one is interested in. This would be unnatural to do anyway.

Let us next formulate the partition function when we have P TFs and N binding sites, one of which is a specific binding site; throughout this section we assume that $P \ll N$; that is, there are many more binding sites than are TFs. With this in mind we can now see that there are only two macro-states to consider: (i) that the specific binding site is occupied; and, (ii) that it is not. The partition function is then the sum of these two configurations. Taking into account the relevant binding free energies as well, we obtain:

$$Z = Z_1 + Z_2$$
$$Z_1 = \binom{N - 1}{P - 1} \exp\left(-G_s - (P - 1)G_n\right)$$
$$Z_2 = \binom{N - 1}{P} \exp\left(-PG_n\right) \tag{6.32}$$

Here Z_1 is the number of micro-states compatible with the specific binding site being bound. There is exactly one way in which a single site can be occupied, hence we need not do anything. To calculate the number of configurations to distribute the remaining $P - 1$ TFs over the remaining $N - 1$ binding sites we again use the binomial coefficient and the exponential terms formulate the relevant weights. The second part of the partition function, Z_2, expresses all the possible ways to distribute P TFs over $N - 1$ non-specific sites. The probability of the specific binding site being occupied is then Z_1/Z.

6.11 Where in the derivation did we use the assumption that the individual TFs are indistinguishable?

If there is more than one specific site then the partition function needs to be extended. The idea is essentially the same as in the case of a single site, but we need to sum over all possible occupation states of the specific sites; that is, we must consider the case of no specific site being occupied, exactly 1 site being occupied, exactly 2 sites, and so on. Continuing with the restricted case that we have only 2 specific sites we obtain for our partition function:

$$Z = Z_0 + Z_1 + Z_2$$

$$Z_2 = \binom{2}{2}\binom{N-2}{P-2}\exp(-2G_s - (P-2)G_n)$$

$$Z_1 = \binom{2}{1}\binom{N-2}{P-1}\exp(-G_s - (P-1)G_n)$$

$$Z_0 = \binom{2}{0}\binom{N-2}{P}\exp(PG_n) \tag{6.33}$$

The case here is analogous to (6.32). The term Z_1 counts all the cases where exactly two specific sites are occupied. Altogether there is only $\binom{2}{2} = 1$ way to occupy two specific sites with two TFs, but there are $\binom{N-2}{P-2}$ ways to distribute the remaining $P-2$ TFs over the non-specific sites. The case for Z_2 and Z_3 can be argued similarly.

6.12 Using the partition function in (6.33) calculate the probability that both specific binding sites are occupied.

6.13 (advanced) Write down the version of the partition function (6.33) for the case when we distinguish between every TF, and show that the probabilities for various scenarios (e.g., "exactly one specific binding site occupied") are unaffected.

The special case of 2 specific sites in (6.33) suggests already how the formula can be extended to an arbitrary number of N_s specific sites, when there are overall N binding sites. (Of course we assume that $N_s < P < N$.) The pattern is the same. We need to sum over all possibilities to occupy the N_s sites. There are always $\binom{N_s}{k}$ possibilities to occupy exactly k of the specific sites. To obtain the total number of configurations compatible with k binding sites being occupied, we need to multiply the binomial coefficient with the number of ways to distribute $P-k$ TFs over $N-N_s$ binding sites. Taking into account the weights as well, we obtain for the partition function:

$$Z = \sum_{k=0}^{N_s}\binom{N_s}{k}\binom{N-N_s}{P-k}\exp(-kG_s - (P-k)G_n) \tag{6.34}$$

As usual, the probability of, for example, exactly one specific binding site being occupied is given by:

$$P(\text{exactly } 1) = \frac{1}{Z}\binom{N_s}{1}\binom{N-N_s}{P-1}\exp(-G_s - (P-1)G_n)$$

The next question to consider is how different types of TFs can be included. Different types of TFs will bind non-specifically to other types' specific sites. To simplify the problem, we will assume two different species of TF and only be interested in specific binding of one type. To formulate the new partition function, we need to extend (6.34) to take into account the additional possibilities of binding. By this we mean that only type-1 TFs have specific binding sites, whereas type-2 TFs bind non-specifically to all sites. Essentially, the approach is the same again. We first distribute the TFs of type 1 to their specific sites. Then we distribute the type-1 TFs over the non-specific sites, just as in (6.34). Finally, we distribute the type-2 TFs over the remaining sites. Since we assume they bind to all sites with the same preference (or rather binding free energy) we do not need to distinguish between specific and non-specific sites there. To obtain to our partition function we assume that there are P_1 TFs of type 1, P_2 of type 2. The latter bind to all sites with a binding free energy of G_{n_2} whereas the former bind to non-specific sites with G_{n_1}:

$$Z = \sum_{k=0}^{N_s} \binom{N_s}{k} \binom{N - N_s}{P_1 - k} \binom{N - P_1}{P_2} \exp -kG_s - (P_1 - k)G_{n_1} - P_2 G_{n_2} \quad (6.35)$$

6.14 Show that the probability of k type-1 TFs binding to specific binding sites does not depend on the number of type-2 TFs.

6.15 (advanced) Formulate conditions that will change the result in the previous exercise; i.e., find the conditions when the probability of binding to specific binding sites does depend on other types of TFs.

6.2.3 Codon Bias in Proteins

Th statistical reasoning used above can also be used to analyze the codon bias in amino acids where the problem there is as follows. Every protein consists of a number of amino acids. Each amino acid is encoded by a sequence of three bases of the genetic code. For example, Met (or Methionine) is encoded by the triplet ATG. Unlike Met, most amino acids are encoded by more than one triplet, because the genetic code is degenerate. As it turns out, however, the different codons are not equivalent. There is a statistically-significant bias in the usage frequency of individual codons for a specific amino-acid, which varies from species to species. The underlying reason for this bias seems to be that the tRNAs specific to a particular triplet of the genetic code are not equally abundant either. Instead, some of the tRNAs are more frequent than others. The more frequent a tRNA, the faster the relevant amino acid can be incorporated into the growing protein by the ribosome. Therefore, the usage of different codons has implications for the time it takes to translate a protein.

It turns out that not all amino-acids are encoded by the fastest codons. As a result, the majority of proteins are expressed with medium speed, while there are a few highly-optimized ones which use mostly very fast/abundant codons. Similarly, very few proteins are coded for by predominantly rare codons. There are many biologically aspects to this, but some initial understanding of the system can be reached by considering simple models based on the partition functions.

To start, let us assume that every protein is under some selection pressure to be expressed rapidly. This selection pressure will vary from protein to protein. Generally, of course, the faster a protein is expressed, the better. However, since proteins are expressed simultaneously there is a competition for tRNA between them. Increasing the speed of one must decrease the speed of the other. Selection pressure itself is not directly measurable, but within the framework of the partition function we can model it as a preference, as in the case of the king and queen above.

Let us consider a protein of some length and consider a single amino acid within the protein. Assume that the protein has N copies of this amino acid. Further assume that this amino acid has n_c different codons. Let us (arbitrarily) designate the first codon as the fastest, i.e., the codon that has the highest number of tRNA and assign to it some preference value $G_1 = a$. To simplify matters, let us now assume that the other codons are much rarer than the first one and we assume that they have, collectively, a preference value of $G_2 = b$. The idea underlying the model is as follows: Over evolutionary time scales, random mutations will lead to a random walk between the individual codons. However, over time, those codons that are faster will be preferred (by how much they will be preferred is expressed as a), which can be interpreted as an evolutionary selection pressure. Given this model we can then ask about the probabilities of observing various possible configurations.

At this level, the model reduces to understanding the probability of various macro-states. Here the macro-states are defined as by the number of amino acids, k, that are encoded by the most frequent codon. We formulate the partition function by considering the number of configurations that are compatible with exactly k amino acids being encoded by the most frequent codon. Formally, this is the same problem as distributing k TFs over N binding sites and is thus given by $\binom{N}{k}$. For the remaining amino acids, we can then choose, at random, one of the remaining $n_c - 1$ codons. Altogether, we thus obtain the partition function:

$$Z = \sum_{k=0}^{N} \binom{N}{k} (n_c - 1)^{N-k} \exp(-ka - (N-k)b) \qquad (6.36)$$

From this we can obtain the distribution of the codons for this particular protein as above.

6.16 Write down the formula that all amino acids are of the optimal type. Find some plausible example values for n_c, k, N and compute the results numerically.

6.3   Markov Chains

In Sect. 6.2 we asked about the probability of finding the king, the queen, and their courtiers in various rooms of their palace. Using methods from statistical physics, we can derive the probability of seeing n people in a room, if we were to take a snapshot of this room at one instant in time. What this approach does not tell us is how long people remain in the rooms. It also does not provide us with any hints as to which room they are likely to go to if they leave a specific room. What is more, the partition function method only tells us about the long-term probabilities. When modeling stochastic processes we often have (or assume) complete information about the state of the system at some time t, which constrains the possible states the system can take in the near future of t.

To make this more concrete, let us concentrate on the king himself in our court example. We could imagine that if the king dwells in one of the dining rooms of the palace then he is likely to remain there for about an hour or so during his lavish meals. After eating he often takes a nap, so it is likely that he will move from the dining room straight into his bedroom. Sometimes, though, he decides to tend to matters of state and will meet the prime minister in the appropriate room instead of taking a nap. Similarly, after getting up in the morning, he enjoys exercising by viewing the paintings of the queen's ancestors around the building before taking breakfast.

If we were to observe the behavior of the king over many days, then we would be able to measure these regularities and tabulate the probabilities of him going to room j given that he is currently in room i. Now, to some extent, these probabilities are constrained by the floor plan of the palace. For example, there is no way to go to the palace kitchen from the bedroom of the king without passing through a few other rooms first; equally the bathroom has only one door, directly into the bedroom of the king. Consequently, the probability of passing from the bathroom into the kitchen in one step must be zero.

Altogether, the probabilities of passing from one room into another can be formalized in a transition matrix, $\mathbf{W}$:

$$\mathbf{W} = \begin{pmatrix} 0.5 & 0.0 & 0.0 & 0.4 & 0.1 \\ 0.1 & 0.1 & 0.1 & 0.4 & 0.3 \\ 0.2 & 0.0 & 0.1 & 0.0 & 0.7 \\ 0.0 & 0.3 & 0.5 & 0.1 & 0.1 \\ 0.0 & 0.0 & 0.0 & 0.1 & 0.9 \end{pmatrix} \tag{6.37}$$

Here we understand the elements of the matrix as follows: The element w_{ij} of the matrix $\mathbf{W}$ gives us the probability that if the king is in room i then he will next go to room j. For example, assuming the king is in room 1 then the probability that he next goes to room 4 is 0.4. Similarly, the probability that he remains in room 1 is 0.5. Consistent with the meaning of the matrix, we must require that all rows sum to 1.

Implicit in the transition matrix of (6.37) is the assumption that time is discrete. By this we mean that decisions as to whether the "state" of the system changes or not (that is whether the king remains in the room or goes to a different one) are taken in fixed time intervals. Time can then be counted by integer numbers, i.e., $t = 0, 1, 2, \ldots$ This is, of course, not a realistic assumption. In reality, the dwelling times of the king in rooms will be drawn from some continuous distribution, not from a discrete one. For example, he may stay in a room for 12.0985 time units. Our model does not allow this, at least not for the moment. However, we can approximate continuous dwelling times to an arbitrary degree of accuracy by making the discrete steps small. For the moment we will not worry about that, but simply accept that this is a *discrete time Markov chain model*; we will consider *continuous time Markov chain models* later.

6.17 Draw a floor plan consistent with the ridiculously small palace implied by (6.37).

6.18 Justify in detail why the rows of a transition matrix of a discrete time Markov chain must sum to 1?

It is crucial to note that the transition matrix $\mathbf{W}$ does not give us the probability of finding the king in a particular room, but just the probabilities that the king moves to a room j given that he is in room i. It is possible to obtain the actual probabilities of dwelling in a room from $\mathbf{W}$, but this requires some work. The good thing about the transition matrix $\mathbf{W}$ is that, unlike the partition function *ansatz*, it can provide us with a much more detailed picture of the stochastic properties of the system we model.

To simplify the language let us now talk more abstractly about systems and their states, and the probabilities of state transitions, rather than about the probability of the king going from one room to the next. We identify the "state" of our stochastic system with the room in which the king is. So, if we say the system is in state 1, then this means the king is in room 1. To describe the system, we need N labels, corresponding to the N different rooms in which the king may be. We could now extend this and include the queen in our model. In this case we would have to talk about the king being in this room and the queen being in that room and the probabilities of them walking from one room to the next. Clearly, this extended system has more states, and requires more labels. A state would no longer be identified with the room in which the king is dwelling, but with the pair of rooms. If there are N rooms, then the state space would be N^2 rather than N in the case of the king alone. The corresponding transition matrix would then be an $N^2 \times N^2$ matrix formulating all the possible transition probabilities of the system. This shows that the size of Markov chain models can grow very quickly as entities are added. Clearly, with every additional person that is included, the state space grows dramatically.

Transition matrices such as these implicitly contain the assumption that the transition from one state to the next does not depend on how the system arrived at a particular state. In other words, the history of the system is irrelevant for the future. All that counts is the current state and the transition probabilities. In the context of our example, this means that the probabilities of the king going from the dining room to the bedroom does not depend on whether he has been in the ballroom before or in the changing room before entering the dining room. This assumption of independence of a stochastic process from the past is the defining property of Markov processes. Luckily for modelers, the Markov property significantly simplifies the mathematics and is also, in most cases, a reasonable assumption to make when modeling natural systems.

Let us now see what we can do with the transition matrix and what information we can extract from it. For this purpose, we return again to the simple example of the king walking through his palace. First we consider the transient probabilities of the system given by the transition matrix (6.37); that is, the probabilities not too long after a time when the state of the system was known. The simplest case is to assume that, at a time $t = 0$, the system is in a specific state, say that the king is in room 1. Where will he be next? In fact, we do not even need to calculate this, because a simple look at the $\mathbf{W}$ reveals that he will be in room 4 and 5 with the respective probabilities 0.4 and 0.1, and in room 1 with probability 0.5. We can also calculate this, if we wish. To do so, we must introduce the state vector. As the name suggests, this is a vector that tells us in which state the system is at a given time. The entries of the vector should all sum to one, so the entries of the vector can be interpreted as giving a probability of being in a particular state. For example, the state that the king is in room 1 can be represented by S_0:

$$S_0 = \begin{pmatrix} 1 & 0 & 0 & 0 & 0 \end{pmatrix}$$

Similarly, the state vector S_x represents that the king is in room 2 or 4 with probability 0.5:

$$S_x = \begin{pmatrix} 0 & 0.5 & 0 & 0.5 & 0 \end{pmatrix}$$

To obtain the state-vector $\mathbf{S}_1$ at the next time step, $t = 1$, we simply compute the product of $\mathbf{S}_0$ and $\mathbf{W}$:

$$\mathbf{S}_1 \doteq \mathbf{S}_0\mathbf{W} = \begin{pmatrix} 1 & 0 & 0 & 0 & 0 \end{pmatrix} \begin{pmatrix} 0.5 & 0.0 & 0.0 & 0.4 & 0.1 \\ 0.1 & 0.1 & 0.1 & 0.4 & 0.3 \\ 0.2 & 0.0 & 0.1 & 0.0 & 0.7 \\ 0.0 & 0.3 & 0.5 & 0.1 & 0.1 \\ 0.0 & 0.0 & 0.0 & 0.1 & 0.9 \end{pmatrix} = \begin{pmatrix} 0.5 & 0 & 0 & 0.4 & 0.1 \end{pmatrix}$$

This yields the expected result and confirms what we have already known: at time $t = 1$, the king is in rooms 4 and 5 with the respective probabilities of 0.4 and 0.1. This result also gives us the state vector for $t = 1$, which we call S_1. We can use this to continue computing the state vector at time $t = 2, 3, \ldots$:

$$S_2 = S_1\mathbf{W} = \begin{pmatrix} 0.25 & 0.12 & 0.2 & 0.25 & 0.18 \end{pmatrix}$$

This tells us that, at time $t = 2$, the king might be anywhere in the palace, although it is most likely that he is in room 1 or room 4, each of which has a probability of 0.25. We can now continue this process indefinitely, and compute various state vectors, such as[2]:

$$S_{10} = \begin{pmatrix} 0.04 & 0.04 & 0.07 & 0.11 & 0.72 \end{pmatrix}$$

---

[2] Here we truncate the values after two decimal places, which is why the state vectors do not sum to exactly 1.

and:

$$S_{20} = (0.03\ 0.03\ 0.06\ 0.11\ 0.74)$$

Looking at these sequences of numbers, the reader might now wonder whether these state vectors will eventually stabilize. Comparing S_1 with S_{10} and S_{20} suggests that the probability of being in state 1 seems to decrease over time. On the other hand, room 5 emerges as a preference for the king. While it can be very interesting to numerically calculate how the probabilities change over time, one of the things we may be interested in are the long-term, or steady state probabilities of the system, if they exist at all. The steady state of the Markov chain model would describe the probabilities of finding the system in a specific state long after the transient effects of initial conditions have been "forgotten" by the system. This steady state would correspond to the quantities we have calculated in Sect. 6.2 using the partition function approach.

The good news is that, at least for the Markov chain model defined by the transition matrix (6.37), there exists a unique steady state probability vector defining the long-term distribution of states. Intuitively, it is very easy to understand how this long-term behavior can be obtained. The idea of steady state behavior is that it does not change when acted upon by the transition matrix. This is very much like the case of the steady state in the case of differential equations (see Chap. 4), where the steady state was defined by the state of the system when the time derivatives vanish. In the case of Markov chains, the change operation is the application of the transition matrix to the state vector. So, if the system is in a state S^* and we apply the transition matrix to it, i.e., if we calculate S^*W, then the probabilities should not have changed. This is perhaps best expressed formally. The steady state probabilities of the system are defined by the vector S^* that fulfills the steady state equation:

$$S^*W = S^* \tag{6.38}$$

An equivalent form of this equation is:

$$S^*(W - I) = 0 \tag{6.39}$$

Here we use I as a shorthand for the identity matrix. Equation (6.39) is essentially an eigenvalue problem. We can either solve it by hand or use a computer algebra system to obtain the solution:

$$S^* = (0.03\ 0.03\ 0.06\ 0.11\ 0.74)$$

This tells us that, based on the transition matrix defined above, the king is most likely to be in room 5, where he spends around 74 % of his time.

6.3.1 Absorbing Markov Chains

Let us now change our matrix $\mathbf{W}$ to make it a so-called *absorbing Markov chain*:

$$\mathbf{W} = \begin{pmatrix} 0.5 & 0.0 & 0.0 & 0.4 & 0.1 \\ 0.1 & 0.1 & 0.1 & 0.4 & 0.3 \\ 0.2 & 0.0 & 0.1 & 0.0 & 0.7 \\ 0.0 & 0.3 & 0.5 & 0.1 & 0.1 \\ 0.0 & 0.0 & 0.0 & 0.0 & 1.0 \end{pmatrix} \tag{6.40}$$

The relevant change with respect to our original transition matrix has taken place in the last line, where we have altered the transition probabilities such that room 5 has only a single transition—to itself. With this modification, our example of the king in the palace becomes a bit unrealistic, in that now room 5 becomes a sink; once the king enters it, he will remain there for ever. Correspondingly, the steady state is always given by the vector, $S^* = (0, 0, 0, 0, 1)$.

Absorbing Markov chains can (and often do) involve more than a single absorbing state. We could have constructed our system such that there is no way from room 4 and 5 to any of the other rooms, but free movement is possible between them and into them; or we could have stipulated that rooms 3 and 1 are connected to each other but that there is no escape from them to any of the other rooms. All the techniques we are going to present over the following pages are perfectly applicable to such more general cases.

6.19 Calculate the steady state vector $\mathbf{S}^*$ for the Markov chain in (6.40).

Our absorbing transition matrix is of the general form:

$$\mathbf{W} = \begin{pmatrix} \mathbf{A} & \mathbf{B} \\ \mathbf{0} & \mathbf{C} \end{pmatrix} \tag{6.41}$$

Here the lower row corresponds to all the absorbing states. In the present case, we have only one such absorbing state, hence in our case the lower row corresponds to $(\mathbf{0}, \mathbf{C}) = (0, 0, 0, 0, 1)$. The sub-matrix $\mathbf{A}$ corresponds to the following:

$$\mathbf{A} = \begin{pmatrix} 0.5 & 0.0 & 0.0 & 0.4 \\ 0.1 & 0.1 & 0.1 & 0.4 \\ 0.2 & 0.0 & 0.1 & 0.0 \\ 0.0 & 0.3 & 0.5 & 0.1 \end{pmatrix}$$

Note that transition matrices of absorbing Markov chains can always be coerced into the shape of (6.41), where the sub-matrix $\mathbf{C}$ summarizes the absorbing states simply by renumbering states. As it turns out, the matrix $\mathbf{A}$ is of fundamental significance. In absorbing Markov chains it can be used to compute the average number of times a specific state is visited before the absorbing state is entered. To be more precise, let us define the so-called fundamental matrix $\mathbf{Q}$:

$$\mathbf{Q} \doteq (\mathbf{I} - \mathbf{A})^{-1} \tag{6.42}$$

Here $\mathbf{I}$ is the identity matrix. The entry q_{ij} of this matrix is the average number of times the system will be in state j given that it started in state i before the absorbing state is entered. To make this concrete, we can compute this matrix for our modified transition matrix $\mathbf{W}$ above:

$$\mathbf{Qw} = \begin{pmatrix} 2.37 & 0.41 & 0.73 & 1.23 \\ 0.53 & 1.39 & 0.63 & 0.85 \\ 0.52 & 0.09 & 1.27 & 0.27 \\ 0.47 & 0.51 & 0.91 & 1.54 \end{pmatrix}$$

This means that the king starting in room 1 will, on average, be expected to be in room 2 ≈ 0.41 times before entering room 5. So, if we repeated a thought experiment where we let the king start from the same room, then in just under every second experiment we would observe him passing through room 2 on his way to the absorbing state. This procedure generalizes to all absorbing Markov chains.

Using this fundamental matrix $\mathbf{Q}$ we can calculate another quantity that is often of interest: the mean time before the absorbing state is entered. In discrete time Markov chains the mean time is the mean number of steps. The fundamental matrix already contains the answer. Say we start in room 1, then the mean time to enter room 5 would be the sum of the mean times to pass through rooms 2, 3, and 4. We can obtain this by multiplying the fundamental matrix with a vector v that contains only 1's:

$$\mathbf{T} = \mathbf{Qv}$$

Using the numbers of our specific example with the king, we obtain the following mean times before the king enters room 5:

$$\mathbf{T} = \begin{pmatrix} 4.76 \\ 3.41 \\ 2.16 \\ 3.45 \end{pmatrix}$$

This tells us that, if we start from room 1 then, on average, it takes 4.7 time steps before we reach the absorbing state.

6.20 Assume the system is in state $(0.5, 0.5, 0, 0, 0)$. What is the mean time to reach the absorbing state?

6.3.2 Continuous-Time Markov Chains

So far we have assumed that our Markov chains are updated in discrete ticks of time. At each time step the system makes at most one transition. The problem of such discrete Markov chains is that they do not represent time very well. In real systems events play out in continuous time. An event may happen at $t = 0.032$ and another 0.327 time units later. Continuous time cannot be modelled exactly by discrete time Markov chains. Discrete time models can approximate continuous time dynamics to an arbitrary degree of precision by decreasing the transition probabilities, that is by making every single time step correspond to smaller and smaller time quanta.

One can compare this to a series of pictures taken at 1 second intervals. From the series of pictures one can obtain a sense of the dynamics of the scenery, although one would miss some faster movement. If one reduces the interval to half a second then more detail will visible. As one increases the frequency with which pictures are taken, at some point the series of pictures will be a very good approximation of continuous motion. In fact, any movie is just a series of still pictures taken at discrete time points.

The same procedure applies with discrete Markov chains. In order to arrive at continuous time Markov chains, we start with a discrete chain and make a time-step correspond to ever smaller units of real time. If we continue this process indefinitely then we will eventually arrive at a continuous model.

Unlike discrete time Markov chains, the continuous version is not defined by a matrix of transition probabilities, but by transition *rates*; that is, transitions per time unit. This transition matrix $\mathbf{Q}$ can be obtained from the discrete matrix of transition probabilities by a limiting process. The idea is to divide the transition probabilities per time step by a time interval, and then let both the probabilities and the time interval go to zero. In the limit one obtains the transition probabilities per unit time for a continuous process. Concretely, for each element of the transition matrix one gets:

$$q_{ij}(t) = \lim_{\Delta t \to 0} \frac{p_{ij}(t + \Delta t)}{\Delta t} \qquad i \neq j \qquad (6.43)$$

Unfortunately, this procedure does not make sense for the diagonal elements. In the discrete case, at every time step the Markov chain makes the transition into a specific state, which could be the state that it was in at the previous time step. This idea does not translate well into a continuous model where transitions can happen at any time; hence, staying in the same state cannot be easily interpreted as a transition any more. We therefore have to slightly re-interpret the diagonal elements. To do this, we derive them from formal considerations.

A formal requirement on transition matrix $\mathbf{W}$ was that the rows sum to one. In the continuous time version this is no longer the case; instead the rows should sum to zero. To see where this requirement comes from it is best to take the limit of small time in a version of (6.43). Assume that P_1 and P_2 are state vectors, and that $P_2 = P_1 \mathbf{W}$:

$$\lim_{\Delta t \to 0} \frac{\mathbf{P_2} - \mathbf{P_1}}{\Delta t} = \dot{\mathbf{P}}_1 = \lim_{\Delta t \to 0} \frac{\mathbf{P_1 W} - \mathbf{P_1}}{\Delta t} = \mathbf{P}_1 \lim_{\Delta t \to 0} \frac{\mathbf{W} - \mathbf{1}}{\Delta t} \qquad (6.44)$$

If we now define the new matrix $\bar{\mathbf{Q}} \doteq \lim_{\Delta t \to 0} ((\mathbf{W} - \mathbf{I})/\Delta t)$ and omit the index to P then (6.44) becomes a differential equation:

$$\dot{\mathbf{P}} = \mathbf{P}\bar{\mathbf{Q}} \qquad (6.45)$$

This is precisely what we would have expected from a continuous version of the transition matrix—that it tells us the change of the probability during an infinitesimal amount of time. Moreover, the off-diagonal elements of $\bar{\mathbf{Q}}$ are identical to the elements of $\mathbf{Q}$. Altogether, this gives us the confidence to simply identify $\mathbf{Q}$ and $\bar{\mathbf{Q}}$, and we now know how to calculate the diagonal components of the matrix:

$$q_{ii} = \lim_{\Delta t \to 0} \frac{w_{ii} - 1}{\Delta t} = \lim_{\Delta t \to 0} \frac{1 - \sum_{i \neq j} w_{ij} - 1}{\Delta t} = -\sum_{i \neq j} q_{ij} \qquad (6.46)$$

This interpretation of the diagonal elements also makes sense given the changed interpretation of the entries of the continuous matrix with respect to $\mathbf{W}$. The transition matrix $\mathbf{Q}$, unlike $\mathbf{W}$, does not contain probabilities as elements, but rates. Every row specifies how, in any infinitesimal time interval, probability flows to other states. Seen from this perspective, it is clear that the sum of flows to states j given by $\sum_{i \neq j} q_{ij}$ must be accompanied by a flow from state i.

If we take this perspective of probability flows then it is also instructive to consider (6.45) in an alternative form:

$$\dot{P}_i = \sum_{j \neq 1} a_j P_j - b_i P_i \qquad (6.47)$$

Here we have temporarily set $a_j \doteq a_{ij}$ and $b_i \doteq -q_{ii}$. The reader will immediately recognize that this is just a master equation for the stochastic process.

Returning to our normal notation, (6.45); in steady state, the left-hand side of the equation will vanish, which immediately leads us to the continuous time equivalent of the steady state (6.38):

$$\mathbf{Q} \cdot \mathbf{P} = 0 \qquad (6.48)$$

6.21 Write down the transition matrix for the Poisson process.

6.3.3 An Example from Gene Activation

We illustrate the use of continuous time Markov chains with an example from gene activation (see [2]). For this example we assume that a gene G is regulated by three regulatory binding sites. Each of these sites can be occupied by a transcription factor. Assuming a binding rate constant of k_b and an unbinding rate constant of k_u, we want to calculate the relative amount of time all three binding sites are occupied as a function of the number of transcription factors N.

To solve this problem we formulate it as a Markov chain whose states correspond to the occupation levels of the binding sites. In order to simplify the problem, we do not distinguish between the three binding sites; that is, we are only interested in the total number of occupied binding sites but we are not worried about which ones are occupied. Altogether our system then has four different states corresponding to 0, 1, 2, or all binding sites being occupied. Consequently, the transition matrix is a 4×4 matrix.

We next need to find the transition rates. We make the (common) assumption that the binding rate to the individual binding sites is proportional to the number of TFs in the cell; the proportionality constant is the rate constant k_b. Hence, altogether the rate of binding is Nk_b. We will assume that binding to the different binding sites is independent. Consequently, if there are 3 unoccupied binding sites then the total binding rate to the sites is $3Nk_b$.

In the case where one site is already occupied, then the binding rate per site is somewhat reduced because the number of free TFs is reduced by one. Moreover, the number of free binding sites has reduced to 2, making the total binding rate $2(N-1)k_b$. Analogously, when there is only a single free remaining binding site, then the transition rate is given by $(N-2)k_b$.

Unbinding of TFs is also a stochastic process, but the total rate from one site only depends on the unbinding rate constant k_u, not on the total number of free TFs. As in the case of binding, the unbinding rate needs to be multiplied by the number of bound sites. So the transition from a state of three bound sites to a state of two bound sites happens with a rate of $3k_u$ where k_u is again the unbinding rate constant.

To get to our Markov chain we assume that there are only transitions to the "neighboring" binding states. For example, from a state corresponding to two occupied states there are transitions to 1 or 3 sites being occupied, but not to the state corresponding to all sites being unoccupied. We can now write down our transition matrix Q whose off-diagonal elements are these transition rates. The diagonal elements are the negative of the sum of the off-diagonal elements in the same row:

$$\mathbf{Q} = \begin{pmatrix} -3Nk_b & 3Nk_b & 0 & 0 \\ k_u C & -k_u C - 2(N-1)k_b C & 2(N-1)k_b C & 0 \\ 0 & 2k_u C & -2k_u C - (N-2)k_b C & (N-2)k_b C \\ 0 & 0 & 3k_u & -3k_u \end{pmatrix} \tag{6.49}$$

The reader will notice the additional factor C in the transition matrix. This is a phenomenological term that takes into account an effect called "co-operativity." In essence, cooperativity is a modification of the assumption that all binding sites are independent. In real systems it has been observed that the binding of TFs to their sites can be increased (or decreased) when other sites are occupied as well. As a result, the binding of a single TF is much weaker when it is bound by itself than when more than one binding site is occupied. The factor $C > 1$ takes this into account by increasing the binding rates once there is at least one particle bound, and it also increases the unbinding rates in the case when not all binding sites are fully occupied.

We can now calculate the steady state probability of finding the system in various states. To do this we introduce the probability row-vector $\mathbf{P}$:

$$P = \begin{pmatrix} p_0 \\ p_1 \\ p_2 \\ p_3 \end{pmatrix}$$

Here p_i corresponds to a state where exactly i TFs are bound. To obtain the probabilities p_i in steady state, we must solve (6.48). We are primarily interested in the probability of all binding sites being occupied:

$$p_3 = \frac{Nk_b^3\left(N^2 - 3N + 2\right)}{k_u^3 + 3k_u^2c^{-1}Nk_b + k_b^3N^3 - 3k_b^3N^2 + 2k_b^3N + 3k_uN^2k_b^2c^{-1} - 3k_uNk_b^2c^{-1}} \quad (6.50)$$

We assume that the co-operativity is infinitely strong; that is, we take the limit $c \to \infty$ and the terms involving c disappear from (6.50) giving:

$$p_3 = \frac{Nk_b^3\left(N^2 - 3N + 2\right)}{k_u^3 + k_b^3N^3 - 3k_b^3N^2 + 2k_b^3N}$$

$$= \frac{Nk_b^3\left(N^2 - 3N + 2\right)}{k_u^3 + Nk_b^3\left(N^2 - 3N + 2\right)} \quad (6.51)$$

This can be simplified by introducing a new parameter $K \doteq \frac{k_u}{k_b}$:

$$\lim_{c \to \infty} p_3 = \frac{N(N-1)(N-2)}{N(N-1)(N-2) + K^3} \quad (6.52)$$

Finally, if the number of TFs in the cell is high, that is if $N \gg 1$ then (6.52) can be approximated by a much neater formula:

$$p_3 \approx \frac{N^3}{N^3 + K^3} \quad (6.53)$$

The reader will undoubtedly recognize (6.53) as the Hill equation for $h = 3$ (see Sect. 4.5.1 on p. 159). For the special case of three binding sites we have shown that, in the limit of infinite cooperativity, the probability of all three binding sites being occupied approaches the Hill function. Note that in Chap. 4 we merely assumed Hill dynamics for gene activation, but here we actually derived it. The derivation shows that the Hill coefficient is equal to the number of binding sites, but only in the case of infinite co-operativity. If co-operativity is lower, then the Hill coefficient will also be lower than the number of binding sites. Note, however, that a crucial assumption here is that the cell acts as a perfectly-mixed compartment. In reality this assumption is not strictly met. The interested reader is referred to [2] and references therein.

6.22 Show that a Hill coefficient of 4 applies for the case of four binding sites.

6.23 In the case of three binding sites, calculate p_0, p_1, p_2.

6.4 Analyzing Markov Chains: Sample Paths

In principle, every Markov chain can be represented by a single matrix, as discussed in the previous sections. In practice, this matrix form may not always be the most intuitive way to formulate models. Markov chain models quickly grow very large as they are extended, and filling in large (or even moderately sized) transition matrices becomes infeasible very rapidly. How serious an issue this is is illustrated by the state space explosion of the example of the king and the queen. If we include the queen as well, then there are not $2N$ but N^2 states. For this reason it is desirable to find alternative ways to formulate Markov chain models.

However, formulating the models is not the only problem. Even if we had a finished model, manipulating large models will stretch computer algebra systems to the limit. Normally, it is therefore only possible to obtain precise results from Markov chain models when the models are of relatively modest size.

In biological modeling, Markov chains find widespread application. As a consequence, many methods and computational tools have been developed to solve them. Many of these methods can be used without actually being aware of the fact that the underlying model is a Markov chain model. One can distinguish two approaches in "solving" Markov chain models.

The first approach is to generate so-called "sample paths" of the model; that is, a particular sequence of random states that is consistent with the specification of the Markov chain model. Taking as an example the case of the king who walks through his palace, such a sample path would be a sequence of rooms each with a time tag that specifies at what time the king entered the room. One normally uses computer programs to solve such Markov chain models using a pseudo-random number generator.

In the case of a discrete time Markov chain, such sample paths are very easy to generate. The algorithm essentially consists of generating a random number to decide which state to enter next. Once this is decided, the time counter is incremented by 1, and the process repeated. For example, assume the system is in state X_0 at time $t = t_1$, and the available transitions from X_0 are to X_0, X_1, X_2, X_3 with probabilities $0.1, 0.4, 0.2, 0.3$ respectively. To determine the next state, we draw a random number between 0 and 1; let us call this number n_0. If $n_0 \in [0, 0.1)$, then the system stays in state X_0; if $n_0 \in [0.1, 0.1 + 0.4)$ it makes the transition into X_1; if $n_0 \in [0.5, 0.5 + 0.2)$ it transitions into X_2, and so on. Repeating this over and over generates a single sample path of the discrete time Markov chain (Table 6.1).

Generating sample paths for continuous time Markov chains is somewhat more involved and requires more sophisticated algorithms. There are a number of highly efficient algorithms; the two best known ones are the so-called Gillespie algorithm and the somewhat more efficient Gibson-Bruck algorithm. Details of these algorithms are discussed in Chap. 7.

Such sample paths are of great value in exploring the possible behaviors of the system, and to obtain a feeling for the kinds of behavior the model can display. They are, so to speak, an example of what the system could do. On the downside, by themselves sample paths do not provide much insight into general properties of the system. For example, to understand how the system behaves in the mean over time, it is necessary to generate a number of paths, starting from the same initial conditions. We can then calculate the *ensemble average* at each time point, thus generating a sequence of averages. Given a large enough number of generated sample paths, this

Table 6.1 Samples path through a Markov chain

$t = 0$	1	1	1	1	1	1	1	1
$t = 1$	2	3	2	4	5	1	1	2
$t = 2$	2	2	1	4	5	2	2	4
$t = 3$	2	3	4	4	5	1	4	3
$t = 4$	3	4	5	4	5	1	5	4
$t = 5$	3	3	3	4	5	3	6	4
$t = 6$	2	3	5	4	5	4	7	7
$t = 7$	4	3	3	3	2	5	6	7
$t = 8$	5	2	4	4	7	6	7	6
$t = 9$	6	4	5	5	5	5	5	5
$t = 10$	6	6	6	5	5	6	7	5
$t = 11$	6	7	6	6	6	6	7	6
$t = 12$	6	7	6	4	5	6	6	6
$t = 13$	7	7	6	3	7	6	6	6
$t = 14$	6	6	7	7	5	6	6	6

The first column shows the time and the remaining 8 columns show specific, randomly-generated paths. All models start in state 1 at time $t = 0$ and only differ in their random seed

sequence of ensemble averages normally reflects more or less accurately the mean behavior of the system over time.

While the generation of sample paths is interesting, important, insightful and, in many cases, the only choice one has of obtaining general insights into a system, the process requires many sample paths to be generated. This is not always either practical or possible. Moreover, sample paths will only provide insight for one specific set of parameters. In order to understand how the properties of the system change as parameters are modified, a new ensemble of sample paths must be generated. Generating sample paths is particularly impractical when one is interested in rare events of the stochastic system. In these cases, one has to spend many cycles of computation exploring transitions one is not really interested in for every event one is really interested in.

6.5   Analyzing Markov Chains: Using PRISM

The second approach to solving Markov chains is to derive general statements from the Markov chains directly. This approach can avoid many of the disadvantages that come with the extensive simulations necessary to generate a large set of sample paths. In the case of very small models, properties of the Markov chain can, of course, be derived directly from the transition matrix. For even slightly larger models, this quickly becomes difficult to handle, and is impractical. To solve those cases there are a number of computational tools available to assist the modeler. A particularly powerful one is PRISM [3,4], developed at the University of Oxford, UK. PRISM was originally developed as a probabilistic model checker—a tool that can be used to explore the properties of stochastic algorithms. In essence, however, it converts the models of algorithms into Markov chain models. Its scope is thus much wider than mere model checking of computer programs and it has been successfully applied to problems in many other fields, including biological modeling.

PRISM has two features that make it a powerful tool for modeling stochastic models in biology. Firstly, it has a very simple specification language that allows a quick and intuitive way of defining Markov chain models, without the need of explicitly specifying the transition matrix (which can be quite a daunting task). Secondly, it has a number of inbuilt tools to explore specific questions about the system at hand and its properties. In particular, it allows the user to ask questions about the probability of a certain event happening at a specific time, or about the expected state at a specific time. It has an easy to use interface to support varying of parameters and an inbuilt graphing capability. PRISM can compute steady state behaviors and can deal with both discrete and continuous time Markov chains.

To find answers to questions, PRISM does not generate sample paths but determines the properties of the system directly from the transition matrix, which it also generates based on user specifications. As such, it can give exact answers. The downside of this is that Markov chain models quickly suffer from a state space explosion. PRISM's model checking facility is thus limited to models of moderate size, which can be a severe limitation in practical modeling applications. There are, however, many cases where it can be used in practical modeling. PRISM includes a simulation facility that enables the user to generate sample paths through the Markov model and extends the scope of the tool to significantly larger systems than the model checking part. While this is less interesting to us, the path generation is implemented in a clever and user-friendly way, so as to retain some of the model checking functionality even for the path generation. This makes PRISM an invaluable tool for stochastic modeling in biology.

6.5.1 The PRISM Modeling Language

At the heart of the PRISM system is the PRISM modeling language, which allows the user to define Markov chains in an intuitive way. It is best to show the structure of the PRISM language by example, rather than to formally introduce it. The following defines a state transition as part of a longer specification in the PRISM specification language:

```
[inca] (a > 1) & (a < 4) & (b = 3) -> 0.3:
        (a' = a+1) & (b' = b-1);
```

We assume here that a and b are state variables that can take various values (the range of values is not specified in this example). For example, $a = 1$ could mean "the king is in room 1" and $b = 2$ could then indicate that the queen is in room 2. Assuming a

and b are the only variables of the system, every state of the Markov chain could then be described in terms of (a, b). The first entry in the line, the word `inca` enclosed in square brackets, is a label for the state transition. It could be left empty, but the square brackets must be written. The label is followed by one or more conditions, known as *guards*. These formulate the circumstances under which the particular transition will take place. So, in this example, the transition will only take place when a is greater than one, but smaller than 4, and when b equals 3. The list of conditions is separated from the transition rate or probability (depending on whether one defines continuous or discrete Markov chains) by the symbol `->`. The transition probability/rate is not necessarily a fixed value, but can itself be a formula that contains the values of state variables; examples below will demonstrate this. Finally, separated from the rest by a colon is the definition of the actual transition. In this case, we specify that a is increased by 1 and b is decreased by 1, and it takes place with a rate of 0.3.

Altogether, the statement `inca` stipulates that, if the Markov chain is in the state $(a = 2, b = 3, \ldots)$ or $(a = 3, b = 3, \ldots)$, then with a rate/probability of 0.3 the system will make the transition to the state $(a = 3, b = 2, \ldots)$ or $(a = 4, b = 2, \ldots)$, respectively. The dots indicate state variables other than a and b; these remain unchanged by the transition. On its own, `inca` would make a very boring transition rule for the Markov chain as it only allows the chain to take one of four different states. In PRISM one can define multiple transition rules. For example, we could add a new transition labeled `bounda`:

```
[bounda]  (b > 3) & (a < maxa)-> 0.1:
          (a' = a+1) & (b' = b-1);
```

This specifies that when $b > 3$ then b will be decreased by 1 and a will be increased by 1. A condition on the state transition is that a must be smaller than the value `maxa`, which we define to be the maximum value a can take. Internally, PRISM generates the Markov chain model from the user-specified transition rules and it requires the range of possible states for every state variable to be specified (we will show in a moment how this can be done). The parameter `maxa` is, in this case, the upper limit for the state variable a. The update rule for `bounda` stipulates that a be increased by 1 and the additional condition on a must be made so as to avoid this state variable increasing beyond its allowed maximum value. Failure to guard against state variables violating their boundaries would lead to an error message in PRISM.

The same applies for decreasing values. If we assume that the minimum state values for a and b are 0, then we need to take care that guards reflect this:

```
[lbounda]  (b = 2) & (a > 0)  -> 0.6:   (a' = a - 1);
[lboundb]  (a = 0) & (b > 0)  -> 0.01:  (b' = b - 1);
```

These two rules stipulate that a is decreased by 1 with a rate of 0.6 whenever b takes the value of 2, provided that the transition does not take a below its minimum value. The transition `lboundb` specifies that b be decremented by 1 when a is 0, provided it is above the minimum value. The guards in transition rules must always be formulated in such a way that the limits of the state variables cannot be violated by the transition.

Let us now add a final rule to specify that *b* should increase with a probability/rate of 0.1:

```
[uboundb] (b < maxb) -> 0.1: (b' = b + 1);
```

So far we only have specified the transitions, but we still have to say whether our model should be interpreted as a continuous time or a discrete time Markov chain. This can be done in PRISM by starting the model definition file with the keywords ctmc or dtmc to specify a continuous and discrete time Markov chain, respectively. We also need to specify the range of possible values the state variables can take. In the present case we choose to let them vary between 0 and limits named as maxa and maxb. Finally, we need to specify a module name in which we enclose our transition rules, which we choose to be pmodel1—modules are simply collections of transition rules that belong together. We then obtain our first workable PRISM model definition (Code 6.1):

```
ctmc

// Upper limits on state variables a and b.
const int maxa = 6;
const int maxb = 10;

module pmodel1

        a : [0..maxa] init 0;
        b : [0..maxb] init 0;

[inca]    (a > 1) & (a < 5) & (b  = 3) -> 0.3:
                        (a' = a + 1) & (b' = b - 1);
[ubounda] (b > 3) & (a < maxa) -> 0.1:
          (a' = a + 1) & (b' = b - 1);
[lbounda] (b = 2) & (a > 0) -> 0.6:  (a' = a - 1);
[lboundb] (a = 0) & (b > 0) -> 0.01: (b' = b - 1);
[uboundb] (b < maxb) -> 0.1: (b' = b + 1);
endmodule
```

Code 6.1 Our first PRISM model, prModel1.pm

Note that we preceded the module with a definition of the constants maxa and maxb which define the maximum values. It is not strictly necessary to define the limits via parameters, but it increases the clarity and maintainability of the model.

The keyword init following the square brackets in the state variable definition specifies the initial value, that is the state of the Markov chain at time $t = 0$. In the present example, PRISM would assume that the state variable *a* is in state 0 at time $t = 0$. If we asked PRISM about the probability of *a* being in state 1 at time $t = 0$ it would return as answer "0" whereas it would return the answer "1" if we asked about the probability of the state at time $t = 0$ being 0. We will describe in the following section how these and similar questions can be put to PRISM.

6.5.2   Running PRISM

Before we can reach a position where we can ask questions about the model, we will need to perform some basic checks of it. For this it is necessary to put the model definition into a normal text file that we choose to call `prModel1.pm`. Then the simplest way to call PRISM is to enter on the command line:

```
prism prModel1.pm
```

This will produce some text containing information about the model. Crucially for our model, it will also generate a warning message stating that it contains a *deadlock state*. This means that the Markov chain has at least one absorbing state. By default, PRISM will "fix" this by adding a self-loop to such states. PRISM will tell us which of the states is absorbing if it is called with the `-nofixdl` flag to ask it not to provide the fix. It then reports that the deadlock state is $(6, 10)$; that is, $a = 6$ and $b = 10$.

6.24 Go through the model definition and confirm that PRISM is correct in pointing out that $(6, 10)$ is a deadlock state.

6.25 What is the steady state behavior of the model?

6.26 Come up with a change to the model to fix the deadlock in `pmodel1` (there are many ways to achieve this).

We now have the option either to fix the deadlock, by changing the model, or to ignore it and continue. We will choose the latter option. Re-running PRISM, the program will give us information about the Markov chain. Specifically, it will tell us that the model defines 64 states. This is a notable result. Given that our parameters allow for 11 values of b and 7 values of a, one might have expected that the model has $7 \times 11 = 77$ states, rather than a mere 64. The reason for the discrepancy is that some states are inaccessible from the initial state we specified.

6.27 Change the PRISM model specification so that all 77 states are accessible.

Looking at the model specification we can guess which states these are. From lines `ubounda` and `lbounda` we might guess that states satisfying $b < 2$ and $a > 1$ cannot be reached. Going through the various transitions of the Markov chain, one can try to determine whether this guess is indeed correct. However, there is a much better option. We can ask PRISM itself whether or not the states can be reached. In order to do this we must use the query language. We will not give a detailed introduction to this query language, but will introduce sufficient for the reader to be easily able to formulate her own questions. We will do so mainly by illustrating the use of PRISM with examples.

The main way to put queries to PRISM is to ask it about the probability of a particular state. If we want to know whether there is any possibility that the Markov chain will be in a state consistent with $b < 2$ and $a > 1$, then what we ask PRISM is whether the probability of ever being in this state is greater than 0:

```
P > 0 [F (b < 2) & (a > 1)]
```

The first part of the expression, `P > 0` formulates the question itself. The letter `P` indicates that we are asking for a probability. The expression can be loosely translated as, "Is the probability greater than 0?" In the square brackets we then formulate the event whose probability we would like to determine. The condition on the Markov states, `(b < 2) & (a > 1)` are self-explanatory. What requires more explanation is the use of the letter `F` in the expression. This *path property* specifies a particular *temporal property*. This might be translated as, "...it is ever the case that ...". So, altogether, the question could be read as saying,

> `P > 0 [F (b < 2) & (a > 1)]` = "Is the probability greater than zero that *it is ever the case that* $b < 2$ and $a > 1$ (assuming we start from the initial state specified in the model specification)?"

In order to put this question to PRISM, we type the following on the command line[3]:

```
prism -csl "P>0 [F (b<2) & (a>1)]" prModel1.pm
```

The question must be enclosed in quotation marks and be preceded by the command line option, "`-csl`". It indicates the particular flavor of the query language used. This is of no further concern to us here, except that the query specification must be preceded by it. This will cause PRISM to produce some output, including the answer to the question, namely, "false":

```
Result: false (property not satisfied in the initial state)
```

We now know that, starting from the initial conditions $a = b = 0$, the Markov chain will never enter any states compatible with `(b < 2) & (a > 1)`.

The initial conditions are not explicitly specified in the PRISM question, but the answer depends on the initial condition. Clearly, initial conditions are important. To see this, consider an initial condition of $b = 0, a = 3$. In this case, at least at time $t = 0$, the condition `(b < 2) & (a > 1)` is certainly fulfilled.

6.28 Perform this model check. Experiment with variations to the question to obtain different answers.

Questions to PRISM are not limited to those with "yes" or "no" answers. The system can also return probabilities of certain states being reached or not. For example, to

---

[3] Depending on the exact details of the reader's working environment, it may be necessary to precede the prism command by a full specification of the path to the prism executable. Exact details of this will depend on the operating system and the installation.

obtain the probability that the state `(b < 2) & (a > 1)` will ever be entered, we enter the following variant of our first PRISM query:[4]

```
P=? [F (b < 2) & (a > 1)]
```

The key difference between this question and the previous one is that now we write `P=? [...]`, which requests PRISM to return the probability of an event. In this particular case, we would expect PRISM to return the probability 0, which it does. To obtain a somewhat more interesting result, we vary the question, and ask now for the probability that the system encounters a state compatible with `(b < 3) & (a > 1)`. The corresponding query is:

```
P=? [F (b < 3) & (a > 1)]
```

For this we obtain as a result that the probability is a little bit less than 0.5.

6.29 We ask about the probability that the relevant states will ever be entered. It is clear that, due to the transition matrix, some states are inaccessible. However, since the Markov chain has only a finite number of accessible states, but the probabilities we ask about really are about what happened after a possibly infinite number of transitions, we would expect that a state is either entered with probability 1, or with probability 0. Yet, the states consistent with $(b < 3)$ and $(a > 1)$ are entered with a probability of 0.5. How can that be explained?

So far we have asked questions about whether or not a specific state or set of states will be entered at some point during the stochastic process defined by the Markov chain. This was indicated by the symbol F in the query specification. PRISM supports other operators to formulate questions about tasks. In the next example we ask for the probability that our Markov chain is *always* in the state $(b = 0)$ and $(a = 0)$. This means simply that the system is in the corresponding state for all times (including at time $t = 0$). We expect of course, in this case, the corresponding probability to be zero. In the query, note the new path property G replacing F:

```
P=? [G (b=0) & (a=0)]
```

We encourage the reader to try this out and convince herself that the answer is indeed 0, as expected.

To obtain a more interesting response, let us now change the query by limiting our question to a specific time; instead of asking about the probability that a state is always true, we ask about the probability that a state is always true during a given time interval. The next example asks for the probability that the system is in state $a = b = 0$ for the first two units of time:

```
P=? [G<=2 (b = 0) & (a = 0)]
```

---

[4]From here, we omit the full command line unless it introduces a new feature.

6.30 Using the path property G, trace by hand the probability $P(t)$ that the Markov chain remains in the initial state for the first t time units?

PRISM allows an alternative way to restrict the scope of the G path property. We can explicitly specify a time-interval. For example, the case above could have been specified by writing:

```
P=? [G[0,2]  (b=0) & (a=0)]
```

This requests the probability that the chain remains in the state $a = b = 0$ between time $t = 0$ and time $t = 2$. The same syntax can also be used with the F path property. So:

```
P=? [F[1,2]  (b=0) & (a=0)]
```

This requests the probability that, at some point between $t = 1$ and $t = 2$, the state of the chain is $a = b = 0$.

Another useful path property to check properties of Markov chains is U, which stands for "Until". Unlike F and G it operates on two states, a left state L and a right state R. It formulates the probability that L is always true until the chain enters the state R; the truth of the statement does not depend on whether the system remains in R or not. Note that, one way to make this true is to have R as an initial condition. If at time $t = 0$ the system is compatible with R then a query specified with U will always be true.

The general syntax of requests involving the U path property is P=? [L U R]. An interesting special case that will be of practical use is when the left state is specified as true, which means that any state satisfies the condition. The path property U can also take a time specification. So, the property P=?[L U[1,2] R] asks for the probability that between the time $t = 1$ and $t = 2$ the chain is first in state L, and then in state R; once R has been entered, it does not matter what comes thereafter and the system may leave R without affecting the truth of the query. A few examples will clarify this.

As a first example, let us ask for the probability that our initial state is followed by the state $a = 0$ and $b = 1$:

```
P=? [(a=0) & (b=0) U (a=0) & (b=1)]
```

PRISM returns as an answer that this is true with probability 1. This means that there is only one possible transition from our initial condition, namely to increase b by one. Taking a look at the definition of the Markov chain confirms that this is correct. The only transition that is compatible with the initial condition is the line labeled uboundb stipulating that b be incremented by 1.

The states L and R can be sets of states. The next example asks for the probability that a always remains zero until b reaches 10. Note that the L state must be compatible with the initial conditions as specified in the model file. The reader is encourage to check that this is indeed the case in the following example:

```
P=? [(a=0) U (a=0) & (b=10)]
```

In this case the left state specification L is compatible with several states; in fact, it is completely independent of the value of the state variable b. The result of this query turns out to be very low (≈ 0.014). If we wish to restrict the time, then we could instead ask the question:

```
P=? [(a=0) U[10,20] (a=0) & (b=10)]
```

which queries the probability that between $t = 10$ and $t = 20$ the state variable b reaches the value 10, and a has always remained 0 until b reached its value.

PRISM allows the user to specify initial conditions in the property specification, which is convenient because it means that the user who wishes to change the initial conditions of her problem does not need to revise the model specification file. This can be done by simply adding the initial conditions in curly brackets before the closing square brackets in the query specification. For example, to ask for the probability that the state ($b = 10$) and ($a = 0$) is reached given that we start in the state ($b = 9$) and ($a = 0$), while a never changes until we reach the state, would be formulated as follows:

```
P=? [(a=0) U (a=0) & (b=10){(b=9) & (a=0)}]
```

6.31 If L and R are states or sets of states, what would PRISM return if asked the following question: P=?[L U R{R}]. Check your answer on some specific examples.

PRISM can also answer questions about the next state of a Markov chain. This is perhaps not as useful in the context of biological modeling, but is mentioned for the sake of completeness. The relevant path property is X and it takes only one argument: the putative next state (say N). If S is our initial state, then we can ask the question about the probability of N following the initial condition Q by asking P=? [X NQ]. For example, to find the probability that the state $a = 0$ and $b = 10$ follows the state $a = 0$ and $b = 9$ we write:

```
P=? [X (a=0) & (b=10){(b=9) & (a=0)}]
```

A very useful feature of PRISM is that it allows the user to ask questions about the long term, or steady state behavior of the system. To do this one needs to exchange the operator P by S. For example, to ask about the probability that in steady state the system is found in state N, we write:

```
S=?[N]
```

As far as the steady state behavior of our particular example model, pmodel1, is concerned, a query about the probability of a steady state can only result in one of two answers. We have previously established that it has deadlock-states; in other words, the underlying Markov chain has absorbing states. Therefore, in the long-run, the system has to end up in this absorbing state. Consequently, in steady state the system is in this absorbing state with probability 1 and in all other states with probability 0. If we ask about the probability of the steady state ($a = 6$) and ($b = 10$), PRISM will return 1 as the answer:

```
S=? [(a=6) & (b=10)]
```

If we queried the steady state probability of any other state, we would obtain the answer 0.

Note that a state can have a steady state probability that is less than 1 and still be an absorbing state. This would be the case when there are several absorbing states.

6.32 Change the model `pmodel1` so as to remove the absorbing state. Use PRISM to prove that the revised model does not have any absorbing state (beyond just trusting the absence of the warning). Then determine the steady state probability of being in various states.

6.33 In the revised model from the previous exercise, check the probability that the system will be in state $(a = 3)$ and $(b = 7)$ between time $t = 10$ and $t = 21$.

6.5.3   Rewards

PRISM has a function for querying how long the system has spent in a given state, or even how often a certain state transition has taken place. To access this functionality PRISM provides so-called *rewards*.

6.5.3.1 Rewards for States

To highlight some of the possible usages of rewards, we introduce our second model (`pmodel2`) which has a very simple specification (Code 6.2).

```
ctmc

const int maxb = 10;
const int maxa = 6;

module pmodel2

        a : [0 .. maxa] init 0;
        b : [0 .. maxb] init 0;

[inca]   (a < maxa) -> 0.3: (a' = a + 1);
[incb]   (b < maxb) -> 2.0: (b' = b + 1);
endmodule

rewards
      b = 1: 1;
endrewards
```

Code 6.2 A PRISM model illustrating rewards (`prModel2.pm`)

In essence, this second model consists of two independent variables, a and b. They are incremented with different rates. By now, the reader should have no difficulty in understanding the new model specification, so we concentrate here on the part containing the rewards. The reward specification always needs to be enclosed by the `rewards ... endrewards` key words. In this first example we specify that a reward of 1 should be given when the system is in a state compatible with $b = 1$ or, to be more precise, a reward of 1 should be given for *every unit of time* that the system spends in this state. The part containing `b=1` specifies the state (or set of states) for which the reward should be given, and the number after the colon specifies the actual reward. This can be any number. A reward structure can contain more than a single reward specification and we will demonstrate this below.

To illustrate rewards, we have designed our Markov chain such that the system can only enter the state $b = 1$ once; once it has left this state it cannot come back. Hence, the reward obtained will be a direct measure of the time the system spends in state $b = 1$. We can now ask PRISM what the expected reward is once we reach $b = 3$ for the first time. Since this system can only enter a particular state once, and the rate of moving b on is 2, we would expect the reward obtained to be $1/2$. This reflects that the system spends on average 0.5 units of time in the state. To formally query PRISM we need to use the R operator to indicate that we are asking about rewards rather than probabilities. Following the operator in square brackets we formulate the condition we are after: that the system has reached $b = 3$:

```
prism -csl "R=?[F b=3]" prModel2.pm
```

Due to the structure of this Markov chain, the answer would have been no different had we asked `R=?[F b=4]` or indeed specified any $b > 1$. To illustrate this further; had the transition rate of b been 4 then the reward obtained would have been $1/4$; had the rate been 2 but the per-unit reward also 2 then the reward obtained would have been 1, i.e., $2/2$, and so on.

We can now modify our question, and ask about the mean time b has spent in the state $b = 1$ by the time a reaches the value 1. The expected time can no longer be easily seen directly from the specification, but can be obtained without any difficulty through a variant of our first reward query:

```
R=?[F a=1]
```

Note that b does not feature directly in the query any longer. The query just tells us when to stop allocating rewards to the state $b = 1$ (namely, once a has reached 1). PRISM tells us that the chain will have spent ≈ 0.38 time units in state $b = 1$ by the time a reaches 1.

PRISM also allows querying of the cumulative rewards gained up to a particular time. In the context of Markov chains this can be used to find how long a system has spent in a particular state, on average. Using the path property C we can ask about the expected total time the system will have spent in the state $b = 1$ up to time $t = 1$:

```
R=?[C<=1]
```

The answer to this turns out to be ≈ 0.3. It is worth pondering for a moment the difference between the F and the C path property in the context of rewards. In the first case, the rewards are counted until the condition specified after the F path property is met. So in our example, once the system fulfills $a = 1$ for the first time, the counting of rewards is turned off. In contrast, cumulative rewards are counted until a specified time, rather than a specified set of states, is reached.

6.34 Answer the following question first without using PRISM and then check your answer on the computer. Suppose we ask for the cumulative reward for `pmodel2` at time $t = T$. As $T \to \infty$, what will be the response?

6.35 Without actually using PRISM, answer the following question: Assume that we modify `pmodel2` by adding an additional state transition from b=10 to b=0. We then query PRISM about the modified system: R=?[F b=3] What will the answer be? Check this in PRISM.

In addition to cumulative rewards, PRISM also allows one to ask for so-called instantaneous rewards. If queried about instantaneous rewards, PRISM returns the expected reward state at a particular time. Instantaneous rewards do not take into account the time the system has spent in the state. In the current simple example where the reward is 1, the result of an instantaneous reward query works out to be the probability of being in a state $b = 1$ at a particular time. In the next example we ask about the reward at time $t = 1$:

```
R=?[I=1]
```

6.36 Check that the instantaneous rewards in the last example gives the probability of being in state $b = 1$. Hint: Use the U path property.

6.37 Change the reward structure in `pmodel2` such that the instantaneous rewards (as in the last example) give the mean value of b at a particular time.

Finally, PRISM also allows the user to ask about rewards in steady state. The syntax for this places the operator S in the square brackets:

```
R=?[S]
```

In our particular model, it will never be the case in the steady state that $b = 1$. Therefore, PRISM will return the value 0 to this particular query. In the case of an *ergodic* Markov chain—that is when all states are visited with a non-zero probability in steady state—the steady state rewards would return the average reward weighted by the probabilities of being in the states to which rewards are assigned.

6.5.3.2 Rewards for Transitions

So far we have tied rewards to the system being in a specific state. PRISM also allows the user to relate rewards to the transition between states, rather than the states themselves. This is a functionality that has many potential applications in biological modeling. For example, bi-stable systems, such as the Cherry and Adler bi-stable switch (see Sect. 4.6), are quite common in biology. In this context one may be interested in the rate of spontaneous transitions between the states, which can be addressed using rewards. It is possible to assign rewards to specific state transitions representing the transition between the two bi-stable states and probe the frequency of these transitions. The syntax to specify transition rewards is very similar to how state rewards are treated.

To explore the use of transition rewards, we create a new PRISM model, pmodel3, by adding the following block to the bottom of our second model (Code 6.2):

```
rewards
 [inca] true: 1;
endrewards
```

The new model contains two reward structures (the new one should come after the original). This new reward structure specifies that a reward should be given whenever the transition inca is made; as we can see from the model specification, this corresponds to incrementing the state variable a. The keyword true in this construct just specifies that we do not worry about what states the system is in when the transition is performed, but only about transition "inca" itself. Instead of true we could have specified a particular state or a set of states, which would then have assigned rewards to the transition only if these states are satisfied. We chose not to do this, in order to clarify the main idea of transition rewards.

By adding a second reward structure to our file, we have created the need to specify which reward we are enquiring about. This is easily done by just adding a number after the R operator in curly brackets to indicate which of the reward structures the query relates to. The reward structures in PRISM are automatically numbered in order of appearance.

We wish now to ask about the number of inca transitions the system has made when the state variable a hits the value 2 for the first time. We would expect this to be 2. To check whether our intuition is correct on this we use the query:

```
prism -csl "R{2}=?[F a=2]" prModel3.pm
```

Note the curly brackets after the R operator to specify which reward structure we refer to. With these transition rewards, we can now also ask questions about the number of transitions of a when b is in a particular state. For example, the query R2=?[F b=2] (which returns $\approx$0.3) asks for the average number of transitions of a before b reaches 2 for the first time. Intuitively, this result makes sense. Transitions in b happen much faster than transitions in a, so we would expect that on average a makes very few transitions compared to b.

The syntax of PRISM also allows the user to make several queries in one go and perform some simple arithmetic on the results. We can, for example, ask for the (rather meaningless) quantity corresponding to the ratio of the average number of transitions of a when b equals 2 and the probability that b equals 3 between times $t = 100$ and $t = 200$:

```
R{2}=?[F b=2]/P=?[true U[100,200] b=3]
```

PRISM accepts the standard arithmetic operators commonly used in computer mathematics, i.e., +, -, /, *, for addition, subtraction, division and multiplication, respectively. To obtain roots and powers, use the `pow` function; for example, to obtain the third power of the respective result one can write:

```
pow(R{2}=? [F b=2], 3).
```

6.5.4   Simulation in PRISM

In practice, models often turn out to be too large for PRISM to handle. The most important (but not the only) limiting factor is explosion of the state space. Since PRISM operates internally on the entire state space of the Markov chain, the size of a Markov chain model can quickly lead to a situation where PRISM queries are no longer feasible. In these situations, one can either try to modify the model or, failing that, use the simulation feature of PRISM.

In addition to model checking PRISM can also generate sample paths of the Markov chain. When used in simulation mode, PRISM no longer analyses the Markov chain as a whole and, consequently, cannot provide exact answers any more, but it can handle much larger models. The path generating capability of PRISM is a useful feature in its own right and, as practice shows, very convenient to use. What makes PRISM an even more essential part of the modeler's toolbox is that it can easily be configured to automatically perform a large number of simulations. A sub-set of the queries we have introduced above can also be used in conjunction with the simulation feature; answers to queries are then derived from a set of sample paths rather than directly calculated from the Markov chain. This vastly extends the scope of feasible models, but comes at the cost of loss of accuracy. In this section, we will show some examples of how this can be used. For the sake of simplicity we continue to use `pmodel3`.

The simplest use of the simulation facility is to generate a single example path. To do this PRISM must be given the option `-simpath` along with the length of the desired path and an output file for the results. As output file we choose `test.txt` to hold our path of length 20. After the following the command, PRISM produces the output in Table 6.2:[5]

```
prism -simpath 20 test.txt prModel3.pm
```

The sample path is stochastic and the reader will almost certainly find different numbers in hers. The first column of the output reports the name of the transition taken, and the second the step number.[6] The third column reports the time when the

---

[5]We have truncated the time values to 3 decimal places.

[6]The astute reader will notice that our sample path is only 16 lines long even though we specified a length of 20. The reason for this is that, after 16 transitions the chain has reached its absorbing state and no more transitions are possible. Hence it halted.

Table 6.2 A sample path for `pmodel3`

Action	Step	Time	a	b
—	0	0.0	0	0
[inca]	1	1.103	1	0
[inca]	2	1.174	1	1
[inca]	3	1.258	1	2
[inca]	4	1.711	1	3
[inca]	5	2.344	1	4
[inca]	6	2.976	1	5
[inca]	7	3.587	1	6
[inca]	8	3.706	1	7
[inca]	9	4.320	1	8
[inca]	10	4.347	1	9
[inca]	11	5.930	1	10
[inca]	12	9.460	2	10
[inca]	13	14.489	3	10
[inca]	14	15.084	4	10
[inca]	15	20.866	5	10
[inca]	16	23.070	6	10

respective state was entered. The fourth and fifth columns report the values of a and b after each transition, enabling the user to reconstruct which transition was chosen.

One of the powerful features of the PRISM simulator is that it allows the user to estimate properties by extensive sampling of paths. Many of the model checking questions can be approximated by simulation. To do this, the user needs to specify the number of samples that should be used to estimate the probability. The next example estimates the probability that the state $b = 3$ will be reached within the

first time unit based on a sample of 2 million sample paths as defined by the option
-simsamples 2000000:

```
prism -sim -simsamples 2000000 -csl "P=?[F<=1 b=3]" prModel3.pm
```

The results for such estimated quantities can vary between requests. In our case we
obtained an estimate for the probability of $0.323622 \pm 8.52149798591374E - 4$,
which is very close to the corresponding result from the model checking version of the
same question, which reports a probability of 0.3233235838169363. The accuracy of
the estimate will crucially depend on the number of sample paths that are generated,
of course.

In many practical applications—particularly when the Markov chain does not have
an absorbing state—it is necessary to limit the number of transitions the Markov chain
performs for each sample path. Depending on the application this can significantly
decrease the time required for the simulations to finish; suitable choices will depend
both on the Markov chain itself and on the particular query. To limit the simulation
time in the above example, one can simply add the simpathlen option to the
command:

```
prism ... -simpathlen 10 "P=?[F<=1 b=3]" prModel3.pm
```

Here we limit the number of transitions to 10.

6.5.5　The PRISM GUI

PRISM can be conveniently used from the command line but also has a graphical
user interface, including an editor and a facility to check properties and plot results
from several model checking queries. The graphical user interface can be called with
the xprism command. The interface is reasonably intuitive to use and the reader
should be able to quickly find her way around without any instruction, so we omit a
detailed description here. Nevertheless, we include a few tips and tricks that may be
helpful.[7]

The PRISM GUI (Fig. 6.4) consists of several tabs at the bottom left: *Model*
containing the editor; *Properties* with the model checking interface where properties
can be specified, analyzed and results plotted; *Simulator* where numerical sample
paths can be created; and *Log* that reports details about the Markov chain.

The advantage of the graphical user interface is that it allows the user to automate
querying of PRISM while varying parameter values. The results can then also be
plotted as a function of the variable parameter. PRISM more or less automatically
varies parameters that are left unspecified in the model definition file. The usage of
this feature is easy but not entirely intuitive. To illustrate the general principle of
interacting with the PRISM GUI we will walk the reader through one example. The
objective of the example is to plot the probability of $b = 1$ as a function of time for
the model prModel3.pm.

---

[7]We are referring here to PRISM version 4.2.1.

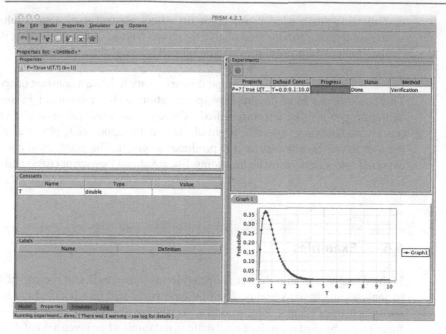

Fig. 6.4 The property-specification interface of *xprism*

The property checking interface of PRISM consists of four different panes, for entering properties, entering constants, defining experiments, and the plot pane that contains all the plots (shown in Fig. 6.4).

1. In the *Model* tab, load the model file prModel3.pm into PRISM and then switch to the *Properties* tab.
2. Move the mouse to the *Properties* pane, right click, choose *Add* and type in the property specification, i.e., P=?[true U[T,T](b=1)]. Note that we leave T unspecified here. PRISM will complain about this but the reader may rest assured that ignoring this complaint will do no harm.
3. Next, right click on the *Contents* pane and choose *Add constant*. In the name field type T, and in the drop down list under *Type* choose double. This means that we specify our constant T to be continuously varying. Leave the value unspecified.
4. Go back to the *Properties* pane and click onto the line containing the previously entered property until the whole line becomes light blue. The pane can list multiple property specification and this step serves to specify which property should be checked. The next step will not work unless this step has been completed successfully and the line containing the query is light-blue.
5. In the *Experiments* pane, right-click and choose *New experiment*. This brings up a window where the range over which T should be varied and the step size can be specified. Click in the right-hand circle to select entry of a *Range*. The graph in Fig. 6.4 was produced by varying T from 0 to 10 using a step size of 0.1. Enter the range and select *Okay*.

6. Following this, a window will come up asking whether to plot the results into a new or existing window and to give the graph a name. After confirming this step the graph of all probabilities will be plotted.

Generally, in order to produce a graph it is necessary to leave a constant unspecified. This could be a part of the property specification itself, or constants in the model definition can also be left unspecified. Through experimentation, the reader will quickly find out how to do this. Right-clicking at the appropriate places also allows the user to see the raw data used to produce the graph. The graph itself can also be exported into various formats, including Encapsulated Postscript (EPS) and Matlab format.

6.6   Examples

In the remainder of this chapter we provide two case studies of the use of PRISM to analyses stochastic systems. The first case study will be a detailed model of the *fim* switch from Chap. 2. The second case study will illustrate how the simulation function can be used to produce stochastic simulations when given a set of differential equations.

6.6.1   Fim Switching

The *fim* switch in *E.coli* is regulated by the invertible genetic element *fimS* (see the discussion in Sect. 2.6.2.2). There are two binding sites to each side of this element. Altogether, there are two types of protein that can bind to these competitively: FimB and FimE. The element *fimS* is inverted when all four binding sites are occupied by the same type of molecule. "Inverting" literally means that the element is cut out of the genome and re-inserted in the opposite orientation. This particular switch regulates the expression of so-called *fimbriae* in *E.coli*; these are small hair-like protrusions from the surface of the cell. Their primary role is to aid attachment to other cells. For *E.coli* cells that live within the gut of other organisms, this means that they are able to hold on to tissue from the host (Fig. 6.5).

Fimbriae are virulence factors. In the context of commensal cells this means that they need to be regulated to avoid the host from triggering its defenses which would lead to an inhospitable environment for the parasites. Interestingly, the way nature chose to regulate fimbriation is not at a global level; for example, via cell-cell communication. Instead, fimbriae are regulated as a stochastic function at the level of the individual cell. Each bacterium switches randomly between the fimbriate and afimbriate states. The levels of fimbriation in the population are controlled indirectly by the environmental conditions through the switching bias from afimbriate to fimbriate. As a result, the population as a whole is a mix of fimbriate and afimbriate cells.

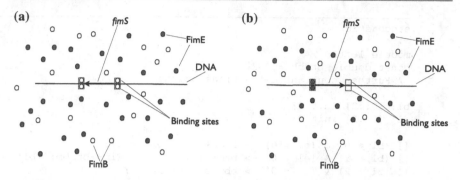

Fig. 6.5 A schematic illustration of the fim switch. FimB and FimE compete for binding sites. If all four binding sites are occupied by one species then the inversion of *fimS* can happen. In the schematic figure on the left, all binding sites are occupied and the switch is ready to invert. The right-hand side shows the inverted switch. The binding sites are not fully occupied so no switch can take place at this stage

Mechanistically, the switching rates to and from the fimbriate state emerge from the competition of two proteins, or recombinases, for the four binding sites near *fimS*. The recombinase FimB can switch *fimS* in both directions, however, it does so with a very low rate. The other recombinase, FimE, only switches *fimS* from the on state (where fimbriae are expressed) to the off state (where they are not). While it switches only in one direction, it does so with a much higher efficiency than FimB. FimE is only expressed when the *fimS* switch is in the on orientation.

We will now present a simplified model of this system to demonstrate the practical use of PRISM. To keep the model transparent we will make a number of key simplifications:

- The concentrations of FimB and FimE are constant over time.
- Binding to each of the binding sites takes place with a rate that is proportional to the concentrations of the FimB and FimE proteins respectively.
- On each side of *fimS*, if one of the two binding sites is occupied by, say FimB, then FimE cannot bind to the other site. There are no such constraints between the double sites on each side of *fimS*.

In the model shown in Code 6.3, we have split the Markov chain over three modules: one to keep track of the switching, and one each to keep track of the binding at each site of *fimS*.

```
stochastic

module leftBS
const double btobb = 0.2;
// Further constant definitions omitted ...

bl: [0..2] init 0;
el: [0..2] init 0;

[] (bl = 0) & (el = 0) -> FimB*tob: (bl' = 1);
[] (bl = 1) & (el = 0) -> bto: (bl' = 0) + FimB*btobb: (bl' = 2);
[] (bl = 2) & (el = 0) -> bbtob: (bl' = 1);

[] (el = 0) & (bl = 0) -> FimE*toe: (el' = 1);
[] (el = 1) & (bl = 0) -> eto: (el' = 0) + FimE*etoee: (el' = 2);
[] (el = 2) & (bl = 0) -> eetoe: (el' = 1);
endmodule

module rightBS
// As leftBS; br replaces bl and er replaces el ...
rightBS=leftBS [bl=br, el=er]
endmodule

module switch1
// We say that state 0 corresponds to off
switch: [0..1] init 0;

[switch1] switch = 0 & ((bl = 2) & (br = 2)) ->
                        bbswtichon: (switch' = 1);
[switch2] switch = 0 & ((el = 2) & (er = 2)) ->
                        eeswtichon: (switch' = 1);

[switch3] switch = 1 & ((bl =2) & (br = 2)) ->
                        bbswtichoff: (switch' = 0);
[switch4] switch = 1 & ((el =2) & (er = 2)) ->
                        eeswtichoff: (switch' = 0);
endmodule
```

Code 6.3 A PRISM model of Fim binding (fim-model.pm)

In the PRISM specification code, we have only explicitly written one of the binding modules (the one for the "left" binding site). The right module is exactly the same as the left module. We can therefore simply copy the existing module and rename the state variables. PRISM provides a function for this, as shown. We have omitted the specification of the constant rate factors here to save space. The reader who wishes to test this model is encouraged to obtain a copy from the book's web site and experiment with her own values.

Biological modeling in PRISM can be very intuitive. In this case, we represent each of the two binding sites in their own modules. The state variable bl counts how many molecules of FimB are bound to the left binding site; el is the analogous state variable for FimE on the left binding site. In the copied modules rightBS we have the corresponding state variables br and er.

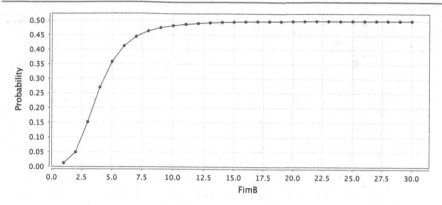

Fig. 6.6 The steady state probability of *fimS* being *on* for various concentrations of FimB

For each of the binding sites there are only three transitions: from no occupation to one molecule being bound; from one molecule bound to two molecules bound; and unbinding of molecules. Note that the state transitions disallow a mixed-bound state where on the left side, say, one FimB and one FimE are bound.

The inversion of *fimS* is represented in the `switching` module. The transitions `switch1` and `switch2` formulate the switching on of the system. The former is the more important, as it controls switching when all four binding sites are occupied by FimB; this is represented by the conditions $bl = br = 2$. The latter represents switching *fimS* from off to on when all four sites are occupied by FimE. We know empirically that this is a very rare event and therefore one should set the rate-constant `eeswitchon` to a very low value. The same module also represents the switching off of *fimS*. Note that the conditions for switching are such that switching only occurs when all binding sites are occupied by the same type of protein.

Using this model (together with the parameters that can be found in the online version of the model file) we asked PRISM about the steady state probability of the switch being on, i.e., in state 1. Using the PRISM GUI we generated a graph (see Fig. 6.6) that shows these probabilities for various concentrations of FimB.

A salient feature of the graph is that, for increasing FimB, the steady state probability rapidly approaches 0.5. This can be easily explained by taking into account that the FimB-mediated switching becomes dominant as the concentration of FimB increases. For high levels of FimB, the importance of FimE for switching becomes negligible. That this is indeed true can be checked using transition rewards. Figure 6.7 shows the proportion of switching events that are mediated by FimB; i.e., that happened when all four binding sites were occupied by FimB. Clearly, as we increase the concentration of FimB this proportion goes to 1. Since FimB switches in both directions with equal rate, in the long run we would expect that the probability of a cell to be on or off approaches 1/2 as the number of FimB increases, which is exactly what we observe in Fig. 6.6.

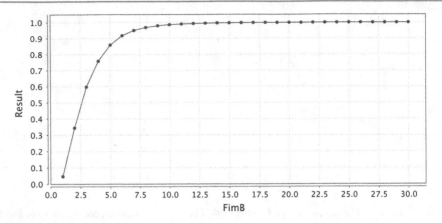

Fig. 6.7 The proportion of FimB-mediated switching events in steady state

6.6.2   Stochastic Versions of a Differential Equation

Differential equation models are normally much more convenient to handle and explore than stochastic models. The formalism of differential calculus is well developed. We have also seen that it is often possible to derive exact equations for the steady state behavior of a system, even if the full set of differential equations cannot be solved. Unfortunately, in biology stochastic effects can often not be ignored. In these situations a dual approach is helpful where a differential equation model of the system is analyzed first. This provides basic insights into the behavior of the system, but tells us nothing about stochastic aspects of the system. PRISM can then be used to simulate a stochastic version of the differential equation system.

In this second case study we give a short example of how this can be done. Assume we have a system of differential equations for x, y, z. These variables could be the number of particles of specific proteins in the cell, for example. We could solve the system numerically, but that would only give us the deterministic behavior. In order to analyze the stochastic behavior of the system we need to translate the system of differential equations into the PRISM specification language, which is relatively straightforward to do. Note, however, that this will only work when the variables of the differential equations represent numbers of entities, such as numbers of molecules or numbers of agents. This will not always be the case. A case in point is the differential equation model of malaria in Sect. 4.4. If the differential equation model represents, say, fractions of quantities, then for obvious reasons the model cannot directly be translated into a stochastic PRISM model. Assuming this caveat is met, we can create PRISM models directly from the differential equation (or the differential equations) by following these steps:

1. Create a new stochastic state variable, $\bar{x}$, for each deterministic variable x in the system of differential equations.
2. For each term in the right hand side of the differential equations, enter a new transition in the Markov chain with the same rate as formulated in the differential equation. For example, if the left hand side of the equation is $\dot{x}$ and the right hand side contains a term $-2azx$, then add the following transition:

```
[] (x > 0): 2 * a * z * x -> (x' = x - 1)
```

3. In the case of differential equations with several variables, there is often a flow from one variable to another, in the sense that one variable is converted into another one. If this is the case, then this must be taken into account. For example, if the right hand side of the equation $\dot{z}$ contains a similar term $+2azx$, then this term must also be taken into account in the PRISM model. The previous transition must be amended as follows:

```
[] (x > 0) & (z < maxz): 2 * a * z * x ->
            (x' = x - 1) & (z' = z + 1)
```

Note here that it was necessary to add an additional guard to prevent z from going over its allowed limit.

Let us now illustrate this using a specific example, using a simple production-decay differential equation:

$$\dot{x} = a - bx \tag{4.19}$$

In Chap. 4 we show how to derive the general solution for this equation. Assuming initial conditions of $x(0) = 0$ this works out to be:

$$x(t) = \frac{a}{b} \left(1 - \exp\left(-bt\right)\right) \tag{6.54}$$

We do not need this general solution to generate our PRISM model, but the solution makes it more convenient to compare the behavior of the differential equation with the model we create. (Instead, we could of course also solve (4.19) numerically, although this seems rather silly if we have the general solution.)

Following our rules of conversion, in our PRISM version of (4.19) we need only a single stochastic state variable, which we call x. The differential equation has a production term and a decay term. The former translates into a transition to increase the value of x by 1 with a rate of a. Similarly, there is a corresponding term for decreasing x with a rate of bx. Altogether we obtain the PRISM model shown in Code 6.4.

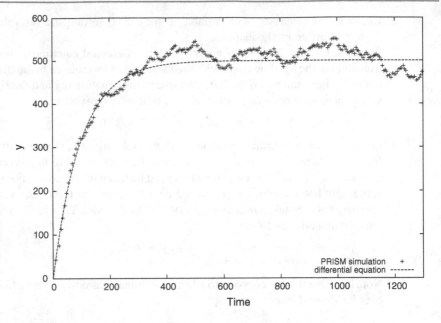

Fig. 6.8 A stochastic simulation in PRISM compared with the differential equation

```
ctmc

const double a = 5;
const int maxx = 4000;
const double b = 0.01;

module drift

x: [0 .. maxx] init 0;

[] (x < maxx) -> a:   (x' = x + 1) ;
[] (x > 0)    -> x*b: (x' = x - 1) ;

endmodule
```

Code 6.4 A production-decay model (drift.pm)

Figure 6.8 shows a comparison of the differential equation model with the stochastic PRISM model. The graph illustrates well how the stochastic system fluctuates around the solution of the deterministic system.

6.38 Make a stochastic version of the bistable switch system (4.17) on p. 165. Compare the behavior of the stochastic and the deterministic system.

6.6.3 Tricks for PRISM Models

PRISM is a very powerful tool that is easy to use for creating and exploring stochastic models in biology. Like any other software tool it does have its idiosyncrasies that can appear somewhat unexpected to a novice user, but they are simple to work around if one is aware of them. We will address some of those in this section.

A common problem that can be quite puzzling at first could be caused by a transition like this one:

```
[]  (x < maxx) ->  x * b: (x' = x + 1)
```

Syntactically this line is correct. However, if a line like this is incorporated into a model PRISM will complain about it during model building, mentioning that transition rates sum to 0. It would not build the model until the problem is fixed. The root of the problem is that the state variable x appears in the transition rate. If $x = 0$ then the rate would also be zero. In order to avoid this error, the guard (the conditions) must be strengthened:

```
[]  (x< maxx) & (x>0)->  x*b: (x'=x+1)
```

This version will allow the model to be built. This type of error is easy to make, but often surprisingly difficult to spot. When writing a model it is helpful to check conditions carefully and to make sure that the probability cannot become zero. Sometimes, there are also situations where the transitions in the model are such that a state variable cannot become zero; in this case PRISM may still complain if the rate formula would formally allow negative rates.

A major limitation of PRISM when model checking is the size of the model. There are no hard and fast rules as to what can and cannot be model checked successfully. Generally, the more states there are, the longer the model checking takes. The speed (and feasibility) of model checking can also be influenced by the rates. During the testing phase it is therefore important to keep the number of states as low as possible. This allows one to obtain a feel for feasible model sizes. This is specifically the case when using the graphical interface, which can lock during model building or model checking, preventing the user from gracefully exiting the program.

Model checking times can sometimes be reduced by choosing the correct options. The GUI version allows the user to select various model checking algorithms interactively. The command line offers a number of switches to adjust how PRISM works internally. The options can be seen by typing prism -help. A detailed explanation of the various algorithms and options available in PRISM is beyond the scope of this book but a summary of basic operator use may be found in Appendix A.3. Further good and practical advice is available from the FAQ section under *Documentation* on the PRISM web site [4].

References

1. Gardiner, C.: Handbook of Stochastic Methods: For Physics, Chemistry and the Natural Sciences. Springer, Berlin (2008)
2. Chu, D., Zabet, N., Mitavskiy, B.: Models of transcription factor binding: sensitivity of activation functions to model assumptions. J. Theor. Biol. **257**(3), 419–429 (2009). doi:10.1016/j.jtbi.2008.11.026
3. Kwiatkowska, M., Norman, G., Parker, D.: PRISM 4.0: verification of probabilistic real-time systems. In: Gopalakrishnan, G., Qadeer, S. (eds.) Proceedings of the 23rd International Conference on Computer Aided Verification (CAV'11). LNCS, vol. 6806, pp. 585–591. Springer, Berlin (2011)
4. PRISM. http://www.prismmodelchecker.org/. Accessed 27 June 2015

Simulating Biochemical Systems

This chapter describes in detail how modeling techniques can be used to simulate biochemical systems. In particular, we look at Gillespie's *stochastic simulation algorithm* (SSA) [1] and some of the variations spawned from this to effect efficient modeling of systems of coupled chemical reactions. Gillespie's work offers a bridge between the continuous models of differential equations (Chap. 4) and the individual representations of agent-based models (Chap. 3). Distinctive of the SSA approach is that it treats individual molecular *species* as individual entities, yet processes all agents of the same species *en masse*. Nevertheless, *the number* of molecules within each species plays a significant role within the evolution of a model. Exploration of the SSA and related approaches includes detailed Java code to illustrate the main elements of their implementation. In addition, we introduce the widely available modeling environments Dizzy [2] and SGNS2 [3], that provide convenient, packaged implementations of these algorithms.

7.1   The Gillespie Algorithms

In the differential equation approach, a model of a chemical reaction system is built as a set of coupled differential equations. The variables of these equations are molecular concentrations. For other than relatively simple sets of equations, analytical solutions are likely to be unavailable and the equations have to be solved numerically. A notable feature of ODE models is that they imply a system of reactions that is both continuous and deterministic, neither of which accurately matches the realities of the physical world. For instance, we know that the inherent stochasticity of nature can produce important distinctive non-deterministic effects, and at the beginning of Chap. 6 we discussed some of the errors that creep in when we ignore the fact of integral particle numbers at low concentrations.

In contrast to ODE models, agent-based models consider each individual molecule along with the discrete interactions occurring between them. Random number

© Springer-Verlag London 2015

D.J. Barnes and D. Chu, *Guide to Simulation and Modeling for Biosciences*,
Simulation Foundations, Methods and Applications,
DOI 10.1007/978-1-4471-6762-4_7

generators support the stochastic aspect but individual-based approaches can be computationally very expensive. Given that the states of individual molecules are effectively indistinguishable from each other at the level of the model, representing each molecule distinctly is questionably detailed.

Gillespie [4] formulated an approach that was designed for systems with relatively low numbers of the individual molecules of interest within *a perfectly mixed* chemical volume, treating different molecular species *en masse* rather than individually. His formulation was derived from an analysis of the "collision probability per time unit" of two reactive molecules within a perfectly mixed volume in thermal equilibrium. Collisions between molecules of interest lead to reactions which then discretely alter the numbers of reactant and product molecular species.

Rather than having a *deterministic reaction rate constant* (k_i) for each reaction, Gillespie associated a *stochastic reaction constant* with each, c_i, representing the probability that a particular pair of reactant molecules would react in the next infinitesimal time interval. By combining c_i with the number of possible combinations of pairs of those reactant molecules (h_i) at a particular time then the probability of that reaction taking place within the next time interval can be calculated. Given a set of reactions, the task of simulation becomes to repeatedly *identify which reaction is most likely to occur next, and when.* The stochastic nature of a reaction system means that these two questions should be answered probabilistically. As numbers of reactant molecules change through reactions occurring, so do the probabilities of the reactions change continually. This probabilistic element means that the results from using Gillespie's approach naturally have the desirable deviations from the smooth results of deterministic approaches, which are unrealistic when relatively small numbers of molecules of the species of interest are involved.

The probability of any particular reaction occurring next is dependent on the stochastic reaction constant values of all the reactions in the system, as well as on the numbers of all of the different reactant molecules. Associated with each reaction, i, is a *probability density function*:

$$P(\tau, i) = a_i exp(-a_0 \tau) \tag{7.1}$$

This is the probability at time t that i will be the next reaction to occur, and that it will occur between time $(t + \tau)$ and time $(t + \tau + \delta\tau)$, where $\delta\tau$ is an infinitesimal time interval, $a_i = h_i c_i$, $a_0 = \sum_{i=1}^{M} a_i$, and M is the number of reactions. a_i is usually referred to as the *propensity* of reaction i. For a reaction involving a single molecule of each reactant species, h_i is calculated as the product of the current number of molecules (N) of each reactant species. If a reaction requires k molecules of a particular species rather than just 1 then the binomial coefficient $\binom{N}{k}$ must be used instead of N to calculate the number of combinations involving that species. If μ is the next reaction to occur then no reaction takes place between the current time t and time $t + \tau$.

Using this formulation, Gillespie described two methods for providing the answers to the two questions posed above: identifying the next reaction, μ and when it would occur, $t + \tau$. These are known as *the direct method* and *the first reaction method*. Both are described in the following sections.

7.1.1 Gillespie's Direct Method

In the direct method (Fig. 7.1), the probability density function (7.1), along with a pair of random numbers from the unit interval, are used to generate the answers to the two questions at the heart of the simulation's iteration: Which is the reaction most likely to occur next and when will it occur? Given two random numbers r_1 and r_2, we calculate τ, the time to the next reaction, as: $\tau = (1/a_0)ln(1/r_1)$ and choose μ such that

$$\sum_{i=1}^{\mu-1} a_i < r_2 a_0 <= \sum_{i=1}^{\mu} a_i \qquad (7.2)$$

This approach is known as the *direct method* because of the way it identifies both τ and μ directly (Fig. 7.1).

In Gillespie's original formulation, all of the a_i values are recalculated after each reaction. However, note that a reaction's propensity value will only change if the number of molecules of one of its reactants is altered by the selected reaction. Therefore, given suitable data structures relating reaction dependencies to each other, some optimization of the calculations on each iteration will be possible. We will explore this aspect in more detail in Sect. 7.2, when we consider Gibson and Bruck's variations on Gillespie's SSA. However, it is worth noting that much of the literature that uses Gillespie's SSA approach for its models tends to omit these improvements for the sake of simplicity. In Sect. 7.3, we will also consider how the search to identify μ, corresponding to (7.2), can be made efficiently.

7.1.2 Gillespie's First Reaction Method

The first reaction method is entirely equivalent to the direct method but works slightly differently. In this method, the τ value for every reaction is calculated and the one

1. Initialization:
   - Set the simulation time to zero, and initialize the random number generator.
   - Set up data for the reactions to be modeled: reaction constants and numbers of each type of molecular species.
   - Calculate the initial propensity values, a_i, for each reaction, and a_0 as the sum of these.

2. Iteration:
   - Generate random numbers r_1 and r_2 from a uniform distribution.
   - Identify the next reaction, μ, and the time until its occurrence, τ_μ, using r_1 and r_2.
   - Increase the current time by τ_μ.
   - Adjust the reactant and product levels according to reaction μ.
   - Output the time and molecular levels, if required.
   - Recalculate the a_i and a_0 values for the next iteration.

Fig. 7.1 Outline of Gillespie's *direct method*

1. Initialization:

   - Set the simulation time to zero, and initialize the random number generator.
   - Set up data for the reactions to be modeled: reaction constants and numbers of each type of molecular species.

2. Iteration:

   - Calculate the propensity value a_i for each reaction.
   - Calculate τ_i (the putative time delta) for each reaction from an exponential distribution with parameter a_i.
   - Choose the reaction, μ, with smallest τ_i.
   - Increase the current time by τ_μ.
   - Adjust the reactant and product levels according to reaction μ.
   - Output the time and molecular levels, if required.

Fig. 7.2 Outline of Gillespie's *first reaction* method

with the smallest value is identified as the next reaction. It is this approach that also forms the basis for Gibson and Bruck's improvements to Gillespie's SSA, which we shall look at in Sect. 7.2. Figure 7.2 shows an outline of the first reaction method. An obvious difference between the direct and first reaction methods is that the former generates two random numbers per iteration whereas the latter generates only one. Where random number generation is a relatively expensive operation, this difference may be significant. However, note that *both* methods take runtime that is proportional to the number of reactions, M. We say, therefore, that these algorithms are of *order* M, written as $O(M)$. In other words, if we were to double the number of reactions in the model then we would expect something like a doubling of the runtime for the simulation component of the model. That the algorithms are $O(M)$ can be observed, for instance, in the way that the selected reaction is identified using $r_2 a_0$ in the direct method and the identification of the smallest τ_i in the first reaction method. Both involve a linear search, whose length increases in direct proportion to the number of reactions. Where the number of reactions is small, this will not be particularly significant, but it does become an issue with large numbers. Figure 7.3 illustrates the scaling effect via our implementation, using sets of randomly generated reactions. This is why other researchers have sought to improve the scalability of the SSA.

7.1.3 Java Implementation of the Direct Method

In order to illustrate some of the practical issues associated with the Gillespie and related algorithms, we will present some Java implementations of them. Even if the reader has no intention of ever implementing these algorithms from scratch, this material will still provide insights into their different characteristics and properties, which will contribute to an overall better basis for making decisions over how to choose between them for a particular simulation task.

Consider a chemical volume containing two distinct molecular species of interest, X and Y. In our models, we use two instances of a `Molecule` class to represent these

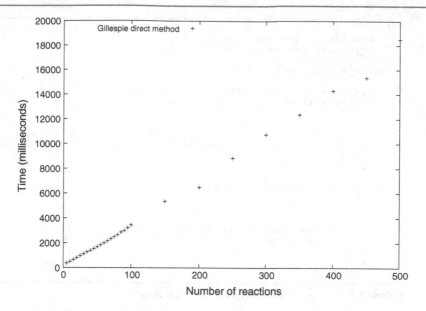

Fig. 7.3 Mean runtime in milliseconds for 1000000 iteration steps of an implementation of Gillespie's direct method. Reaction sets were randomly generated

species, with each instance storing a count of the number of molecules of that species present in the volume at a particular time. These counts will be adjusted whenever reactions that affect them take place, reflecting both consumption and production.

In our implementation, a chemical reaction of the form

$$X \rightarrow Y \tag{7.3}$$

is represented as an instance of the `Reaction` class. Instances of this class store separate lists of `Reactant` and `Product` objects (Code 7.1). `Reactant` and `Product` objects are wrappers for `Molecule` objects. As part of their wrapper function they include a `multiplicity` attribute, indicating how many molecules of the wrapped species is involved as a reactant or product in a particular reaction. Multiplicity cannot be represented at the level of a Molecule because a species may well occur with different multiplicities in different reactions within a single reaction set. So, the following reaction

$$X + 2Y \rightarrow X + Z \tag{7.4}$$

would be represented as a Reaction object with a list containing two Reactant objects wrapping Molecules X and Y, and another list containing two Product objects wrapping Molecules X and Z. The Reactant object wrapping the Y Molecule would have a field recording a multiplicity of two. Note that the same Molecule may appear as both a Reactant and a Product in a single reaction, for instance when it acts catalytically. Reaction's `react` method adjusts molecular counts by consuming reactants and producing products.

```
/**
 * Model a single reaction in terms of
 * molecular reactants and products.
 */
public class Reaction
{
    private List<Reactant> reactants;
    private List<Product> products;
    ...

    /**
     * React by adjusting quantities of reactants and products.
     */
    public void react()
    {
        for(Reactant r : reactants) {
            r.consume();
        }

        for(Product p : products) {
            p.produce();
        }
    }
}
```

Code 7.1 The react method of the Reaction class

Code 7.2 shows how we have mapped the outline of Fig. 7.1 to the code of a Model class's run method. The initModel method creates the model's Molecule, Reactant, Product and Reaction objects and passes them to a ReactionSet object, which stores all reactions in a list. Code 7.3 shows an unoptimized ReactionSet with its findReaction method, which implements the search for μ from (7.2).

7.1.4   Example Reactions

As well as outlining the theoretical basis for his approach, Gillespie illustrated its efficacy with a number of interesting sets of reaction sets, and his original papers are well worth reading as superbly elucidated combinations of theory and practice in this area. We have based a couple of example reactions sets on his, in order to illustrate our Java implementation.

7.1.4.1 A Single Reaction
Consider a reaction set containing the single reaction

$$X \xrightarrow{k} Y \tag{7.5}$$

which represents the isomerization of molecule X to the form Y with a rate of k. In the stochastic model we replace the reaction rate k with the stochastic reaction constant c. Code 7.4 shows the way in which the model is set up in terms of Molecule and Reaction objects, reactants and products. We assume that, at the start of the simulation, there are 1000 molecules of X and the stochastic constant is 0.5. Figure 7.4

```
/**
 * Run the model until the stop time,
 * or no more reactions are possible.
 */
public void run()
{
    int step = 0;
    ReactionSet system = new ReactionSet();

    // Initialize the particular model.
    initModel(system);

    // Obtain the initial cumulative propensity.
    double aZero = system.calculateAZero();

    time = 0;
    while(time < stopTime && aZero != 0.0) {
        // Calculate the time to the next reaction.
        double deltaT = (1.0 / aZero *
                            Math.log(1.0 / Math.random()));
        time += deltaT;
        Reaction r = system.findReaction(Math.random() * aZero);
        r.react();
        step++;
        if(step % samplingRate == 0) {
            system.showStatus();
        }
        aZero = system.calculateAZero();
    }
}
```

Code 7.2 Main iteration of Gillespie's direct method

```
/**
 * Model a set of reactions.
 */
public class ReactionSet
{
    private List<Reaction> reactions;
    ...

    /**
     * Find the required reaction via the given propensity.
     */
    public Reaction findReaction(double aValue)
    {
        double sum = 0;
        for(Reaction r : reactions) {
            sum += r.getA();
            if(sum >= aValue) {
                return r;
            }
        }
        throw new RuntimeException("Internal error.");
    }
}
```

Code 7.3 Reaction selection in an unoptimized implementation of Gillespie's direct method

```
/**
 * Set up the model with a single isomerization reaction.
 */
private void initModel(ReactionSet system)
{
    Molecule x = new Molecule("X", 1000);
    Molecule y = new Molecule("Y", 0);

    // Specify the reaction constant.
    Reaction r = new Reaction(0.5);
    r.addReactant(x);
    r.addProduct(y);
    system.addReactionToSet(r);
}
```

Code 7.4 Setting up the model for the reaction $X \rightarrow Y$

contains a plot of the number of molecules of X against time from a single run of this model. Note that the curve only roughly follows the shape of $X_0 exp(-ct)$, and it is exactly this sort of variation from the continuous function, due to integral particle numbers and stochasticity, that we are looking for from the SSA.

7.1.4.2 Multiple Reactions
It is obviously of greater interest to investigate reaction sets with more than a single reaction. Consider the following set of three arbitrary reactions involving five

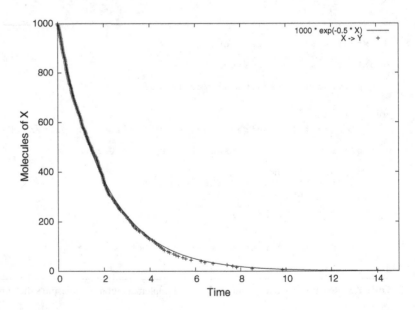

Fig. 7.4 Isomerization of $X \rightarrow Y$ with reaction constant $c = 0.5$ and $X_0 = 1000$. One point plotted every 5 iterations

molecules:

$$A + \overline{B} \xrightarrow{c_1} C \tag{7.6}$$

$$A + D \xrightarrow{c_2} E$$

$$C + E \xrightarrow{c_3} 2A + D$$

Following Gillespie, we use the over bar to indicate a *boundary species*, $\overline{B}$, meaning that the size of the species remains constant. The use of boundary species could be relevant, when, for example, the level of a molecule is constantly replenished. Figure 7.5 shows the variations in molecular numbers of A, C, D and E over a single run of about 2 million iteration steps. Under the conditions of this particular run, the system is reasonably stable after an initial settling down period. In general, the steady-state ranges of A, C, D and E will vary depending on the reaction constants used, and some combinations will lead to the exhaustion of some species.

7.1.4.3 The Lotka–Volterra Equation
An interesting example of dynamic multi-species interaction to study via stochastic modeling is that described by the Lotka–Volterra equation, whose general form is

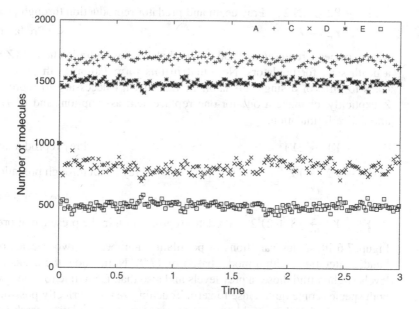

Fig. 7.5 A single run of the reaction set in (7.6) with reaction constants $c_1 = 0.12$, $c_2 = 0.08$, $c_3 = 0.5$ and initial levels of 1000 molecules for all species. One point plotted every 10 000 iterations

given in (7.7), where the $b_{i,j}$ are real numbers (positive or negative).

$$\dot{y}_i = y_i \left(r_i + \sum_{j=1}^{n} b_{i,j} y_j \right), i = 1, \ldots, n \tag{7.7}$$

It describes the interaction between n species whose population numbers are described by the y_i values. Species i grows at rate r_i and interactions between pairs of species are determined by the values of $b_{i,j}$. The simplest case of two species is often characterized as a model of predator and prey interaction (7.8).

$$\dot{y}_1 = y_1(r_1 + b_{1,2} y_2), \quad r_1 > 0 \quad b_{1,2} < 0$$
$$\dot{y}_2 = y_2(r_2 + b_{2,1} y_1), \quad r_2 < 0 \quad b_{2,1} > 0 \tag{7.8}$$

Species y_1 is the prey. The value of r_1 represents population growth in the presence of sufficient food supply and $b_{1,2} y_2$ represents predation proportional to predator numbers. Conversely, species y_2 is the predator, with r_2 standing for death rate and $b_{2,1} y_1$ for population growth proportional to prey numbers.

SSA is an ideal technique for modeling these interactions and there are various ways in which the two-species version could be encoded, for instance:

$$\overline{X} + Y1 \rightarrow 2Y1 \qquad \text{Prey reproduction from undiminishing food supplies}$$
$$Y1 + Y2 \rightarrow 2Y2 \qquad \text{Prey death and predator reproduction through predation}$$
$$Y2 \rightarrow Z \qquad \text{Predator death}$$

where species X represents a food supply that remains undiminished and Z represents a death state for predators. Note that in this formulation death of a single prey leads to birth of a single predator. In fact, it isn't necessary to represent X and Z explicitly, or make a one-for-one replacement assumption, and (7.9) offers an alternative formulation.

$$Y1 \xrightarrow{c_1} 2Y1 \qquad \qquad \text{Prey reproduction} \quad (7.9)$$

$$Y1 + Y2 \xrightarrow{c_2} Y2 \qquad \qquad \text{Prey death through predation}$$

$$Y2 \xrightarrow{c_3} \qquad \qquad \text{Predator death}$$

$$Y1 + Y2 \xrightarrow{c_4} Y1 + 2Y2 \qquad \text{Predator reproduction in the presence of prey}$$

Figure 7.6 illustrates variations in population numbers of two species over a few hundred iterations with a model based on (7.9). Notice how the peaks in predator levels slightly trail those in prey levels and also that, from time to time, the levels of both species come quite close to zero. Touching zero is perfectly possible with the stochastic model; this would amount to a permanent population crash from which the model cannot recover.

Having looked at Gillespie's SSA approach, and a few small example models, we will now look at some variations that seek to improve the implementational efficiency and scalability of the original approach.

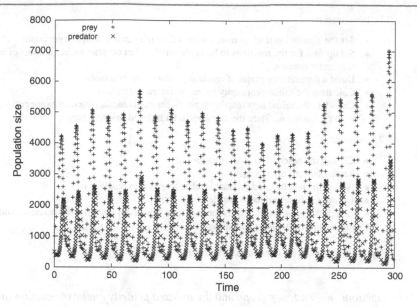

Fig. 7.6 Predator-prey population numbers from (7.9) with $Y1_0 = 1000$, $Y2_0 = 500$, $c_1 = 0.5$, $c_2 = 0.0005$, $c_3 = 0.5$, $c_4 = 0.00025$

7.2   The Gibson–Bruck Algorithm

Gibson and Bruck [5] developed a variation of Gillespie's first reaction method which they called the *next reaction method*. They sought to reduce the $O(M)$ complexity of the SSA by lowering the number of calculations on each iteration. In particular, they looked at the areas that produced the linear dependency on the number of reactions, M:

- Recalculating the a_i values.
- Recalculating the τ_i values.
- Identifying the smallest τ_i value.

It has already been noted in Sect. 7.1.1 that it is only necessary to recalculate the a_i values of reactions whose reactant molecule levels were affected by the immediately preceding reaction. As a result, the next reaction method is able to limit recalculation of τ_i values to only those reactions whose a_i were changed. This optimization is made possible by using *absolute time* for τ values rather than relative time.

Figure 7.7 shows an outline of how the next reaction method works. While very similar in outline to the first reaction method, this more efficient version is often overlooked in favor of the Gillespie methods. The reason for this appears to be primarily because it requires additional data structures to support it; in particular, the

1. Initialization:

   - Set the simulation time to zero, and initialize the random number generator.
   - Set up data for the reactions to be modeled: reaction constants and numbers of each type of molecular species.
   - Build a dependency graph of reaction products and reactants.
   - Calculate the initial propensity value, a_i, for each reaction.
   - Calculate the initial next reaction time, τ_i, for each reaction from an exponential distribution with parameter a_i. Store the τ_i values in an indexed priority queue.

2. Iteration:

   - Choose the reaction, μ, with smallest τ_i.
   - Set the current time to τ_μ.
   - Adjust the reactant and product levels according to reaction μ.
   - Output the time and molecular levels, if required.
   - Generate new a_i and τ_i values for reactions dependent upon the last reaction, and update the indexed priority queue accordingly.

Fig. 7.7 Outline of the Gibson–Bruck *next reaction method*

reactions' *dependency graph* and the *indexed priority queue* of reaction times. There is little rationale for avoiding Gibson–Bruck, if this is the reason, given the ready availability of modern programming libraries of data structures. We will explore these data structures in the following sections.

7.2.1 The Dependency Graph

The dependency graph is used to identify relationships between reactions. For instance, the propensity of a reaction having molecule X as a reactant will be reduced when any reaction takes place that consumes X. Conversely, its propensity will be increased when any reaction takes place that produces X. This reaction is said to be dependent upon those other reactions. In Java, we can easily represent the dependency graph as a *map* between a reaction and the set of reactions it affects:

```
Map<Reaction, Set<Reaction>> dependencyGraph
```

Whenever a reaction takes place, the set of dependent reactions can easily be identified from the map, and their propensities updated appropriately.

The dependency graph has to be built just once during the initialization stage. For each reaction, it is necessary to identify which molecular species are affected by that reaction. This is usually the set of all reactant and product molecules except (as in the case of a catalytic reaction) where a reactant appears with equal multiplicity as a product. In addition, we identify which species a reaction depends upon—typically simply its reactants.[1] Code 7.5 shows the way in which a dependency graph can

---

[1] We ignore here the indirect dependencies introduced through non-constant reaction probabilities that are affected by non-reacting molecular species, although this is handled in the full version of the code provided for this chapter.

```
/**
 * For each reaction, identify which molecular species it affects
 * and which reactions depend upon those species.
 */
void buildDependencyGraph()
{
    // Iterate over the list of reactions.
    for(Reaction source : reactions) {
        // Create a set for the reactions affected by
        // the source reaction.
        Set<Reaction> affectedReactions = new HashSet<>();
        // Add the self edge.
        affectedReactions.add(source);
        // Get the species whose numbers are changed by
        // this reaction.
        Set<Molecule> affects = source.getAffects();
        // Identify all reactions affected.
        for(Reaction target : reactions) {
            // Determine whether target depends on one of
            // the species affected by source.
            Set<Molecule> dependsOn = target.getDependsOn();
            for(Molecule m : dependsOn) {
                if(affects.contains(m)) {
                    affectedReactions.add(target);
                }
            }
        }
        // Store the set in the dependency graph.
        dependencyGraph.put(source, affectedReactions);
    }
}
```

Code 7.5 Building a reaction dependency graph (unoptimized)

be built by iterating over the full set of reactions and making the link between the molecular species affected by each reaction and the reactions dependent on those species. We shall come back to this task in Sect. 7.4 when we look further at wider efficiency considerations.

7.2.2 The Indexed Priority Queue

The algorithm uses an indexed priority queue to record the next reaction time of each reaction in the model. Its structure makes it very easy to identify the reaction with the earliest reaction time, which is exactly what we want. A priority queue can be represented as a *binary tree* data structure whose nodes each consist of a reaction paired with its putative next reaction time. A binary tree is a recursive data structure consisting of parent nodes linked to up to two child nodes. The child nodes are, themselves, parents of up to two children, and so on. A node with no children is called *a leaf node*.

The tree is ordered such that the time value of any node is always less than or equal to the time values of both of its children. However, the children are not necessarily ordered. Such a data structure is also called a *heap*. The advantage of maintaining this structure is that no search is required to identify the next reaction—it is necessarily stored at the root of the tree.

```
// The indexed priority queue.
private Reaction[] timeOrderedQueue;
...

/**
 * Initialize the indexed priority queue.
 */
private void initQueue()
{
    int numberOfReactions = reactions.size();
    timeOrderedQueue = new Reaction[numberOfReactions];
    for(int index = 0; index < numberOfReactions; index++) {
        insertIntoQueue(index, reactions.get(index));
    }
}

/**
 * Insert reaction into the priority queue at the given index
 * and bubble it upwards to its correct position.
 */
private void insertIntoQueue(int index, Reaction reaction)
{
    timeOrderedQueue[index] = reaction;
    // Make sure that the order is properly maintained.
    int parentIndex = (index - 1) / 2;
    Reaction parent = timeOrderedQueue[parentIndex];
    // Bubble upwards, if necessary.
    while(index > 0 && reaction.getNextTime() <
                           parent.getNextTime()) {
        swap(parentIndex, index, reaction);
        index = parentIndex;
        parentIndex = (index - 1) / 2;
        parent = timeOrderedQueue[parentIndex];
    }
}
```

Code 7.6 Building the initial state of the priority queue

The binary tree can be implemented as an array with the property that a node stored at index i has its parent node at index $(i - 1)/2$ and child nodes at $2i + 1$ and $2i + 2$. The length of the array is equal to the number of the reactions. Code 7.6 illustrates the initial building of the tree during model setup. Note that the nodes are Reaction objects, which we have modified from the version used with Gillespie to store their putative next reaction time.

The ordering of the nodes in the tree will obviously change as reaction times are updated. This is the most complex aspect of the next reaction method, yet it is still straightforward to program. As long as the number of changes to the tree is relatively modest compared with the number of reactions then this is one step towards reducing the runtime speed complexity from $O(M)$.

In addition to the tree that relates reactions to their time ordering, we need a data structure to relate a reaction to its node in the tree. This is so that a node can be updated with its new time when a reaction is affected by the occurrence of another. In our case, we use a map relating a reaction to its integer index in timeOrderedQueue:

```
Map<Reaction, Integer> queueIndex;
```

Code for updating the tree when a reaction's next time is changed can be seen in Code 7.7.

```
void updateTree(Node n)
{
    Node parent = n.getParent();
    if(n.getNextTime() < parent.getNextTime()) {
        // Swap position with the parent node.
        swap(n, parent);
        // Continue upwards, recursively.
        updateTree(n);
    }
    else {
        Node child = child with smallest tau;
        if(n.getNextTime() > child.getNextTime()) {
            // Swap position with the child node.
            swap(n, child);
            // Continue downwards, recursively.
            updateTree(n);
        }
    }
}
```

Code 7.7 Recursively updating a node in the indexed priority queue

7.2.3 Updating the τ Values

What remains to be described is the way in which the τ_i values of the dependent reactions are updated, as this is a key difference between the first reaction and next reaction methods. Whereas the first reaction method requires new τ_i values to be generated afresh from an exponential distribution with the new a_i values, the next reaction method only requires this for the reaction that has just taken place. For all other affected reactions, their old τ values (τ_{old}) are *adjusted* according to the current time and the ratio between their old and new a_i value in the way shown in (7.10), where t is the (absolute) time of the most recent reaction.[2]

$$\tau_{new} = t + (a_{old}/a_{new})(\tau_{old} - t) \tag{7.10}$$

In other words, if the propensity is reduced then the next reaction time is pushed further into the future, but if the propensity is increased then the next reaction time is brought forward. The amount of the change is proportional to the change in propensity. The new τ values are then used to update the nodes in the priority queue, which will be reordered as a consequence, leaving the next reaction at the root of the tree ready for the following iteration.

7.2.4 Analysis

There are several ways in which the next reaction method is faster for large numbers of reactions than Gillespie's original first reaction method:

- Identification of the next reaction is immediate, given the indexed priority queue.

---

[2]Note that we ignore here the possibility that a_{new} is zero, although this is handled in the full version of the code provided for this chapter.

- Only one random number is generated per iteration—to calculate the new τ value for the most recent reaction—rather than one per reaction.
- Only those reactions that were dependent on the one that has just take place must be considered for update, and these are readily identified from the dependency graph.

The limiting factor for the complexity of the next reaction method is the dependencies between the reactions. Where these are relatively few per reaction compared with the overall number of reactions, M, then this method scales at the much slower rate of $O(log_2 M)$ compared with $O(M)$ for the first reaction method. Figure 7.8 provides illustrative timings from our implementations of both methods up to a few hundred reactions. The difference is clear to see. Gibson and Bruck's method is generally to be preferred over Gillespie's, therefore, where the number of reactions is large enough to be a consideration.

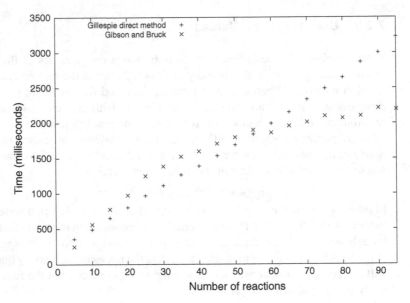

Fig. 7.8 Mean runtime in milliseconds for 1000000 iteration steps for our unoptimized implementations of Gillespie's direct method and the Gibson–Bruck algorithm, with random reaction sets

In their paper [5], Gibson and Bruck also discussed improvements to Gillespie's direct method that also improved its complexity to $O(log_2 M)$. These involve the use of a dependency graph, as in the next reaction method, and a binary tree to store cumulative propensity values, which makes both the calculation of a_0 and selection of the next reaction more efficient than in the original version. We will look at the use of the binary tree in connection with another version of the SSA in Sect. 7.3.

What we see in Gibson and Bruck's approach is a reduction in runtime through the use of more memory, in the form of the dependency tree and priority queue. While the data structures bring slightly increased implementational complexity, they make practical both larger models and more runs of a model for increased accuracy of results. This kind of tradeoff between speed and memory is a recurring theme in algorithm design and implementation in the field of computer science.

7.3 A Constant Time Method

In 2008, Slepoy et al. [6] provided a further variation in the implementation of the stochastic simulation algorithm. In ideal circumstances, their algorithm scales as $O(1)$—in other words, independently of the number of reactions. Where the number of reactions is very large, this has the potential to offer significant speed advantages over both the Gillespie and Gibson–Bruck algorithms. As one might expect, however, the speed improvements require a more complex implementation.

Slepoy et al noted that the SSA is a particular application from a general area known as *random variate generation* (RVG) [7]. They took an existing RVG algorithm and adapted it to apply to the SSA. Their approach is based on a technique called *composition and rejection*, leading them to use the shorthand SSA-CR for it. The essence of the approach is to *group together reactions with similar propensities* and select a group before selecting an individual reaction from within that group. Each group has a lower-bound for the propensities of the reactions it contains. If $lwb(g)$ is the lower-bound propensity for group g, then a reaction with propensity a_i is placed in group g_n, such that, $lwb(g_n) <= a_i < lwb(g_{n+1})$. Successive powers of 2 times the minimum possible non-zero propensity value (a_{min}) are used as the boundary values for groups. The boundary values remain fixed for a particular simulation run, but a reaction may move between groups as its propensity varies. By grouping reactions in this way, much of the dependence of the algorithm on the number of reactions, M, is removed, and hence a scaling independent of M becomes possible. However, there are two assumptions required:

- The ratio of maximum to minimum probability for any two reactions must be bounded.
- The average number of dependencies between reactions must remain fairly constant and not grow in proportion to the total number of reactions.

Figure 7.9 shows an outline of the SSA-CR method.

7.3.1 Selection Procedure

The selection of the right group is based on the approach outlined in (7.2) for reaction selection in Gillespie's first reaction method. However, the implementation illustrated in the `findReaction` method in Code 7.3 is a linear summation and search, which

1. Initialization:

   - Set the simulation time to zero, and initialize the random number generator.
   - Set up data for the reactions to be modeled: reaction constants and numbers of each type of molecular species.
   - Build a dependency graph of reactions.
   - Calculate the initial propensity value a_i for each reaction and the cumulative propensity a_0.
   - Calculate the minimum propensity value a_{min}.
   - Assign each reaction to group n such that $a_{min}2^{n-1} < a_i <= a_{min}2^n$.

2. Iteration:

   - Generate uniform random numbers, r_1, r_2, from the unit range.
   - Set the time increment, $\tau = (1/a_0)log_e(1/r_1)$.
   - Use r_2a_0 to select a group, g.
   - Repeatedly generate random numbers r_3 and r_4 until a reaction, μ, is selected from group g.
   - Adjust the reactant and product levels according to reaction μ.
   - Output the time and molecular levels, if required.
   - Generate new a_i values for reactions affected by the last reaction, adjusting group membership accordingly.

Fig. 7.9 Outline of the *SSA-CR method*

is not particularly efficient—indeed, it is one of the elements that contributes to the unoptimized first reaction method being $O(M)$. An alternative, which is applicable to both the first reaction method and the SSA-CR method, is to store cumulative propensity values in a binary tree. In SSA-CR, the leaves correspond to groups and store the sums of the reaction propensities within each group, whereas in the first reaction method the leaves would store the individual reaction propensities. Each parent node stores the sum of the values in its two child nodes. As a result, the root of the tree stores the sum of all group (and, therefore, all reaction) propensities: a_0. We can use the same style of tree implementation as that for the indexed priority queue discussed in Sect. 7.2.2. An important difference is that the propensity tree is not sorted; the leaves can be stored in group-number order, and changes to propensities only result in node values changing rather than the tree being reordered. Code 7.8 illustrates how the tree is constructed. For the sake of simplicity, we assume that the number of groups, G, is a power of two, rounding up and padding with empty groups at the upper end if necessary. The depth of the tree is log_2G, requiring $2^{depth+1} - 1$ nodes. The initial state of the tree is calculated by storing the group propensity values in the leaves, and then recursively summing the child values up to the root in the `calculatePropensityChildren` method.

On each iteration of SSA-CR, the selection of the group using r_2a_0 is performed by working down from the root of the tree. If r_2a_0 is less than the value in the left child of the root then the left child is descended, otherwise the right child is descended, subtracting the value in the child node from r_2a_0. This process is repeated until a leaf is reached, corresponding to the selected group. The fact that the tree has depth log_2G means that this search is typically considerably faster than the unoptimized linear search illustrated in Code 7.3, and is not directly dependent on the number of reactions.

```java
// The tree of cumulative propensity values.
private double[] propensityTree;
// The offset in propensityTree of the first leaf
// -- i.e. propensityTree[propensityTreeLeafOffset] is
// the value for groups[0].
private int propensityTreeLeafOffset;

/**
 * Build the tree of cumulative propensity values.
 * The sum for all groups will be at the root of the tree.
 */
private void buildPropensityTree()
{
    // Calculate how much space is required for the tree,
    // storing the group propensity values in its leaves.
    int numGroups = groups.size();
    int depth  = (int) Math.ceil(logBase2(numGroups));
    int nodes  = (int) (Math.pow(2.0, depth + 1) - 1);
    int leaves = (int) Math.pow(2.0, depth);

    this.propensityTreeLeafOffset = nodes - leaves;
    this.propensityTree = new double[nodes];

    // Fill in the leaf values of the tree.
    int leafNode = propensityTreeLeafOffset;
    for (PropensityGroup group : groups) {
        propensityTree[leafNode] =
            group.getGroupPropensitySum();
        leafNode++;
    }

    // Pull the leaf values cumulatively up to the root.
    calculatePropensityChildren(0);
}

/**
 * Sum the child values of node to give the value for node.
 * @param node The node to be computed.
 * @return The computed value of the node.
 */
private double calculatePropensityChildren(int node)
{
    int leftChild = 2 * node + 1;
    if (leftChild < propensityTree.length) {
        return propensityTree[node] =
            calculatePropensityChildren(leftChild) +
            calculatePropensityChildren(leftChild + 1);
    }
    else {
        return propensityTree[node]; // Leaf
    }
}
```

Code 7.8 Building the binary tree of cumulative propensity values for SSA-CR

7.3.2   Reaction Selection

Selection of a reaction from within the chosen group involves the rejection element of the algorithm: a reaction is selected and a random propensity generated. Two random numbers are used for this, r_3 and r_4 in Fig. 7.9. If the random propensity is larger than the selected reaction's propensity then that particular reaction is rejected, and both parts of the selection within the group are repeated until a match is found.

Clearly, it is important to minimize the chance of rejection once a group has been selected, otherwise reaction selection could take an arbitrary length of time. Minimization is achieved by setting the upper limits of group membership to be successive powers of 2 times the minimum possible non-zero propensity value (a_{min}) for the set of reactions. When selecting a reaction from within a group, a uniform random number is generated between zero and the upper boundary of the group. Because every reaction in the group has a propensity greater than half the upper bound, there is a less than a 50 % chance that a reaction will be rejected by the random number, significantly limiting the number of iterations of the rejection cycle.[3]

The number of groups required will be $log_2(a_{max}/a_{min})$, where a_{max} is the maximum possible propensity for the set of reactions. Notice that the number of groups does not depend on the number of reactions, M, but bounding the number of groups does depend on being able to calculate the maximum propensity, a_{max}. We will discuss below the potential problems where it is difficult or impossible to determine a_{max} for a set of reactions.

SSA-CR uses the same dependency tree as described in Sect. 7.2 to determine the impact of reaction selection on other reactions' propensities. Once a reaction has been selected, both reaction and group propensities are adjusted accordingly. The adjustment of a reaction's propensity may, of course, result in having to move it from one group to another if its new propensity has crossed either the lower or upper bound of the group's range. Because updating the propensity tree is a relatively expensive component of the iteration, the tree update is best done after *all* the propensity revisions have been made, rather than as each reaction is adjusted. This suggests maintaining a record of changed the groups, in the form of a *set* as part of the revision step. This will save updating the tree twice for a group if two or more of the group's reactions have revised propensities. We will consider this step further in Sect. 7.4 because its efficient implementation is important to the scalability of the algorithm.

If it is possible to determine accurately a_{max} at the start of the model then moving groups does not present a problem because all possible groups will have been provided for in the propensity tree. However, if a_{max} could not be determined in advance, and the number of groups expands into the range of the next power of two, then the propensity tree will require rebuilding with an additional level in the middle of simulating the reactions. This is not disastrous, but the possibility needs to be taken account of in the implementation and will have a small effect on the runtime, as it is a relatively expensive operation.

Figure 7.10 displays a comparison between the runtime performances of implementations of the Gibson–Bruck and SSA-CR methods. As can be observed, the averaged runtime for the SSA-CR implementation rises less steeply than for the Gibson–Bruck algorithm as the number of reactions increases. However, it is still

---

[3]In measurements with our implementation, using a large set of random reactions, the average number of iterations was around 1.4.

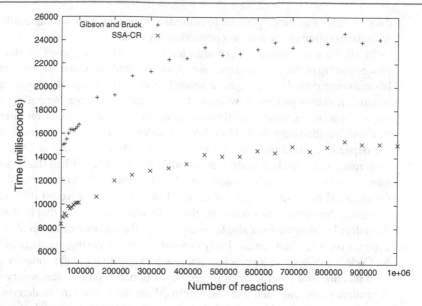

Fig. 7.10 Mean run times in milliseconds for 1000000 iteration steps for our implementations of the Gibson–Bruck and SSA-CR algorithms, with random reaction sets

some way off being the $O(1)$ we were looking for and we will discuss some of the reasons for this in Sect. 7.4.

7.4   Practical Implementation Considerations

One of the reasons we have considered a number of different approaches to the implementation of stochastic simulations is that, aside from the accuracy of the modeling, our primary concern is likely to be having a fast runtime for our models. We have discussed the theoretical runtime scaling complexity of the different methods but, in practice, we may find that our programs are unable to match up with these. Why is this? The two principle reasons are: the way we have implemented the algorithms and the environment in which we run the programs. In this section we will look at the effect of both of these issues.

7.4.1   Data Structures—the Dependency Tree

In Sect. 7.2 we illustrated that Gibson and Bruck were able to significantly reduce the time complexity of Gillespie's method, in part through the use of additional data structures for storing reaction dependencies and putative reaction times. Appropriate data structure selection is crucial to runtime performance, as we can easily illustrate with the building of the reaction dependency graph.

In all these algorithms, there are clearly two distinct phases of the modeling process: setup of the reaction structures prior to simulation, and the simulation itself. In analyzing complexity, we have tended to completely ignore the setup element because it occurs just once, whereas the simulation involves repeating a series of steps an enormous number of times—suggesting that the simulation time will tend to dominate the setup time. However, consider the approach we used in building the dependency tree in Code 7.5 and note that its complexity is $O(M^2)$ because each reaction is checked against every other for its effect. When dealing with large numbers of reactions, this version would add significantly to the overall runtime— for practical purposes it does not scale. How can we improve on it? The key is to recognize that dependency arises via the molecular species, and that both the number of molecules affected by a single reaction, and the number of molecules a reaction depends on will often remain fairly constant with increasing numbers of reactions. So Code 7.9 offers a more efficient approach to building the dependency graph. For each reaction, we record which molecules it depends on in a temporary molecule-dependency list, and then use that intermediate data structure to determine which reactions are affected by each other. It is worth noting that we end up with exactly the same data structure for the dependency graph as in the original version, but the complexity of the building process is more like $O(M)$ rather than $O(M^2)$.

7.4.2 Programming Techniques—Tree Updating

We can further illustrate the impact of our implementation choices on runtime with a look at the way in which the propensity tree of the SSA-CR method is updated. The tree holds the group propensity values in its leaves and the cumulative propensity value, a_0, in its root. Each time a reaction's propensity is affected by the occurrence of another reaction, its group's propensity must be adjusted. In Sect. 7.3.2 we noted that it is best to defer updating the propensity tree until all the group changes are known at the end of a single step, rather than updating it each time a reaction's propensity is adjusted. That way we can avoid multiple updates to the tree for the same group. We suggested storing the changed groups' identities in a set to avoid updating the tree multiple times for the same group. The idea is that, once all the individual reaction changes have been implemented within the groups, each leaf value for the changed groups is updated and trickled up to the root. However, the process of updating the tree deserves further consideration because it still potentially involves some duplication of effort. Consider the case where two groups' leaves have the same parent node in the tree and both group's propensities have changed following a reaction. If the leaf-to-root update is made separately for each leaf, then most of the work of the first update will be overwritten by the second. The update process can be streamlined a little, therefore, by actually storing the set of *parent nodes* of the group leaves that have been adjusted and recalculating the parent values as the sum of their two child values, regardless of whether one or both child values has

```
/**
 * Build the dependency graph for the reactions.
 */
private void buildDependencyGraph()
{
    // Map of molecules to the list of reactions depending on them.
    Map<Molecule, List<Reaction>> moleculeDependencyList =
            new HashMap<>(molecules.size());

    // Build a map of which reactions depend on which molecules.
    for(Reaction target : reactions) {
        Set<Molecule> dependsOn = target.getDependsOn();
        for(Molecule m : dependsOn) {
            List<Reaction> dependency =
                moleculeDependencyList.get(m);
            if(dependency == null) {
                dependency = new ArrayList<>();
                moleculeDependencyList.put(m, dependency);
            }
            dependency.add(target);
        }
    }

    // For each reaction, identify which molecules it affects
    // and which reactions depend on those molecules.
    for(Reaction source : reactions) {
        // Create the set for affected reactions.
        Set<Reaction> affectedReactions = new HashSet<>();

        // Add the self dependency.
        affectedReactions.add(source);

        // Identify the reactions affected.
        Set<Molecule> affects = source.getAffects();
        for(Molecule m : affects) {
            List<Reaction> dependency =
                moleculeDependencyList.get(m);
            if(dependency != null) {
                affectedReactions.addAll(dependency);
            }
        }
        dependencyGraph.put(source, affectedReactions);
    }
}
```

Code 7.9 Improved building of a reaction dependency graph (see Code 7.5 for comparison)

been adjusted. However, we can take this even further by recognizing that updating from two leaves with different parent nodes but a common grandparent node will also involve duplication of update from the grandparent node upwards, and the argument can be continued for different ancestry levels within the tree. What we have actually done in our implementation, therefore, is to trickle each update only to the next level above in the tree, and then repeated this process for each level, until all the changes eventually accumulate in the root. Each changed node is only updated once with this approach.

7.4.3 Runtime Environment

Aside from implementational effects on overall runtime, there will be potentially significant effects from the hardware on which a program is run, the operating system environment and the programming language. An obvious factor is the amount of memory available to the program, and the way it is managed by the runtime environment and operating system. When discussing the comparison in Fig. 7.10, we noted that our implementation of SSA-CR did not appear to be giving the $O(1)$ behavior for increasing numbers of reactions we had anticipated. One of the potential reasons for this is the effect of memory management within a Java runtime environment. The *garbage collector*, responsible for recycling out-of-date objects, is an ever-present element in Java programs. Depending on how the Java runtime is configured, this can result in either periodic pausing of the model for garbage collection, or a continual underlying overhead in the runtime. While this might be seen as a disadvantage of Java, the corresponding advantage of memory security is hard to dismiss. Furthermore, just because memory management is not always automatically managed in other languages does not mean that the implementor is thereby freed from having to worry about it. The need to manage memory carefully will always be an issue with large simulations and is left entirely in the hands of the programmer in many programming languages. Pitfalls for the unwary are legion!

In addition, memory and processor contention from other processes running on the same machine may be hard to avoid without exclusive access to the available resources, and will apply regardless of the programming language used.

7.5 Reaction Equation Systems for Biological Models

Up to this point, the example reaction systems we have used to illustrate the SSA have been rather abstract and, apart from the Lotka–Volterra equations described in Sect. 7.1.4.3, not directly related to biological systems. Here we will redress the balance by illustrating how the reaction equation notation can be used to describe a small biological model whose dynamics might be the subject of investigation. The material in this section is based on that found in an analysis of reaction-parameter values in noisy biological systems [8].

In bacteria, uptake of nutrient is often mediated by *porins* that are inserted into the cell surface. These porins can be specific to the uptake of only one particular type of nutrient from the environment. This specificity allows the cell to be selective over which types of nutrient it takes up. Through a system of tight regulation, a specific porin is only expressed when the corresponding nutrient is available. A common regulatory mechanisms is that the nutrient displaces a repressor of the gene coding for the nutrient-specific porin, thus activating its expression. However, this creates a Catch-22 situation because nutrient needs to be in the cell before the nutrient uptake system can be activated by it. As a consequence, the cell must exhibit a degree of "leakiness" of porin-expression in order to be able to activate the uptake switch.

Here we will outline a model, in the form of a series of reaction equations, that might be used to represent the fundamental steps in porin-mediated nutrient uptake and conversion to growth. As always, our primary driving force from the beginning will be simplicity of the model.

The external nutrient is converted into energy via a number of steps to eventually fuel production of ATP and biomass. To keep the model simple, the complexity of real-world metabolism will be reduced to a 3 step process:

- Nutrient is taken up;
- The nutrient is converted into ATP;
- The ATP is converted into BIOMASS.

We assume that the first two conversions are catalysed reactions and require the enzymes $ENZ1$ and $ENZ2$ respectively. $ENZ1$ could be thought of as a porin that enables the cell to take up nutrient. $ENZ1$ and $ENZ2$ are continuously broken down and need to be produced by the cell. Protein expression involves a cost to organisms and needs to be represented in the model. For the sake of simplicity it is assumed that the synthesis of one enzyme molecule requires one molecule of ATP. If $e1$ is the gene for enzyme 1, then this can be represented formally by the following chemical equation (analogously for $ENZ2$):

$$e1 + ATP \xrightarrow{k_1} ENZ1 + e1$$

ATP is also converted into biomass:

$$ATP \xrightarrow{k_2} BIOMASS$$

The enzyme-mediated conversion of a substrate is modelled as a two step process. During the first step an enzyme binds a substrate (for example an external nutrient (SUBS1) or the internal nutrient (SUBS2)) and forms an intermediate compound. The intermediate compound then decays into an enzyme particle and a converted substrate (either SUBS2 or ATP):

$$SUBS1 + ENZ1 \xrightarrow{k_3} TMP1$$

$$TMP1 \xrightarrow{k_4} SUBS2 + ENZ1$$

$$SUBS2 + ENZ2 \xrightarrow{k_5} TMP2$$

$$TMP2 \xrightarrow{k_6} ATP + ENZ2$$

Under certain conditions this dynamics could be approximated by a Michaelis-Menten dynamics (Sect. 4.5.1), although here the unpacked model is used instead. Biomolecules are assumed to be diluted away with a rate proportional to the growth of biomass. In addition to this, even in the absence of growth, enzymes are "lost" with a constant rate. This ensures that even if there is zero activity, the cell loses enzymes continuously. If X is a place-holder for the biomolecules, then dilution is represented as the reaction:

$$X \xrightarrow{k_7}$$

While we won't take this model any further, it does serve to illustrate the way in which biological processes can be represented readily in the form of chemical reaction equations. Given a model description such as this, multiple runs of an SSA using varying values for the reaction constants $k_1 \cdots k_7$ could be used to explore the solution space of values that lead to significant production of *BIOMASS*, for instance. While the size of the parameter space would likely make exhaustive exploration impractical, techniques such as *genetic algorithms* do provide an effective and efficient approach. Unfortunately, these are beyond the scope of this book.

7.6   The Tau-Leap Method

Aside from the work of Gibson and Bruck and Slepoy et al, there have been a number of variations of the SSA proposed for the purpose of improving performance. The tau-leap method [9] is one of these. At its heart is the recognition that one of the fundamental costs of the SSA is that the exact effect of every single reaction taking place is explicitly modeled. With the tau-leap approach, a time frame, *tau*, is chosen and the multiple reactions that occur within it are all acted on together. This approach necessarily represents an approximation in comparison to the exact SSA, and the balance between simulation accuracy and speed up lies in the appropriate selection of the leap distance.

The number of times that a given reaction with propensity a_i will take place within the τ period is calculated from a Poisson random distribution with mean $\lambda = a_i\tau$. There is clearly a gotcha in this method in that there is potential for the number of molecules in a species to become negative if τ is set too high; on the other hand, setting it too low reduces the potential efficiency gains. The right value is likely to vary with the dynamic state of the system.

While we do not discuss this method in any further detail, it is worth being aware of it because it is often offered in modeling toolkits, such as Dizzy (which we look at in Sect. 7.8), and it can offer a significant reduction in execution time if appropriate for the model.

7.7   Delayed Stochastic Models

A characteristic of all of the modeling algorithms we have looked at so far is that reactions are assumed to take place instantaneously: reactants are consumed and products produced with no passage of time and no intervening reactions taking place. While this assumption may work satisfactorily for many of the situations we wish to model, there are also many scenarios where taking account of reaction duration may have a significant effect on the overall outcome. It is for this reason that some effort has gone into the idea of introducing delay formulations into the specifications of reaction systems.

A specific use for delays can be found in the technique of collapsing chains of similar reactions, such as the following, taken from Gibson and Bruck [5]:

$$A + B \rightarrow S_0$$
$$S_0 \rightarrow S_1$$
$$\ldots$$
$$S_{n-1} \rightarrow S_n$$
$$S_n \rightarrow C + D$$

Where the intermediate products, $S_0 \ldots S_n$, play no independent role in other reactions, there are clearly advantages to not having to represent the intermediate reactions separately—the number of intermediate reactions may be very large and this will have a negative impact on the runtime of the model. Instead, we can conceptually represent the chain above in the following way:

$$A + B \rightarrow P_{int}$$
$$P_{int} \rightarrow C + D$$

in which a single module of an artificial intermediate product, P_{int}, is produced. P_{int} is then converted, after a calculated delay, into the final products of the chain. The length of the delay and its determination depend, of course, on the characteristics of the chain being collapsed. In Gibson and Bruck's example of a chain of n identical first-order intermediate reactions the delay is based on a probability drawn from a *gamma distribution*. However, where there is implementational support for the principle of delays, they could be based on anything, including absolute time intervals.

Delayed production of products can be introduced relatively easily into the Gillespie direct method [10,11], as partially illustrated in Code 7.10. The idea is that a queue of delayed products is maintained in addition to the reaction set. The queue is ordered by the time at which the products should be added to the system. Before a selected reaction is acted on at time τ, the head of the queue is checked for a pending product. If the next pending product is due to be produced prior to time τ then the product is produced and the pending reaction canceled. Cancellation is necessary because reaction propensity values are likely to change as a result of the product being added to the system. In our implementation, the queue is populated through a change to the behavior of the `produce` method (see Code 7.1) for delayed products. Delayed product objects calculate the time of their production, based on the current simulation time, and place a `PendingProduct` object into the queue.

We have already identified that the Gillespie algorithms have the worst scaling complexity of all of the algorithms we have considered in this chapter; the addition of the queue adds further overheads to the basic iteration. Every delayed product that is added to the queue will result in the work done to identify a future reaction being wasted, for instance. It follows that, for large reaction systems, the unmodified SSA is unlikely to be the most efficient choice where large numbers of delays are involved.

```
public void run()
{
    int step = 0;
    ReactionSet system = new ReactionSet();
    initModel(system);

    // Obtain the initial cumulative propensity.
    double aZero = system.calculateAZero();
    time = 0;

    // Stop when the simulation time has passed or
    // no change possible.
    while(time < stopTime &&
            !(aZero == 0.0 && delayedProducts.size() == 0)) {
        double deltaT = (1.0 / aZero *
                          Math.log(1.0 / Math.random())));

        // See if a delayed product should be produced first.
        PendingProduct p = delayedProducts.peek();
        if(p != null && p.getProductionTime() <= time + deltaT) {
            delayedProducts.remove();
            p.produce();
            time = p.getProductionTime();
        }
        else {
            Reaction r =
                system.findReaction(Math.random() * aZero);
            time += deltaT;
            r.react();
        }
        aZero = system.calculateAZero();
    }
}
```

Code 7.10 Gillespie's direct method with support for delayed reaction products (compare Code 7.2)

The inefficiency in adding delayed reactions to the Gillespie approach is due, in part, to the iteration loop dealing with two distinct entities—reactions and products. Since both are time-dependent entities, it should be possible to manage both in a consistent way. Gibson and Bruck's next reaction approach is particularly suited to this task because of the way it orders reactions by their next reaction time. If delayed products can be included somehow in the time-ordered queue then they will naturally arise interleaved with selected reactions, without the need for a separate decision-making process. Our implementation is based around the idea of introducing an artificial `DelayedProductReaction` type which has no reactants and a single delayed product. Each delayed product occurrence in a normal reaction is made the responsibility of a delayed reaction object which maintains a time-ordered queue of when the product should next be released. When a normal reaction with a delayed product takes place, the product is not produced but its production time is passed to the corresponding delayed reaction object. The delayed reaction appears in the normal time-ordered queue with its next reaction time corresponding to the head of its delayed product queue. In this way, a delayed product naturally works its way to the head of the reaction queue as simulation time passes. Delayed reactions also fit naturally into the dependency graph so that reactions that are dependent on the delayed product are updated when production occurs, rather than when the originating ordinary reaction occurs.

7.8 Dizzy

So far in this chapter, we have introduced a succession of approaches to modeling reactions stochastically. We have illustrated how they might be implemented in the form of Java class definitions in order to provide insights into some of the practicalities of using the different algorithms. Although these sample programs could be used to model reaction sets in the way we have illustrated, we have not sought to provide a full programming environment to the reader, primarily because a number of modeling environments already exist. One of these is Dizzy [2,12].

Dizzy provides implementations of the Gillespie and Gibson–Bruck algorithms as well as the tau-leap approximation (Sect. 7.6), for instance. It has a GUI that incorporates features for displaying results, but its code can also be accessed in the form of libraries via an application programmer interface (API) for incorporation into other programs. Dizzy is freely distributed under the terms of the GNU Lesser General Public License (LGPL) [13]. In this section, we will briefly introduce the features of Dizzy.

When Dizzy is started it brings up an editor window (Fig. 7.11). A model can be created directly within this window, or prepared externally in a text file and loaded via the *File* menu option. The preferred suffix for such files is `.cmdl` to match *Chemical Model Definition Language* (CMDL), the language used to define the reaction sets. Figure 7.11 illustrates the editor window with the following CMDL statements for the isomerization reaction of (7.5):

```
// Reaction parameters
X = 1000;
Y = 0;

re1, X -> Y, 0.5;
```

Each statement is terminated by a semicolon. Symbols for the molecular species are defined in assignment statements containing the initial species size. The single *reaction definition* has been given a name, *re*1, and it includes the value of the reaction constant parameter, separated by a comma from the reactants and products.

Once a model has been defined completely it can be run by selecting the *Tools* → *Simulate* menu item.[4] Figure 7.12 shows the simulator controller dialog window where the simulator type (Gillespie, Gibson–Bruck, etc.) is selected, along with the species to be monitored in the output, and the stop time. We have selected a stop time of 20 and to view the population numbers of both species against time in the output. At the end of the run, a visualization of the output is displayed (Fig. 7.13).

---

[4]If the model has been defined directly in the editor window rather than loaded from file, it may be necessary to specify a parser; select *command-language*.

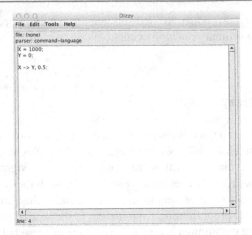

Fig. 7.11 The Dizzy editor window with CMDL commands for an isomerization reaction

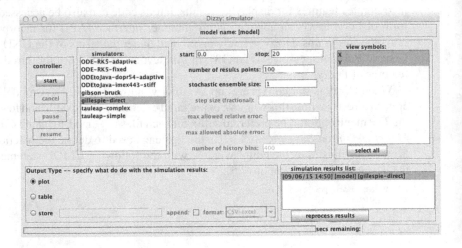

Fig. 7.12 The Dizzy simulator controller dialog window

Defining species' names and initial population levels is a particular example of the ability to associate names with values and expressions. Code 7.11 is a CMDL version of the reactions shown in (7.6) which illustrates this.

Fig. 7.13 Display of results within Dizzy

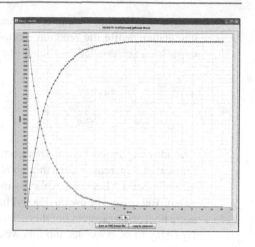

```
// The species and initial population levels.
A = 1000; B = 1000; C = 1000; D = 1000; E = 1000;

// Reaction parameters.
ab_rate = 0.12;
ad_rate = 2 * ab_rate / 3;
ce_rate = 0.5;

// Reactions.
ab, A + $B -> C, ab_rate;
ad, A + D -> E, ad_rate;
ce, C + E -> A + A + D, ce_rate;
```

Code 7.11 CMDL implementation of (7.6)

In the ab reaction, species *B* is prefixed with a dollar symbol to show that it is a boundary species. Notice that we have used comments to make the model more readable, and named the reaction constants. The definition of the constant ad_rate involves an expression based on the value of ab_rate. Expressions written like this are always evaluated immediately but there is also a syntax to delay evaluation until runtime. This is particularly useful for reaction rates that are not constant but species-dependent. An expression written between square brackets is a *deferred evaluation expression*. Each time it is used the reaction rate is required the expression It will be re-evaluated each time its value is required. A particular use for deferred evaluation expressions is in the calculation of *custom reaction rates* rather than calculation using the *built-in method*. With the built-in method, a reaction's rate is determined from a numeric reaction parameter and the number of distinct combinations of the reactant molecules. In contrast, a custom reaction rate specifies a rate expression that is not necessarily directly dependent upon the number of combinations—it is simply a general expression that may or may not include species numbers within it.

In the following example, the reaction ab uses the built-in method in the rate calculation while the reaction ac uses the custom rate method in the form of an expression surrounded by square brackets.

```
A = 100.0; B = 0.0; C = 0.0;
// built-in method rate.
ab, A -> B, 0.1;
// custom reaction rate.
ac, A -> C, [A * 0.3 * B];
```

The idea expressed here is that conversion of A to C becomes more likely as the amount of B increases, even though B is not a direct reactant. Note that the population size of species A has been explicitly included in the rate expression, while it is implicit in ab. You should be aware that if A had not been included in the expression and there were no A in the system but there was some B, then reaction ac would be considered by Dizzy to be possible, and this would result in the value of A becoming negative.

The Dizzy environment has some support for delayed reactions using the following syntax:

```
X -> Y, probability, delay : d
```

However, at the time of writing, delayed reactions are restricted to having only one reactant and one product. Better support for delayed reactions is available in the SGNS2 environment, which we describe in Sect. 7.9.

7.9   The Stochastic Genetic Networks Simulator

The Stochastic Genetic Networks Simulator (SGNS2) [3, 14] is a simulator for modeling reactions stochastically. It offers a "no-frills" user interface (i.e., it has no GUI!) but, in some areas, provides support for more sophisticated models than Dizzy, such as reaction dynamics. Implementation is based around Gibson and Bruck's next reaction method (see Sect. 7.2). Here we will provide a brief summary of its basic usage and the interested reader is referred to the full user manual that is available with the software from the SGSN2 web site [14].

```
// Basic isomerization of X to Y with reaction constant c.

// Simulation control.
time 0;
stop_time 15;

// Output control.
readout_interval 0.1;
output_file isomerization.txt;

// The reaction constant.
parameter {
    c = 0.5;
}

population {
    X = 1000;
    Y = 0;
}

reaction {
    X --[c]--> Y;
}
```

Code 7.12 The isomerization reaction of (7.5) in SGNS2

Reaction scripts are created using an ordinary text editor and passed to the simulator, sgns2, at the command line using a command such as:

```
sgns2 script.g
```

where script.g is the name of the file containing a script. Code 7.12 shows a script for the basic isomerization reaction of (7.5) in the syntax of SGNS2.

The syntax of SGNS2 is fairly fussy, reflecting the fact that quite a lot of programmatic control is available to the modeler. In particular, Lua code [15] may be included in a script for computational purposes.

Notice that there is a clear identification of parameterization values, molecular species and reactions in named blocks within the script. Rather confusingly, setting values for simulator parameters, such as time and output_file, does not require an assignment symbol, =, whereas other setting situations do. It is important to set an explicit value for stop_time as it is zero by default.

The reactant or product list may be left empty in a reaction and multiplicity may be indicated by using a numerical prefix immediately in front of a species name (there must be no intervening space), for instance:

```
X --[1]--> 2Y;
```

Delayed reactions are supported, per-product, by adding the delay time in parentheses after the product name, for instance:

```
Pro + RNAp --[kt]--> Pro(d1) + RBS(d1) + RNAp(d2) + R(d2);
```

where d1 and d2 have been defined in the parameter block and could be more complex than constant values. The syntax for delays also includes built-in support for random distributions, such as Gaussian and Exponential; for instance:

```
Rib + RBS --[ktr]--> RBS(d3) + Rib(d4) + P(gaussian:mean,sd);
```

Here the names `mean` and `sd` are parameter values for the mean and standard distribution of the normal distribution.

The queue in which delayed products are held pending their production is accessible from the script as a way to seed additional molecular species into the system part way through the simulation. For instance:

```
queue 100X(25);
```

would release an additional 100 molecules of species X into the system at time 25.

7.10 Summary

In this chapter we have looked in detail at Gillespie's stochastic simulation algorithm for modeling systems of coupled chemical reactions. Such an approach is appropriate when the stochasticity of the system plays an important role in the outcomes. This is often the case when the interacting entities are discrete individuals and their numbers are relatively low. Following on from Gillespie, Gibson and Bruck developed an approach with better scalability for use with large numbers of reactions. This work was then taken further by Slepoy, Thompson and Plimpton who developed an even more efficient approach suitable for very large numbers of reactions. We have also touched on some of additional developments, such as tau-leap and the use of delays in reactions, that are increasingly being discussed in the research literature.

As well as providing illustrative implementations of all of the main approaches, for the sake of helping the reader to appreciate the differences between them, we have also briefly introduced two widely-available simulation environments, Dizzy and SGNS2, that can be easily used 'off-the-shelf' to build models. It is our expectation that typical readers will prefer to use one of these rather than build their own.

References

1. Gillespie, D.: Exact stochastic simulation of coupled chemical reactions. J. Phys. Chem. **81**(25), 2340–2361 (1977)
2. Ramsey, S., Orrell, D., Bolouri, H.: Dizzy: stochastic simulation of large-scale genetic regulatory networks. J. Bioinform. Comput. Biol. **3**, 415–436 (2005)
3. Lloyd-Price, J., Gupta, A., Ribeiro, A.S.: SGNS2: a compartmentalized stochastic chemical kinetics simulator for dynamic cell populations. Bioinformatics **28**(22), 3004–3005 (2012). doi:10.1093/bioinformatics/bts556
4. Gillespie, D.: A general method for numerically simulating the stochastic time evolution of coupled chemical reactions. J. Comput. Phys. **22**, 403 (1976)
5. Gibson, M., Bruck, J.: Efficient exact stochastic simulation of chemical systems with many species and many channels. J. Phys. Chem. **104**, 1876–1889 (2000)
6. Slepoy, A., Thompson, A., Plimpton, S.: A constant-time kinetic Monte Carlo algorithm for simulation of large biochemical reaction networks. J. Chem. Phys. **128**, 205101 (2008)

7. Devroye, L.: Non-uniform Random Variate Generation. Springer, New York (1986)
8. Chu, D.: Evolving parameters for a noisy bio-system. In: IEEE Symposiom Series on Computational Intelligence (2013)
9. Gillespie, D.: Approximate accelerated stochastic simulation of chemically reacting systems. J. Chem. Phys. **115**(4), 1716–1733 (2001)
10. Bratsun, D., Volfson, D., Tsimring, L.S., Hasty, J.: Delay-induced stochastic oscillations in gene regulation. PNAS **102**(41), 14593–14598 (2005)
11. Roussel, M.R., Zhu, R.: Validation of an algorithm for delay stochastic simulation of transcription and translation in prokaryotic gene expression. Phys. Biol. **3**, 274–284 (2006)
12. Ramsey, S., Orrell, D., Bolouri, H.: Dizzy home page. http://magnet.systemsbiology.net/software/Dizzy/ (2010)
13. Foundation, F.S.: GNU general public license. http://www.gnu.org/licenses/gpl.html (2007)
14. Lloyd-Price, J., Gupta, A., Ribeiro, A.S.: SGNS2 stochastic simulator. http://www.cs.tut.fi/%7Elloydpri/sgns2/index.html (2015)
15. Ierusalimschy, R., de Figueiredo, L.H., Celes, W.: Lua—an extensible extension language. Softw.: Pract. Exp. **26**(6), 635–652 (1996)

Biochemical Models Beyond the Perfect Mixing Assumption

Biological cells are spatially extended systems. This means that any biochemical reaction that happens in the cell occurs at a particular place at a particular time. However, particles move around in the cell volume via diffusion and in many instances this acts as an equalizer and the cell offers a good approximation to a perfectly-mixed system. Nevertheless, there are situations where the assumption of perfect mixing ceases to be a good one and systems must be treated as spatially extended, where the variation in concentration of chemicals across the volume must be considered. One class of problems where this becomes important are models of reaction bio-polymers such as mRNA or DNA. Each typically has a large number of binding sites and, for many scenarios, the spatial arrangement of those binding sites matters.

Take, as an example, *translation* [1]. The mRNA molecules are long polymers containing the genetic message as a string of codons. The molecules are scanned by ribosomes. A ribosome is a complex molecular machine that reads each codon in sequence. Each reading event takes a certain amount of time to be processed. The average time to read a codon can be calculated but the period of any individual decoding event fluctuates strongly. In this sense, a ribosome performs a directed 1-dimensional random walk along the mRNA. At any particular point in time, an mRNA may be being read by multiple ribosomes.

Modeling this as a perfectly-mixed system is not meaningful. Since the individual decoding times fluctuate very much, individual ribosomes move at different speeds. This means that a slow ribosome may block a faster one coming from behind, not unlike slow vehicles on a narrow road blocking faster cars. In order to understand what is happening in the model we need to know: where the individual ribosomes are located relative to one another; how they are moving over the mRNA; and how and when they are interacting. In such a case, one needs to represent the space of the system. A perfectly-mixed model is not useful here.

Another example where spatial organization cannot be ignored is cellular sensing systems. Many single-celled organisms have the ability to detect gradients of nutrients or toxins in the environment. When detected, such gradients allow the cell to seek the area of higher nutrient concentration (by moving up the gradient and closer

© Springer-Verlag London 2015
D.J. Barnes and D. Chu, *Guide to Simulation and Modeling for Biosciences*,
Simulation Foundations, Methods and Applications,
DOI 10.1007/978-1-4471-6762-4_8

to the source) or to avoid the toxin (by moving down the gradient) in a process called chemotaxis. Clearly, this system of detection of gradients relies on the detection of differences in the concentration of certain molecules across the cell. Therefore, in order to understand and analyse this problem it is necessary to model the gradient in the environment, for example in three dimensional space. The detection of concentration differences across the cell is often achieved by a local clustering of receptors. The idea is that the probability of clustering of a receptor is larger where the concentration of a nutrient is higher. This is one way for the cell to polarise itself and detect differences in external conditions. Again, to model this the position of receptors on the surface must be represented.

The traditional stochastic simulation algorithms (SSAs) (Chap. 7) cannot be used to model such systems in an unmodified way because they assume that all reactions take place in a perfectly-mixed reaction vessel. There is no notion of space for these algorithms. Spatial models need different approaches. Either extensions to the SSAs, or completely different modeling techniques altogether.

An exception arises when the number of particles is very large and space can be described as a continuous system. In such cases the system can be modeled using differential equations. However, unlike the ODE models of the non-spatial systems, reaction diffusion systems are systems of partial differential equations. Solving them is much more difficult than ODEs and it would go well beyond the scope of this book to describe methods to do this. Moreover, even formulating such models is a challenge for many realistically sized biological problems. None of the required conditions for this method applies to the ribosome random walk described above.

The simplest way to model spatial systems is when space and time are discrete. In this case it is possible to use agent-based modeling approaches (Chap. 3). Space would then be represented as a square lattice, for 2 dimensions, or as a 3-D grid. Diffusion of molecules is then reduced to discrete random walks. Moreover, in discrete grids detection of molecules in neighborhoods is computationally more efficient than on a 3-D grid. Consequently, agent-based modeling has been a popular choice for spatial modeling in biology. It is particularly useful when the underlying biological system lends itself to discrete interpretations, as would be the case in a model of the ribosome random walk during translation. An mRNA molecule consists of a discrete number of binding sites. A model of translation does not necessarily need to represent the continuous embedding space of the mRNA and it may well be sufficient to model the translation process as a random walk in discrete space.

However, there are also downsides in assuming discrete space and time. Unless they are very well resolved spatially, discrete systems can be difficult to interpret physically. They can support overall conclusions, but it is often difficult to map discrete simulations to a specific system. Agent-based models in discrete space and time will therefore often not be precise enough to support strong conclusions.

On the other hand, discrete-space models can be made arbitrarily close to a continuous space simply by making the individual cells smaller. By the same token, discrete time can be made more like continuous time when the length of the time-step is decreased. The problem is that, as time-step size and cell size become smaller, the advantages of discreteness gradually disappear.

8.1   Conceptual Differences Between Perfectly Mixed and Spatial Model Systems

There are a number of conceptual problems that appear when space is introduced. In well-mixed systems the location of molecules does not matter. This means that the modeler can simply take a measured value for a reaction rate constant, specify it in the model and simulate it to reproduce the experiment. For example, assume a reaction of the form:

$$X + Y \xrightarrow{k} Z$$

This can be interpreted as a simple reaction that proceeds with mass action kinetics. Concretely, this means that the propensity of the reaction occurring is given by $k[X][Y]$. The product of the particle numbers $[X]$ and $[Y]$ reflects how likely it is that a molecule of X and a molecule of Y meet, and the rate constant k indicates the likelihood that any such meeting does result in a reaction. What is elegant and convenient about this formulation is that the rate constant can be obtained from experiments where the reaction rate can be measured.

In spatial simulations things are more complicated and there is more than a single parameter to consider. In order for X and Y to react, they first must meet; no meeting, no reaction. Yet, it is not clear what precisely it means for X and Y to meet. A first idea is that they meet when they actually collide. The idea is of course that X and Y have a certain size and may bump into one another. The bigger they are the more likely they are to hit one another.

This sounds like a good approach at first but it is not. Firstly, we normally simply do not know how big molecules are. This information might be available but it is not easy to find unless you know where to look for it. More importantly, however, the physical sizes may not be what actually determines the reaction rate—it is often irrelevant to the rate value. The best way to model molecules is to simply record their position in continuous space without concern for their size, which is also what nearly all available spatial simulation programs do. However, a consequence of this simplification is that under these circumstances the probability of collision between two particles vanishes.

Clearly, it is not possible to do away with the concept of collision altogether. After all, the whole point of a spatial simulation is that particles interact when they meet. So, in order to retain the meaning of space, an abstract concept of particle size is re-introduced: a reaction-specific so-called *binding radius*. Then for the reaction given above, two molecules of X and Y will react to produce one molecule of Z whenever they are within the binding radius for that reaction. It is important to note that a binding radius is specific to a reaction rather than the molecular types involved in the reaction. In practice, it will not be chosen so as to reflect the actual size of a molecules, but in order to match the reaction rate that has been measured.

In summary: One method of organising reactions is to define a reaction-specific binding radius. The reaction takes place when two reactants are closer than the binding radius.

There is another complication. Having established that the molecules of type X and Y only react when they are within a binding radius, we can now wonder how often this actually will be the case. This clearly depends on the size of the binding radius, on the number of molecules in the system and the system size, but it also depends on the speed with which the molecules move through the system.

In the simplest case, molecules perform a random walk through the aqueous solution of the cell, driven by random forces–they are diffusing. The speed of this random movement is given by a constant D, the diffusion constant. Normally, the larger the molecule, the smaller the diffusion constant and hence the speed. To understand the connection between the diffusion constant and the reaction rate, consider the extreme case of immobile particles, that is the molecules X and Y have a diffusion constant $D = 0$; if we further assume that the binding radius is small then a moment's thought shows that there will be no reactions between X and Y during the course of the a simulation. As the diffusion constant increases (while keeping the binding radius fixed) the reaction rate will increase as well.

The upshot of this is that the concept of a reaction rate constant that formed the backbone of the simulations of well-mixed systems is no longer directly applicable for spatial systems. Nonetheless, many of the simulation software packages for spatial systems still allow the user to define a reaction rate. This is convenient for the user— because it allows her to input experimentally known rates directly–but such experimentally observed rates should be reproduced by choosing a suitable combination of diffusion constants and reaction radii.

In the well-mixed models and simulations it is often convenient to use special kinetic laws, such as the Hill kinetics (Sect. 4.5.1). These can coalesce a series of reaction steps into one reaction step. This is very convenient as it maps well onto experimentally measured values and avoids multiplication of unknown parameters. A common example is that of enzyme-catalyzed reactions. Conceptually, a catalytic reaction is a series of steps involving binding of enzyme and substrate into an intermediate compound and the subsequent decay into the unaltered enzyme and the end product. The enzyme is temporarily unavailable as the reaction proceeds and thus limits the speed of the catalytic conversion. In practice it is often possible to write this multi-step process as a single reaction from substrate and catalyst to product, proceeding with a rate determined by the Hill equation. However, in the realm of spatial models such nice short-hands no longer work and the equations must be expanded into their individual steps consisting of elementary reactions determined by simple rate constants alone.

8.2   Spatial Modeling with Smoldyn

While there is a vast choice of high quality modeling software for perfectly-mixed systems, spatial modeling software is much rarer. Among the few choices available, Smoldyn [2,3] is perhaps the most respected spatial simulator in the field. Smoldyn models are defined in a simple text file. Hence it is not necessary to install and learn any additional graphical interfaces. At the same time, Smoldyn offers a usable visualization which is basic but sufficient and does not get in the way. Most importantly the visualization can be turned off to accelerate simulations. Indeed, it turns out that this is not even necessary. Hiding the visualization behind other windows is sufficient to benefit from the speed increases. Smoldyn comes with many example models that illustrate its use. There is also a good documentation available. However, the available tutorials are a bit difficult to access and sometimes lack the clarity that a novice would enjoy. Another disadvantage of Smoldyn is the way spatial topologies are created. These need to be specified by hand using geometrical primitives such as spheres or boxes. In principle, the level of complexity of what can be created is arbitrary, but in practice it can become very cumbersome to specify non-trivial geometries.

Nevertheless, Smoldyn is a reliable and powerful tool for those who want to investigate the spatial properties of a biological system. In the remainder of the chapter, therefore, we will use it to illustrate some of the concepts of spatial modeling we have outlined so far. What follows is not a manual or exhaustive tutorial for Smoldyn—which is work better undertaken by its creators. Rather, we present two case-studies via Smoldyn in order to illustrate basic usage of it for the novice, as well as to highlight some of its pitfalls that are not apparent from the documentation alone.

8.3   Basic Concepts of Spatial Modeling

We have already hinted that introducing space in biochemical models is not just about introducing the dimensions as additional variables, but requires a re-thinking of the simulation. Much of this complication is hidden from the user of Smoldyn (and similar packages), but some of it directly affects model design.

A fundamental concept of spatial models is that of a *compartment*. This is a bounded space in which the molecules diffuse. Let us call this basic compartment simply the *world*. The shape of this world is not very important, but the modeler must take some care choosing its size carefully. If the size of the world is too small, then this can lead to artefacts and edge effects during the simulation. At the same time, it is problematic choosing the world too large. Increasing the size of the world while keeping the total numbers of particles constant implies a reduction in particle concentration; or equivalently the larger the world the more particles are needed in order to maintain a constant reaction rate, which entails higher simulation costs. Edge effects can be alleviated to some extent by specifying so-called *periodic boundary conditions*, which Smoldyn supports. This means that a particle exiting the world at one position will re-enter at an opposite position.

In order to define the world in Smoldyn use the `boundaries` command:

```
dim 3
boundaries  0  -50 50
boundaries  1  -50 50
boundaries  2  -50 50
```

The first line of the code defines the dimensionality of the world to be simulated. In this case, we choose the maximum allowed value of 3; this means that the simulation happens in 3D space. For each of the three axes in the world, `boundaries` defines the extension of the world in this direction. In this particular case, the 0th axis goes from -50 to 50, and the same for the 1st and 2nd. We can think of axes 0, 1 and 2 as the x, y and z axes. Smoldyn does not have any intrinsic units. So, whether the dimensions are expressed in terms of metres, millimetres or yards and miles is for the user to decide by ensuring that the numbers and values used in the simulation are all compatible with one another.

Apart from the world itself it is usually necessary to define additional subcompartments that model specific biological spaces. For example, if the world models the environment of a cell, then the cell itself may be modeled as a sub-compartment. Or perhaps the world is the cell and one may wish to model organelles. Whatever it is, sub-compartments will feature in nearly every model.

In Smoldyn defining compartments is conceptually a two-step process. In step one, the user needs to define surfaces that will act as the boundary of the compartment. The second step defines the compartment itself in relation to previously-defined surfaces that act as enclosing boundaries. For example, if we have previously defined a surface of a sphere then we could specify that everything that is within the sphere is our compartment. Defining the compartment can be done in two ways. The simple way is for the user to define so-called *inner points*—points that lie within the compartment. The compartment is then defined by all points that lie in a straight line between the inner defining point and the surface of the the compartment, without crossing a boundary surface of this compartment. Note however, that the surfaces of other compartments are considered invisible.

To illustrate this, we make a square box of dimension $100 \times 100 \times 100$ along the boundaries of the world and centered on the point (0, 0, 0). (Defining such a boundary is not really necessary, because we have already defined our world. However, we do it anyway here, because it illustrates well how to make square compartments.) The Smoldyn directive for our first compartment is:

```
start_surface walls
action both all reflect
color both 1 0 0
polygon both edge
panel rect +0 -50 -50 -50 100 100
panel rect +1 -50 -50 -50 100 100
panel rect +2 -50 -50 -50 100 100
panel rect -1 -50 50 -50 100 100
panel rect -0 50 -50 -50 100 100
panel rect -2 -50 -50 50 100 100
end_surface
```

The definition of any surface must start with start_surface and end with end_-surface. At the start of the definition a name must be supplied for the particular surface. In this case, we chose it to be walls. The action command specifies that all particles reflect from both sides of the surface. This makes sense for the outer wall, but we will discuss different choices in more detail below. The color command specifies the color for the wall. Again, the keyword both indicates that both sides of the surface have the same color. The color is chosen by adjusting the three numerical parameters of the command specifying the mix of red (1), green (0) and blue (0) (RGB) components. The particular color used here is red; white would be achieved with 1 1 1. Specifying numbers in RGB code gives great flexibility, but Smoldyn would also accept common color names instead of their code (i.e., "black", "red", etc.). The line containing the keyword polygon specifies how to draw the walls. As it is now, it stipulates that only the edges of the box are drawn. This is a good choice, because it makes it possible to see inside the box.

So far the commands were only relevant for the visualization of the world. The actual definition of the box is given by the last six commands of the section. Each of these specifies one side of the cube. The keyword rect specifies the position and size of one of the walls. To specify a rectangle, three pieces of information must be given. These are:

- The number of the axis to which the rectangle is perpendicular;
- The position of the corner of the rectangle;
- The size of the rectangle expressed as the length of its sides.

So, the first rectangle is perpendicular to axis 0, a corner at position (-50, -50, -50) and sides 100 units long.

The reader will also have noticed that the axis numbers are preceded by plus or minus signs. Smoldyn requires this to indicate the orientation of the surface panel. This is used to distinguish (and specify) the inside and outside of the box. For the world enclosure this is not crucial, but it will become important later on for additional compartments. The minus sign indicates that the side facing the negative axis is the inside of the panel. When building panels it is very important to be consistent with the definition of inside and outside in all panels. Otherwise, it will become very difficult to consistently define the interaction of molecules near and with surfaces.

So far we have only defined the walls of a compartment, but not the compartment itself. We will defer this step and define another surface at the inside of our world. We will call it cell:

```
start_surface cell
action both all reflect
color both 0.6 0 0.6
polygon front face
polygon back edge
panel sph 0 0 0 30 20 20
end_surface
```

The syntax for the specification of the cell surface is analogous to the above, only simpler. Instead of defining a box we define a sphere, using the keyword sph. We assume that the sphere is centered at the origin, i.e., the point (0, 0, 0) with a radius of 30. These are supplied as arguments after the sph keyword. Following this we specify the drawing quality which will only affect the graphical output. The values 20 20 are suitable.

As with all surfaces, it is also necessary to specify the inside and the outside of the sphere. This is done by giving a sign to the radius. The positive radius we have chosen here means that the front of the surface is at the outside and the back of the radius is on the inside. Had we specified a negative radius, then this would have been the other way round.

Next we want to define the cell compartment as the volume enclosed by cell. To do this, we must choose an inner defining point. The good thing about spheres is that all points of the surface are "visible" from all inner points. Therefore we could choose any point as the compartment inner defining point. For simplicity though, we may as well choose the center point given by (0, 0, 0):

```
start_compartment center
surface cell
point 0 0 0
end_compartment
```

In close analogy to the definition of the surface, compartment definitions are also enclosed by start and end directives. In this particular case the definition of a compartment, let us call it center, is straightforward. All that needs to be done is to define the surface that is the boundary of the compartment—in this case the cell—and an inner defining point—in this case the origin.

What if we now wanted to define the outside of the sphere as a compartment? This is a bit more complicated if we used compartment inner defining points. We could choose one of the corners of the box as the compartment defining point. This is insufficient, however, because the space behind the sphere is not visible from the corner point (see Fig. 8.1 for illustration). Depending on the distance between the inner defining point and the sphere this may be a substantial part of the compartment to be defined.

In Smoldyn, there are two solutions for this. Either one introduces additional points, so that together these "see" the entire space outside of the sphere. For this to work one needs at least one more point at the opposite end, but perhaps many more. Alternatively, much more elegantly one can use logic composition of spaces to define compartments. To do this, it is first necessary to define the inner surface of the cell. We have already done that above when we defined center. Next we simply define the outer surface, let us call it outer as the space that is not equal to the center surface or equalnot center:

```
start_compartment outside
compartment equalnot center
end_compartment
```

With these lines of code we have now set up the topology of the Smoldyn world. The world, plus a spherical cell in the middle. The next step is to populate this world with molecules.

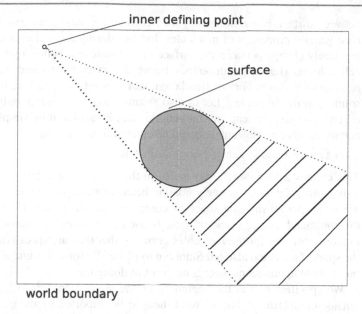

Fig. 8.1 Defining compartments. Assume we wish to define the area outside the *circle* as a compartment, using the *circle* as a boundary. We define the inner point indicated at the *top left*. The *shaded area* behind the *circle* is invisible, so we would need to define a second inner point to specify the whole outer space as our compartment

Molecule definitions are somewhat more complicated in spatial simulations than in perfectly-mixed simulations. In spatial simulations molecules may be freely floating in solution or they may be bound to a surface. When they are surface bound it is necessary to specify whether or not they are bound to the outside of the surface, the inside or whether they straddle the surface and react with particles on both sides. Moreover, for both bound and free molecules it is necessary to define how fast they diffuse through space.

For our first example model we will define three types of molecule with the names freeFL, surfM1 and porin. First, we need to declare them:

```
species freeFL surfM1 porin
```

Following the declaration of the molecules, Smoldyn then allows the user to define the basic characteristics. Here we define the diffusion coefficient difc, the color with which the molecule is drawn and its size on the graphical display:

```
difc freeFL 3
color freeFL red
display_size freeFL 3
```

Note that the directives color and display_size do not influence the simulation itself and are only relevant for the graphical display.

Having defined the diffusion constants and display properties of all the molecules, one can then start to specify reactions. Up to this point there has been no difference

between surface-bound molecules and free-floating ones because we have only specified generic properties of molecules. Indeed, during a simulation molecular types
can freely change between the surface bound state and being freely floating in the
cell volume. Hence, the difference between surface bound and free molecules is
just one of location. For our simulation, let us specify freeFL as freely floating in
solution at the beginning. Let us also assume that it is initially only present in the
outside compartment. In Smoldyn this can be achieved by simply stating that a
certain number of molecules should be created in outside:

```
compartment_mol 200 freeFL outside
```

This creates 200 freeFL molecules in the compartment outside. This implicitly defines freeFL as freely floating because we stipulated that these molecules
are to be placed into the space of compartment outside. Conveniently, when
the command compartment_mol is used then Smoldyn chooses automatically
random positions for the molecules ensuring that they are spread out nicely across
the space. One could also tell Smoldyn to place all molecules into a particular point
but in most applications there is no point in doing this.

   We specified that at the beginning of the simulation freeFL are only in the
outside compartment, but we want them to be imported by the porin molecules
into the center compartment that we have defined above. This means that porin
molecules need to be surface bound. In Smoldyn this can be achieved by placing
them onto a surface, although there is one additional complication. Surface molecules
need to have a state indicating what sort of surface molecule it is. There are four
allowed states in Smoldyn—front, back, up and down—indicating whether the
molecule is bound to the front or back of the surface or whether it is in an up or
down orientation. State does not influence the behavior at all. This is good news
because it means that they can be used/abused by the modeler for her own purposes,
which adds flexibility to the modeling. Molecules in different states should almost be
treated as a different molecular species. Reactions are specific to the molecule state
and in reactions the states must always be specified even though the choice of the
state itself does not influence the reaction dynamics. Formally, states are indicated
in parenthesis after the name of the molecule.

   Let us now place 100 molecules of porin onto the surface cell assigning to
them the state down:

```
surface_mol 100 porin(down) cell all all
```

We will leave it as an exercise for the reader to define the surface molecule surfM1
and create 100 of surfM1 in state down embedded into the surface cell. We will
need those below.

   With all the infrastructure in place, we can now start to think about the reactions.
For the moment, let us only implement importation of particles from the outside
compartment to the center compartment via porin molecules. There are several
ways to do this. The simplest and fastest way—which we choose here—is to model
import as a reaction of an outside freeFL with a surface bound porin resulting in
an unmodified porin and a freeFL inside the cell. This is somewhat unrealistic
biologically because it means that the transport takes no time at all, or alternatively

that porins have unlimited transport capacity. This could be easily remedied by introducing an intermediate compound of `freeFL` and `porin`, but this is not our concern here. Instead, we will now focus on a different question: How to encode in Smoldyn the idea of `freeFL` transport from the `outside` compartment to the `center` compartment?

When compartments are modeled in non-spatial simulation systems then it is necessary to represent the same molecular type inside and outside of a compartment as a separate species. This can be very cumbersome. In spatial systems this is no longer necessary. Since every molecule has a position, it is immediately clear in which compartment a molecule is. Yet, a question remains. When a surface molecule unbinds from the surface, how to specify on which side of the surface it is to be released? One way to do this is to specify the coordinates of release place. Clearly, this is not practical. Instead, Smoldyn has a very general and elegant solution for this. It uses states to define the release relative to surface orientation. We illustrate the idea using our import reactions:

```
reaction ttt1   freeFL(fsoln) + porin(down) ->
                porin(down) + freeFL(bsoln) 10
```

The keyword `reaction` specifies that what follows is a reaction. The first argument of the command is the name of the reaction, in this case we have chosen it to be `ttt1`, but it could be anything else. The second part is the reaction itself followed by the macroscopic rate constant.

A closer examination of the reaction shows that both `porin` and `freeFL` are assigned states. As a surface-bound molecule `porin` clearly must be assigned a state, so no surprise there. In the above example, however, the free-floating molecules `freeFL` also have states, even though they are at no time bound to a surface. Unlike the states of surface-bound molecules, states of free-floating molecules do have a fixed meaning that relates to the geometry of the surface and there are only two allowed ones. The state `(bsoln)` indicates that the free molecule should be at the back of the surface and correspondingly `(fsoln)` indicates that it should be at its front. The meaning of `back` and `front` here relates to the orientation of the surfaces which were discussed above. In the case of the sphere `front` and `back` are defined by the sign of the radius in the surface specification. Since reactions depend on the orientation, it is very important for the modeler to keep track of what is front and what is back in surfaces. Above, we specified that `cell` has a positive radius, which means that the front is outside. Therefore, reaction `ttt1` is the import from the outside to the inside of the cell. We can now add the reverse export, whereby `porin` reacts with `freeFL` inside the cell and exports it. Pay attention to how states are used:

```
reaction ttt2   freeFL(bsoln) + porin(down) ->
       porin(down) + freeFL(fsoln) 10
```

It is important to note that the states `bsoln` and `fsoln` only make sense for freely floating molecules and that it is only ever necessary to specify the state when they react with surfaces or surface-bound molecules. Otherwise, it is not necessary to specify their states.

A final comment about the reaction rate. We have discussed above that the concept of reaction rate that was developed in the context of spatial simulations does not translate well to spatial simulations. Yet in the above reactions we specified a reaction rate constant of 10. In Smoldyn, the value of 10 corresponds to the experimentally observed value for the rate constant and the simulator automatically calculates a binding radius to achieve this rate depending on the diffusion constant of the molecule. The number can be considered a *macroscopic rate constant*. In most cases it will be sufficient to specify it, but there are circumstances where it is necessary to use more advanced features of Smoldyn. One such case will be discussed below.

What remains to do before we can start and run the simulation is to specify the output. To do this, first specify the name of an output file using the `output_files` command:

```
output_files   outfile.txt
cmd N 100 molcountincmpt center outfile.txt
end_file
```

Here we choose the name of the output file to be `outfile.txt`, but any other choice will do as well. The actual output is generated by using the `cmd` directive. This executes a Smoldyn command either once or every so and so many steps. In this case, we specify `cmd N 100 molcountincmpt`, which means that the command `molcountincmpt` should be executed every 100 time steps. This command counts the number of all particles in a given compartment (in this case `center`) and outputs the result into the file specified as a last argument (in this case `outfile.txt`). Rather inconveniently, Smoldyn does not label the names of the molecules in its output, but it does follow the order of declaration with the `species` command, as described above. The output command is ended with `end_file`.

Over the following sections we will further develop this model in the context of concrete applications. Specifically, we will take a closer look at two simple models of *change detection* in biological systems.

8.4   Case Study: Change Detector I

The idea of change detection is closely related to the ability of biological systems to adapt to situations and remain sensitive over a wide range of conditions. For example, human vision is such a system. When fully adapted to dark light humans can detect a small number of photons (individual light particles), but equally they can remain sensitive to differences in light intensities under very bright conditions. *Adaptation* is the process that allows the system to adjust to a variety of conditions. Impressive levels of adaption are not limited to higher organisms, but can be seen even in bacteria. One famous example of adaption is bacterial *chemotaxis* [4]. Bacteria have sensory machinery that allows them to detect a gradient and to swim up it. What makes this sensing mechanism so interesting is that it remains sensitive to small differences in concentration over a large range of absolute concentrations. Consequently, there has been considerable interest from researchers in this phenomenon.

As a case study for spatial modeling we will use a minimal model of adaptation whereby the biochemical system can sense a change in environmental conditions over a wide dynamical range. Such a change would typically be the concentration of some chemical in the environment. The important feature of a change detector is that it resets itself after detecting a change. This means that the detector indicates that a change has happened. This could be changing its own state into an active state, but it should then revert into its ground or base state once the change is finished. There are two quality features for detectors. The first one is its ability to indicate changes over a large range of values (i.e., concentrations) and its ability to reset itself accurately to a ground state once the change is over.

Here, we will consider a minimal implementation of a change detector proposed by De Palo and Endres [5]. They reported two very basic schemes where a cell detects changes in the external concentration of some molecule, translates this into an internal change and relates this via some signalling network within the cell. We will not model this internal signalling pathway but just concentrate on how the external change can be translated into an internal change. The discussion will also remain abstract in that we will not be concerned very much with the meaning of this system and its detailed properties. The main point of the presentation is to illustrate the use of spatial modeling (and Smoldyn).

The idea of the minimal detector is very simple and elegant. It involves a trans-membrane molecule that can be in either an active or inactive state. The level of activity—to be more specific, the number of molecules in the active state across the cell—indicates the amount of change of the external signal. In the first scheme, the transmembrane molecule has two binding sites, one external, and one internal, both binding to the same molecule. The transmembrane can therefore be in four configurations altogether: (i) bound inside, (ii) bound outside, (iii) bound both inside and outside, and (iv) not bound at all. Additionally, it can be in an active state and an inactive state. For simplicity, we will ignore the active and inactive states and assume that the transmembrane molecule is active whenever it is in the second state.

Following [5] we can now implement a functioning detector as follows. Let us assume that the free-floating molecules bind to the transmembrane protein with some rate, both inside and outside. Let us further assume that the particles are able to enter and leave the cell with a certain rate. If import and export happen with the same rate, then the concentration of particles inside and outside of the cell will tend towards the same value in the long run. This can be exploited to detect changes in the environment. The performance (and indeed functionality) of the detector crucially depends on the rates of particle binding and unbinding being set correctly.

The simplest way to obtain a functioning detector is to assume that configuration (iii) (both binding sites occupied) is unstable in the sense that once both sites are bound, the binding sites release molecules with a high rate. Similarly, we stipulate that configurations (i) and (ii) are stable in the sense that the rate of unbinding in these two configurations is low, although new particles may still bind. The effect of such a configuration can be understood without simulation. If the concentration inside and outside of the cell is the same, then half of the transmembrane molecules will be bound by an external ligand and the other half by an internal ligand. More generally, the ratio of transmembrane particles bound inside and outside is an indicator for the difference in concentration of ligand inside and outside the cell volume.

To implement this model in Smoldyn, we can simply extend the model we already have. We use `freeFL` as the ligand whose concentration is to be detected. The molecule `porin` imports the external molecule into the cell and `surfM1` is the transmembrane molecule. In order to specify this model we need to assign `surfM1` two binding sites. In Smoldyn (and most other simulators) it is not possible to specify directly binding sites or to indicate that a molecule is in a bound state. Instead, one needs to work around that and create a specific new molecular species for each compound type. So, if a molecule A binds to a ligand L then the compound should be a new species B. In Smoldyn, however, this task is made a bit easier because states can be abused for this purpose and a particular binding state associated with each state. For a membrane-bound molecule with two receptors there are just enough states available to make it unnecessary to define new species for each compound. The only downside is that we need to keep track of precisely what each state means. This can be a bit confusing and it is best to write this down. For our model here we define the meaning of the states in Table 8.1 and the model is illustrated in Fig. 8.2.

Table 8.1 States in our model and what they mean

surfM1(down)	Unbound
surfM1(up)	Bound from inside
surfM1(front)	Bound from outside
surfM1(back)	Both sides bound

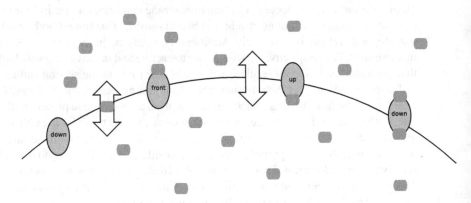

Fig. 8.2 The change detector. The free-floating ligands are taken up via the porin (*double-headed arrows*). The surface-bound molecules `surfM1` (*ovals*) are labeled with their states

```
#binding from inside
reaction r1 freeFL(bsoln) + surfM1(down) -> surfM1(up) 20

#binding from outside when inside is bound
reaction r2 freeFL(fsoln) + surfM1(up) -> surfM1(back) 20
reaction rr2 surfM1(back) -> freeFL(fsoln) + surfM1(up) 20

#first binding from outside
reaction r3 freeFL(fsoln) + surfM1(down) -> surfM1(front) 20

#binding from inside when outside is bound
reaction r4 freeFL(bsoln) + surfM1(front) -> surfM1(back) 20
reaction rr4 surfM1(back) -> freeFL(bsoln) + surfM1(front) 20
```

Code 8.1 Reactions for the binding dynamics of the change detector

Having specified the meaning of states, we can now continue defining the model. We need to write down the chemical reactions that are consistent with the states we have defined. It is very important now to be aware of the orientation of the surfaces, so we can correctly assign the binding dynamics with the external and internal ligands (Code 8.1).

The first reaction is the binding of freeFL from the inside of the cell to the cytoplasmic receptor of surfM1. Note that the reaction uses the state freeFL(bsoln). This together with the specification of the radius in the definition of the cell surface specifies that reaction r1 should only occur with free floating freeFL that are on the inside. The reaction takes surfM1(down) to the state (up) and removes a freeFL. We stress again here that the states (up) and (down) do not carry any intrinsic meaning in Smoldyn. The meanings of the states of the transmembrane molecule arise solely from this reaction. On the contrary, the states of the free-floating molecules— (bsoln) and (fsoln)—have intrinsic meanings and we are not free to assign the states of free-floating molecules.

In order for our scheme to work well, we need the bound states of surfM1 to be stable. Here we realise this by simply not specifying any reverse reaction. That is, once a particle is bound it remains bound. However, surfM1 has a second receptor at the outside of the cell. This receptor can bind another free floating freeFL, but now only if it is in the state (fsoln). The second binding from the outside is specified in the second reaction. The resulting compound is surfM1(up), corresponding to both receptors bound. Unlike the single state, it is not stable at all and decays in reaction rr2. Reactions r3, r4 and rr4 are the exact counterparts of the first three reactions, only that they implement the scenario that the outside receptor is bound first.

The key to the dynamics of the detector is the combination of reactions rr2 and rr4. Whenever surfM1 is in the state back both of those reactions occur with equal probability. In practice this means that a doubly-bound transmembrane protein is equally likely to decay into one that is bound at the outside or at the inside.

Consequently, when the concentration of molecules is the same inside and outside of the cell then—on average—the states front and up should occur equally frequently. Moreover, given the particular reaction scheme we have, the state down corresponding to unbound receptors, should not occur at all. Altogether, one expects that about half of the receptors are bound at the outside and half of them at the inside.

Now, imagine that the system is equilibrated and the number of surfM1 in states front and up is roughly the same. Imagine further, that suddenly all freeFL inside the cell are taken away. The question is now: What will happen in this case to the proportion of surfM1 in the different states?

Naively, one may be tempted to say that nothing happens. The doubly-bound state decays with equal probability into either up or down. Therefore, the proportion of these states should be independent of the external conditions. However, there is a slight twist to the dynamics. While all doubly-bound molecules decay equally likely into either of both configurations, in the special circumstances we describe, only those surfM1 with a ligand bound at the inside will be lifted temporarily into the doubly-bound state back. From there, they may decay again into either up or front. After the decay only half of those (roughly) will again have a free binding site outside and be lifted into the double state again. Therefore, the number of surfM1(up) goes down gradually and over a short period of time all of the surface bound molecules will be in the front state.

This reasoning can be extended to in-between cases where the concentration at the outside is a bit higher than at the inside, or the concentration at the inside is a bit higher than at the outside. In those cases, the concentration difference will be reflected in the composition of states up and front among the freeFL.

8.4.1   Simulations of the System

The property that, the number of surfM1 reflects the difference in concentration between the inside and the outside, can be used as a change detector. Conceptually, this can work as long as any change in concentration of surfM1 at the inside or the outside equilibrates, but equilibrates slower than the state changes in the freeFL. In this case, changes in the outside concentration can be translated into changes in the binding states of the surface proteins and, hence, changing activity levels of freeFL.

In our particular model, the time scale to equilibrate is set by the number of porin molecules and the rate with which reactions ttt1 and ttt2 proceed. In the extreme case, when this rate falls to 0, no equilibration takes place. On the other hand, when the rate is high then equilibration is very fast and the change detection may not work.

In our model we start with 200 freeFL introduced at the outside compartment (see above). This means that we need to wait for the system to equilibrate; that is, we need to wait until the concentration inside and outside the cell have become the same. Import and export of freeFL is mediated via the porin catalyzed input/output reactions (ttt1 and ttt2). Once the equilibration has been reached, the number of surfM1 in states up and front should be approximately the same and we can then test the detector by injecting more particles of freeFL at the outside. This should temporarily raise the number of surfM1 that are bound at the outside. In the model, we want to see this reflected in a transient change of the number of surfM1(front).

In many simulation systems, changing particles suddenly in the middle of the simulation is difficult to achieve. This is not the case in Smoldyn whose command facility can be used for this purpose. We insert into our model the following line:

```
cmd @ 10000 fixmolcountincmpt freeFL 500 outside
```

The keyword cmd specifies that what follows is a command for the Smoldyn simulator. It then stipulates that at time 10000 Smoldyn should adjust the particle number of freeFL to 500 in compartment outside. There are a number of other system manipulation commands available. For example, Smoldyn can be instructed to kill (i.e., remove) molecules of a particular type and even molecules in a particular state either globally or in a particular compartment. For our purposes here, fixmolcountincmpt is the most suitable one.

Figure 8.3 shows the results of simulating the detector. The graph clearly shows the first period of equilibration. At the beginning none of the surfM1 is bound at the outside, but the number of surfM1(front) rises quickly, overshoots a bit and then

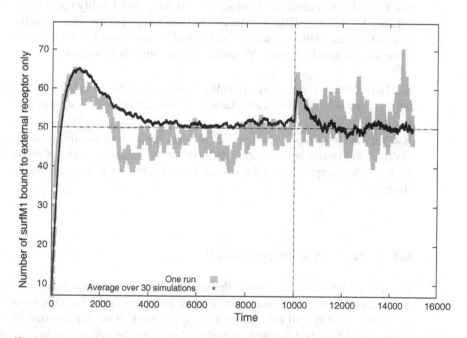

Fig. 8.3 A simulation of the first change-detector model. A single simulation (*lighter plot*) results in noisy output, but when the average is taken over 30 runs (*darker plot*), the signal is clearly discernible. The *vertical line* indicates where the additional freeFL were injected at the outside of the cell. Following this injection, we observe a spike of activity

falls back to the expected 50 molecules corresponding to half of all the `surfM1`. The initial overshoot is connected with the time delay of importing the `freeFL` into the `cell` volume. This process is slower than the binding to the outside of the receptors and is the effect that is also exploited for the change detection.

The injection of `freeFL` at the outside of the cell occurs at time 10000. The graph shows that, shortly after that time there is a rise in the number of `surfM1(front)`, as expected. Our detector works!

There is a problem, however. If we look at just a single simulation, such as the gray line in Fig. 8.3, then this spike happens in a sea of many other spikes. If we had not known that something special happens at 10000, then we would have missed or dismissed it as just another of the random flukes rather than a signal indicating a change of the external concentration. Indeed, in that particular simulation there is a spurious "signal" after time 14000 that is higher than the actual signal, but this is not an indication of significance at this point. To obtain a clean signal, we need to take the average over many runs. The black line in Fig. 8.3 shows an average over 30 simulations. Random fluctuations are very small then and the signal is clearly visible and stands alone.

8.1 Download the Smoldyn code from this book's web site and run this model.

8.2 Consider the definition of the square box on p. 306. Modify the code so that the outward-facing sides of the box are red and the inward facing sides are yellow. Then change the code of the box so that the front of the enclosing box on the positive side of the second axis is missing. Visualize your solution in Smoldyn.

8.3 Define the the space corresponding to the `outside` compartment in model 1, but use inner defining points and the `walls` surface as described on p. 305.

8.4 Vary the parameters of the model and investigate how the ability of the system to detect an external concentration depends on the parameter settings. In particular, explore what happens when the rate of `freeFL` import is increased or decreased strongly.

8.5   Free-Change Detector II

Our first model relied crucially on the ability of the cell to import nutrient. This may not always be practical. Space within the cell volume is limited and importing chemicals unnecessarily will only congest the space more. Also, the external molecules to be sensed may be toxic and importing them into the cell would be detrimental. The question is now whether or not there is a way to detect changes in external concentrations of some chemical without letting it into the cell. Indeed, De Palo and Endres [5] suggested a variation of the first scheme that supposedly achieves

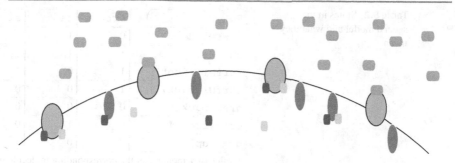

Fig. 8.4 The second change detector. The free-floating ligands bind at the extra-cellular receptor. This leads to `slow` and `fast` being released inside the cell; they can then diffuse to `surfM2` (*smaller ovals*) and rebind

this. We will notice further down that the stochastic implementation of the scheme suggested here does not actually work very well. Yet, this is not the issue here. By exploring how to model this system, we will be able to illustrate important features of Smoldyn.

This second scheme again involves a transmembrane protein (`surfM1`) with binding sites at the inside and outside. Only this time there are two cytoplasmic sites and a single extra-cellular one. This latter binding site does the sensing proper and binds `freeFL`. The intracellular sites can bind to the cytoplasmic ligands `slow` and `fast`. In addition to `surfM1` we assume a second type of transmembrane protein—`surfM2`. It has two cytoplasmic binding sites as well, one for `slow` and one `fast` respectively, but no extra-cellular site (Fig. 8.4).

At the beginning of the simulation we assume that the sites of `surfM2` are unoccupied, that all intra-cellular binding sites of `surfM1` are occupied and that the extra-cellular ones are empty. Once a molecule of `freeFL` binds at the outside, both `fast` and `slow` unbind from the cytoplasmic binding sites and start to diffuse through the cell. The difference between them is that `slow` diffuses at a lower rate than `fast`. We do not really need to worry here about why it diffuses slowly, but we could speculate that `slow` is somewhat bigger than `fast` and hence the difference in diffusion constant. The main point is that the difference in speed exists.

Once the two ligands start to diffuse in the cell volume, they will eventually collide with the, as yet, unoccupied binding sites of `surfM2` and bind to them. The idea of the detector design is that the difference in travelling time of the two ligands from `surfM1` to `surfM2` can be used to indicate changes in external concentrations. To see this, imagine that at first the external concentration is zero. Then all `fast` and `slow` will be bound to `surfM1`. Then, suddenly, the concentration of `freeFL` is increased to some value. Following this, some of the receptors will bind `freeFL` which leads to `fast` and `slow` being released at the inside of the cell. For some time both of the ligands will be in transit. Then, so the idea, the `fast` molecules will start to reach the receptor and bind. Due to the difference in diffusion time, this will happen on average before `slow` binds. Only then after some time the `slow` molecules will arrive and bind as well. At this point then, the system can reset, both `fast` and `slow` can unbind and return to bind to `surfM1`.

Table 8.2 States in the second model and what they mean

State	freeFL	fast	slow
surfM1(down)	0	1	1
surfM1(up)	1	1	1
surfM1(front)	1	0	1
surfM1(back)	1	0	0
gray(back)	0	0	0
gray(front)	0	1	0
gray(up)	0	0	1

A value of 1 means that the corresponding molecule is bound

In this scheme, the detection is associated with the period between fast being bound to surfM2 but slow still being in transit. In this case, the receptor can be considered to be in an active state. In reality, things are not that simple. Even if the concentration of the external freeFL remains constant, there will be continuous binding and unbinding events of slow and fast to and from both receptors because freeFL particles keep binding and unbinding to and from the outside receptors. This entails that at any time there may be a significant proportion of the surfM2 bound to only fast. Yet when there is an increase in the number of freeFL then, transiently, there are many more of the slow in transit before, with some delay, equilibrium is reached and all molecules bind again to surfM2.

To model this, we need to first specify the possible binding states of surfM1. We will use again the trick to represent the various binding states with surface states. Unfortunately, surfM1 now has three binding sites and Smoldyn does not provide enough states to represent all binding configurations with states. Hence, we need to create a helper species to solve this; we call it gray.

Table 8.2 lists the meaning of the states of the molecules. The idea of the model is that, upon binding of freeFL at the outside, both fast and slow are simultaneously released at the inside. Within Smoldyn such simultaneous release cannot be modeled, but it can be approximated by assigning a very high rate to the transition from surfM1(up) to the states (front) and (back). Here, we disallow rebinding of slow and fast to surfM1(back), that is to the receptor that still binds freeFL. We need to remember to include a reaction for unbinding of freeFL from the outer receptor gray(back) and then allow the intra-cellular ligands to rebind surfM1. The rebinding of fast and slow needs to be specified in four steps. It could be that slow binds first to gray(back) to form grayup, followed by fast binding to form surfM1(down); or it could be that fast binds first to form gray(front) followed by slow. We leave it to the reader to formulate the reaction set as an exercise.

More interesting from a technical point of view is binding at the second receptor surfM2. To understand the issue, it is worth keeping in mind that we are interested in diffusion. We want to model the race of slow and fast to the second receptor. Part of the solution to this we get (almost) for free in Smoldyn. The different diffusion speeds can be simply encoded by specifying a suitable difc value for the molecules.

Say, we assign `fast` a diffusion constant of 5 and `slow` is a tenth of this. Once this is specified, so one may think, all that needs to be done is to formulate the reaction set for binding `slow` and `fast` as before.

Not quite! There is one difficulty. To see this, we need to note that Smoldyn executes reactions when the reactants are within a certain binding radius. The binding radius, in turn, depends on the user-defined "macroscopic" reaction rate constant and the diffusion constant. Given a fixed "macroscopic" reaction rate, decreasing the diffusion constant will tend to lead to an increased binding radius.

In our case this is a problem. The molecules `fast` and `slow` have very different diffusion constants. At the same time, we are interested in their difference in expected time difference for them to hit the target `surfM2` when diffusing from their original site `surfM1`. More precisely we are interested in the average time delay between `fast` binding at `surfM2` and `slow` binding there ...and here is the problem. Given the difference in diffusion constants for the two molecules, the binding radii for `slow` and `fast` are not the same. This means that the two molecule types travel different distances before they bind. This is clearly not what we want, because the time taken to travel a given distance is precisely what we want to measure. Variable binding radii undermine this.

There is a solution for this in Smoldyn. Instead of specifying the "macroscopic" reaction rate directly, Smoldyn does allow the user to specify the binding radius instead. Doing this, we can create an even playing field (or rather running track) for both ligands and determine the difference before they bind to `surfM2`. To do this we simply specify that `slow` and `fast` should bind to the receptors of `surfM2` once they are within a given distance—or rather that they should react once they are within a given binding radius. So, we bypass the automatic calculation of binding radii in Smoldyn and specify them manually. The syntax to do this in Smoldyn is shown in Code 8.2.

The syntax of this is not very different from the standard way of specifying chemical reactions. However, note that we omitted the macroscopic reaction rate constants in the reaction definitions. Instead, we defined the binding radius directly by adding another line starting with `binding_radius`. This keyword must be followed by the name of the reaction and the actual binding radius. Note that in the example above we have used a convenient feature of Smoldyn that allows placeholders. Rather than specifying the binding radius directly, we used the placeholder `brad` (for **binding radius**). When the model is interpreted by Smoldyn the word "brad" will simply be replaced by the number specified in the `define` statement. In our case this is 0.5.

The use of the placeholder value `brad` makes it much easier to explore the effect of the binding radius on the behavior of the model. In our model the same binding radius is used in four reactions. Any modification of the model would require all those to be changed. Placeholders combine all of them into one place. The use of placeholders is highly recommended because, apart from speed and convenience, it introduces a level of additional documentation and reduces the risk of inconsistency.

```
define brad 0.5
define reWa 1

#bind fast first, then slow
reaction act1 surfM2(up) + fast(bsoln) -> surfM2(down)

binding_radius act1 brad
reaction_probability act1 reWa

reaction act2 surfM2(down) + slow(bsoln) ->
    surfM2(front) + fast(bsoln)
binding_radius act2 brad

#bind slow first then fast
reaction act3 surfM2(up) + slow(bsoln) ->
    surfM2(front)

binding_radius act3 brad
reaction ract12 surfM2(front) +  fast(bsoln) ->
    surfM2(down) + slow(bsoln)
binding_radius ract12 brad
reaction_probability ract12 reWa
```

Code 8.2 Manual specification of binding radius

In the case when the user determines the binding radius directly, she can also specify a reaction probability. The syntax is the same as for the binding radius, only that `reaction_probability` is the keyword used. By default, Smoldyn reactions are deterministic in the sense that two products always react if they are within each others' binding radii. This default behavior can be changed by setting the reaction probability to a value <1. This makes reactions probabilistic, which may sometimes be useful. However, for our model we have decided to leave the reaction probability at 1 (and we may as well not have specified it).

Simulating the model, we notice that it does not work very well as a detector. While the idea of the detector based on the difference in travel time is sound, there are some implementational details that make the design of the detection hard. At the heart of the matter is the structure of the surface. Receptors of type `surfM1` and `surfM2` are spread over the surface at random positions. Consider now a specific `surfM2`. It will have a number of `surfM1` in its immediate neighborhood. Moreover, depending on the density of receptors on the surface, the distance to the nearest receptor may not be very long. This means that the expected time-difference between two binding events is low and the noise is high. This noise is further amplified by the fact that this specific `surfM2` will receive ligands not only from its immediate neighbor, but from any one of the `surfM1` on the surface. This tends to amplify the noise even more.

To understand this a bit better, we simplified the model by removing the `surfM2` and all its associated reactions. Then, we introduced new compartments in order to determine the particle densities over time in a small volume on the `cell` surface

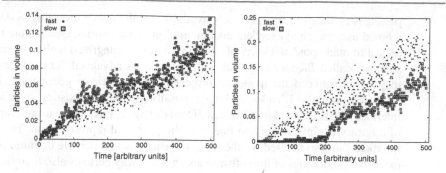

Fig. 8.5 Probing the number of particles over time in a small volume (*left*) around a point on the cell surface and (*right*) around the volume. In the small volume there is clearly a difference in density due to the difference in initial conditions. However, around the cell surface, there is no such difference. From the beginning, fast and slow particles are about equally likely to be in this area. See Exercise 8.6

and a small volume around the origin. Figure 8.5 shows that on the surface there is virtually no difference over time in the concentration of the fast and slow. It is therefore not surprising that the second detector does not work very well. Figure 8.5 shows, however, that the scheme could work if the second receptor were to be placed far from the surface, for example in the center of the cell. There, a substantial signal was detectable.

8.5 Write the reaction rates for the Smoldyn model corresponding to Table 8.2. Compare your version with the model equations available from the web-page.

8.6 Determine in simulation the expected density of fast and slow at an arbitrary point at the cell surface and in the center. To do this, modify the second model by removing surfM2 altogether. Then create two spherical compartments, C1 and C2 with radius 1. The first compartment should be centered around the origin. The second compartment should be centered around a point of the surface of the cell compartment. Then use the molcountincmpt keyword to determine the number of particles.

8.6   Alternatives to Smoldyn

There is a large number of choices for the modeler of perfectly-mixed chemical systems. For spatial modeling much less is on offer, and what exists tends to be more difficult to use. That said, there is a small choice of programs available that can be used to implement spatial models.

STEPS [6,7] is an extension of the Gillespie SSA. It uses a Python interface to interact with models. It is maintained and well documented. One notable feature of STEPS is its support for SBML model definitions. One of the best known software packages is *MCell* [8,9]. It is freely available for the major operating systems, has a broad user base and excellent documentation and a tutorial. One feature that will appeal to many potential users of MCell is that it is integrated with a 3D modeling software called *Blender* [10]. This software allows advanced 3D modeling which may be attractive if the modeling project requires intricate geometries. There is an MCell plugin to Blender, making it possible to use the existing graphical user interface of Blender and its advanced 3D modeling features to create cell topologies of arbitrary complexity, and to fine-tune the graphical representation. The blender interface can also be used for the input of reactions and particle definitions. These extensive capabilities of the software are in some way perhaps also their limitation. It is complex to learn and the graphical environment somewhat cumbersome to use. The complexities of the Blender interface can be an obstacle for impatient modelers who do not like clicking through menus in order to make changes to their model.

References

1. Chu, D., von der Haar, T.: The architecture of eukaryotic translation. Nucleic Acids Res. **40**(20) (2012). doi:10.1093/nar/gks825
2. Andrews, S.S.: Spatial and stochastic cellular modeling with the Smoldyn simulator. In: Bacterial Molecular Networks, pp. 519–542. Springer, Berlin (2012)
3. Andrews, S.S.: A spatial stochastic simulator for chemical reaction networks. http://www.smoldyn.org/. Accessed 19 June 2015
4. Yi, T., Huang, Y., Simon, M., Doyle, J.: Robust perfect adaptation in bacterial chemotaxis through integral feedback control. Proc. Natl. Acad. Sci. USA **97**, 4649–4653 (2000)
5. De Palo, G., Endres, R.G.: Unraveling adaptation in eukaryotic pathways: lessons from protocells. PLoS Comput. Biol. **9**(10), e1003,300 (2013)
6. Hepburn, I., Chen, W., Wils, S., De Schutter, E.: Steps: efficient simulation of stochastic reaction-diffusion models in realistic morphologies. BMC Syst. Biol. **6**(1), 36 (2012). doi:10.1186/1752-0509-6-36
7. STochastic Engine for Pathway Simulation. http://steps.sourceforge.net/STEPS/default.php. Accessed 19 June 2015
8. Kerr, R., Bartol, T., Kaminsky, B., Dittrich, M., Chang, J., Baden, S., Sejnowski, T., Stiles, J.: Fast Monte Carlo simulation methods for biological reaction-diffusion systems in solution and on surfaces. SIAM J. Sci. Comput. **30**(6), 3126–3149 (2008)
9. MCell: Monte Carlo cell. http://mcell.org. Accessed 19 June 2015
10. Foundation, B.: Blender. https://www.blender.org/. Accessed 19 June 2015

Reference Material

A

A.1 Some Common Rules of Differentiation and Integration

In this section we provide a short reference to some differentials and integrals of basic functions. We use the notation to $f(x)$ to represent a function of x and $f'(x)$ to represent the first derivative of $f(x)$, i.e., $\frac{df(x)}{dx}$.

The differential of a sum of two functions of x is the sum of the two differentials:

$$\frac{d}{dx}(f(x) + g(x)) = f'(x) + g'(x) \tag{A.1}$$

whereas the differential of a product is a sum of products:

$$\frac{d}{dx}(f(x)g(x)) = f'(x)g(x) + f(x)g'(x) \tag{A.2}$$

A.1.1 Common Differentials

Table A.1 illustrates the differentials of some common functions.

A.1.2 Common Integrals

The notation $\int_a^b f(x)dx$ represents the *definite integral* of $f(x)$ over the interval $[a, b]$. The interval limits are omitted for indefinite integrals. Common integrals can usually be inferred from the corresponding differential (see Table A.1), with the addition of an integration constant, C. Table A.2 illustrates the integrals of some common functions.

© Springer-Verlag London 2015
D.J. Barnes and D. Chu, *Guide to Simulation and Modeling for Biosciences*,
Simulation Foundations, Methods and Applications,
DOI 10.1007/978-1-4471-6762-4

Table A.1 Common simple differentials

$f(x)$	$f'(x)$
a	0
x^a	ax^{a-1}
a^x	$ln(a)a^x$
e^x	e^x (from above)
$log_a(x)$	$\frac{1}{x ln(a)}$
$ln(x)$	$\frac{1}{x}$ (from above)
$sin(x)$	$cos(x)$
$cos(x)$	$-sin(x)$
$tan(x)$	$\frac{1}{cos^2(x)}$

Table A.2 Common simple integrals

$f(x)$	$\int f(x)dx$
a	$ax + C$
a^x	$\frac{a^x}{ln(a)} + C$
x^a	$\frac{x^{a+1}}{a+1} + C$
$\frac{1}{x}$	$ln(|x|) + C$
$log_a(x)$	$x log_a(x) - \frac{x}{ln(a)} + C$
$sin(x)$	$-cos(x) + C$
$cos(x)$	$sin(x) + C$
$tan(x)$	$-ln(|cos(x)|) + C$

A.2 Maxima Notation

Full documentation on Maxima can be found in the documentation section of the Maxima web site [1]. Table A.3 illustrates some commonly-used Maxima notation.

Table A.3 Commonly-used notation in Maxima

`%e`	Euler's e number
`%i`	Imaginary unit (i.e., %i ^ 2 = -1)
`%pi`	Pi
`log(x)`	Natural logarithm of x (i.e., not ln(x))
`inf`	Infinity
`realpart(c)`	Returns the real part of c
`imagpart(c)`	Returns the imaginary part of c
`binom(a,b)`	$\binom{a}{b}$

Table A.4 PRISM query summary

`P=?[A U[T1,T2] B]`	Probability that between time $t = T_1$ and $t = T_2$ first A is true and then B is true
`P=?[true U B{IC}]`	Starting from initial conditions IC, what is the probability that B will be true
`P=?[F B{IC}]`	Equivalent to the above
`P>=0[G B]`	Returns true if B is always true (independent of random choices)
`S>=0[b=4]`	Returns true if b has a non-zero probability of taking a value of 4 in steady state
`R=?[S]`	Returns the reward in steady state
`R=?[C<=10]`	Returns the cumulative reward before time $T = 10$
`R=?[I=10]`	Returns the instantaneous reward

A.3   PRISM Notation Summary

Full documentation on the PRISM model checker may be found at the PRISM web site [2]. Table A.4 illustrates some of the most common notation used in model queries.

A.4   Some Mathematical Concepts

This section provides a brief introduction to some of the basic mathematics that has been assumed of the reader in this book. The exposition here cannot replace a proper textbook and is intended purely as an *aide memoire*. The reader who has not met these concepts in detail before is strongly advised to read a dedicated textbook.

A.4.1   Vectors and Matrices

A vector in n-dimensional space is an n-tuple of numbers. The following are examples:

$$\mathbf{v_1} \doteq \begin{pmatrix} 1 \\ 3.42 \\ 12 \\ i \end{pmatrix} \qquad \mathbf{v_2} \doteq (1, 3.42, 12, i)$$

v_1 and v_2 are column and row vectors, respectively. We will normally denote the variable names of vectors in boldface, in order to distinguish them from simple

Fig. A.1 A graphical
representation of the two
vectors $(1, 1)$ and $(0, -1)$

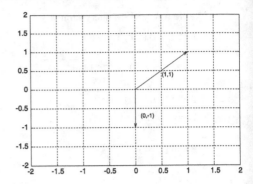

scalars. Our two example vectors contain the complex unit, i. Real vectors—that is vectors that contain only real numbers—can be interpreted graphically (Fig. A.1).

There are a number of operations that can be performed on vectors. The most important is the dot-product. If $\mathbf{v}$, $\mathbf{w}$ are vectors with the elements $(v_1, v_2, \ldots)$ and $(w_1, w_2, \ldots)$, respectively, then $v \cdot w$ is defined as:

$$v \cdot w = \sum_i v_i \cdot w_i$$

An extension of the concept of a vector is that of a matrix. Matrices are arrays of vectors:

$$\mathbf{A} \doteq \begin{bmatrix} 29 & 52 & 42 \\ 70 & -13 & 18 \\ -32 & 82 & -59 \\ -1 & 72 & 12 \end{bmatrix} \qquad \mathbf{A}^T \doteq \begin{bmatrix} 29 & 70 & -32 & -1 \\ 52 & -13 & 82 & 72 \\ 42 & 18 & -59 & 12 \end{bmatrix}$$

Here $\mathbf{A}^T$ is the transposed matrix of $\mathbf{A}$, which means that the columns and rows are exchanged. More conveniently, one can represent a matrix in terms of its elements: A_{ij} is the element in row i and column j. The element A_{12} would be 52 in the example above. Given two matrices, $\mathbf{A}$ and $\mathbf{B}$, we can define matrix multiplication. The product matrix $(\mathbf{A} \cdot \mathbf{B})$ is the matrix of all products of the row vectors of $\mathbf{A}$ and column vectors of $\mathbf{B}$.

$$(\mathbf{A} \cdot \mathbf{B})_{ij} = \sum_k A_{ik} B_{kj}$$

A numerical example might be helpful. Let us define the matrix $\mathbf{B}$:

$$B \doteq \begin{bmatrix} -26 & -94 & -36 & -15 \\ -86 & -97 & -69 & 2 \\ 50 & -38 & 69 & -88 \end{bmatrix}$$

We can now explicitly calculate the product of $\mathbf{A} \cdot \mathbf{B}$:

$$\mathbf{A} \cdot \mathbf{B} = \begin{bmatrix} -3126 & -9366 & -1734 & -4027 \\ 198 & -6003 & -381 & -2660 \\ -9170 & -2704 & -8577 & 5836 \\ -5566 & -7346 & -4104 & -897 \end{bmatrix}$$

In matrix multiplication, the order of the operands is significant. The number of rows in the left operand must be equal to the number of columns in the right operand. Therefore, with the current data, the product of $\mathbf{B} \cdot \mathbf{A}$ is not defined for matrix multiplication, because of the mismatch of the dimension of column and row vectors.

A.1 Confirm the results of the matrix products by explicitly calculating the products by hand.

A.2 Square matrices are matrices that have the same number of rows and columns. Define two square matrices $\mathbf{A}$ and $\mathbf{B}$ and calculate the products $\mathbf{A} \cdot \mathbf{B}$ and $\mathbf{B} \cdot \mathbf{A}$. Compare the results.

A vector can be seen as a special case of a matrix. If $\mathbf{v}$ is a row vector of dimension n and $\mathbf{A}$ is a matrix with n columns then we can calculate the product $\mathbf{v}_2 = \mathbf{v} \cdot \mathbf{A}$ following the same rules as above. Similarly, the matrix product $\mathbf{A}^T \cdot \mathbf{v}^T$ is also well defined. On the other hand, given our definition of matrix product, constructs such as, $\mathbf{A} \cdot \mathbf{v}$ are *not* defined. Henceforth, we will always assume that the dimensions of matrices and vectors match to make sense of the products. It is easy to see that the product of a vector and a matrix is itself a vector. For example, $\mathbf{v}_2$ is again a row vector of dimension n.

Assume now a matrix, $\mathbf{A}$, given in terms of some elements A_{ij}, which we do not need to specify. We can ask whether there is a vector, $\mathbf{v}$, and a scalar value, λ, such that the product of the vector and the matrix equals the vector scaled up by a factor λ. This amounts to solving a linear equation:

$$\mathbf{v} \cdot \mathbf{A} = \lambda \mathbf{v}$$

The vector, $\mathbf{v}$, and the scalar, λ, that solve this equation for $\mathbf{A}$ are called the *eigenvector* and *eigenvalue* of $\mathbf{A}$, respectively. For any specific matrix, there may be more than one solution. Eigenvectors and eigenvalues are important in mathematics and there exists a large body of theory on this topic.[1]

---

[1] The interested reader is encouraged to read a textbook on *linear algebra* for a better idea of the theoretical underpinnings of this concept.

A.4.2 Probability

In order to talk about the probability of something (an "event"), one first needs to define the range of possible events under consideration. The standard example is the rolling of a die. There are six possible outcomes, which we can simply label 1, 2, 3, 4, 5 and 6. Before rolling the die, we do not know what the outcome will be, yet we can assign to each outcome a probability. If the die is fair then we can assume that each outcome is equally likely, occurring with a probability of $1/6$; or $P(x) = 1/6$ if $x \in \{1, 2, 3, 4, 5, 6\}$.

We can now also think about multiple experiments. If we consider two rolls of a single die then the space of possible outcomes is extended in that there are now 36 possible outcomes instead of 6. They are the familiar pairings: $(1, 1), (1, 2), (1, 3)$, ..., $(6, 6)$, where the first entry in the parenthesis refers to the first roll of the die (the result of the first experiment) and the second entry to the result of the second roll. Each pair of numbers is equally likely and, hence, the probability for each is $1/36$ (which is $\frac{1}{6} \cdot \frac{1}{6}$). Correspondingly, one can extend the space of events by introducing further rolls of the die.

If we perform n experiments then we know that there are $1/6^n$ possible outcomes altogether—each equally likely. We might be interested in certain subsets of the possible outcomes. For example, we may wish to ask about the probability that the result of each trial is an even number. This can be formulated in a different way by asking for the probability that the first trial resulted in an even number *and* the second *and* the third ... *and* the nth. In probability theory, whenever we ask about the probability of several events happening jointly, then we need to multiply the corresponding probabilities. Clearly, for each individual trial the probability of an even number is $1/2$, simply because there are equally many even numbers as odd numbers on the die. Hence, the probability of all outcomes of n trials being even is obtained by multiplication:

$$P(\text{all even}) = \prod_{trials} \frac{1}{2} = \frac{1}{2^n}$$

We might now ask for the probability that all rolls of the die yield the result 1. We could calculate this by simply multiplying the individual probabilities, as above. However, this is not really necessary. Since all possible outcomes of our die rolling experiment are equally likely, we already know that a result of only 1's has a probability of $1/6^n$. Having calculated this result, we can now ask for the probability that we *either* have only even outcomes *or* all outcomes are 1. Since these two possibilities exclude one another, we can calculate the desired answer by summing the separate probabilities:

$$P(\text{all even or all 1}) = \frac{1}{2^n} + \frac{1}{6^n}$$

We can also ask about *conditional probabilities*—the probability of an event given that another has occurred. For instance, the probability that we have an outcome of all 2's given that we know that all outcomes were even. Formally, this is normally written using a vertical bar; so $P(\text{all } 2|\text{all even})$. In some contexts conditional probabilities can be quite confusing yet, in essence, the concept is quite straightforward. Instead of considering the space of all possible outcomes, the conditional probability asks for the probability of an event based on a restricted subset of all possible outcomes. So in the example above, we are already given the information that the outcome was even and, based on that, we ask for the probability that the outcome was 2. Another way to look at it is to ask for the probability of obtaining the outcome 2 when all outcomes with odd numbers are discarded from consideration. Formally, it can be calculated as follows:

$$P(\text{all } 2|\text{all even}) = \frac{P(\text{all } 2 \cap \text{all even})}{P(\text{all even})}$$

Here, the symbol $\cap$ means, essentially, that both events need to be true. Since "all 2" is a subset of "all even" the enumerator simply equals $P(\text{all } 2)$. Altogether we thus obtain the solution:

$$P(\text{all } 2|\text{all even}) = \frac{P(\text{all } 2)}{P(\text{all even})} = (2/6)^n = \frac{1}{3^n}$$

This result makes sense. If we assume a die consisting of 3 possible sides only (only even numbers) then the probability for each outcome of the n trials would be exactly $1/3^n$, just as we calculated above.

A.4.3 Probability Distributions

If we roll the die n times, what is the probability that we get an even number exactly $0 \leq k \leq n$ times? Here we do not care about whether the die gives us a 4 or a 6, and we thus reduce the outcomes to purely binary values. As it turns out, we do not need to worry about calculating this as there is a pre-computed answer available: The *binomial distribution*. If we roll the die exactly n times then the probability that we get an even number exactly k times is given by:

$$P(k) = \binom{n}{k} p^k (1 - p)^{n-k}$$

The term $\binom{n}{k}$ is called the *binomial coefficient* and is defined as:

$$\binom{n}{k} \doteq \frac{n!}{k!(n-k)!}$$

The exclamation mark refers to the factorial function, i.e., $k! \doteq k \cdot (k - 1) \cdot (k - 2) \ldots 1$. The binomial coefficient has a specific meaning independent of the binomial distribution. It gives the number of different ways to choose k items out of a collection of n items. For example, assume we have a container with n distinct balls and we randomly pick k of them. The binomial coefficient then tells us the number of possible combinations of balls we end up with. For example, if each ball has a number on it, how many possible combinations of numbers are there when we randomly select k balls (assuming we disregard the order in which the balls are selected). This finds an immediate application in lotteries. If there are $n = 45$ balls in the container and each week $k = 6$ balls are drawn, then according to the binomial coefficient there are altogether 8,145,060 possible combinations; the probability of winning with a single ticket is the inverse of this number.

The binomial distribution is an example of a discrete probability distribution. It can be applied whenever there is a series of events, each of which has just two possible outcomes. An example is the flipping of a coin where the outcomes are "head" or "tail". In this case, both outcomes are equally likely and have a probability of $p = 0.5$. In general, the probabilities of the two outcomes do not need to be equal, although clearly they always need to sum to 1.

Our even/odd example of n rolls of the die is equivalent to n coin tosses. Each roll event has a probability of $p = 0.5$ of returning an even number. Hence, if we assume $n = 20$ and we want to know the probability that there will be exactly $k = 18$ outcomes with an even number, then we can calculate this as follows:

$$P(18) = \binom{20}{18} \left(\frac{1}{2}\right)^{18} \left(\frac{1}{2}\right)^2 \approx 0.000181198$$

Another distribution that is very important in biological modeling is the *Poisson distribution* given by:

$$P(k) = \frac{m^k exp(-m)}{k!}.$$

Assume we have an event that happens, on average, m times per time unit. The Poisson distribution describes the probability of *actually* observing exactly k events per time unit. For example, if somebody receives 10 emails per hour, on average, then we can use the Poisson distribution to calculate the probability that this person receives exactly 2 emails within the next hour by setting $m = 10$ and $k = 2$.

The best known probability distribution of all is the Gaussian distribution. Most readers will be familiar with it, so we will only write it down for reference:

$$g(x) = \frac{1}{\sqrt{2\pi\sigma^2}} \exp\left(-\frac{(x-m)^2}{2\sigma^2}\right)$$

Here σ^2 is the variance of the distribution and m is the mean. This is one of the iconic and ubiquitous functions in science. The classical example of a Gaussian distributed variable is the size of children in a school class. One of the reasons for the omnipresence of the Gaussian is the *central limit theorem* which states that, under a wide range of conditions, random processes consisting of a large number of random variables will show Gaussian distributions.

Fig. A.2 A Taylor expansion of the function $f(x) = \exp(x)$ to second order around $x_0 = 0$. The approximation is compared with the exact result

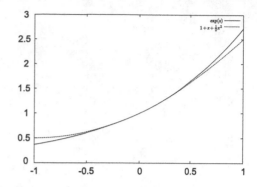

A.4.4 Taylor Expansion

A technique that may be less well known to the reader, but is of high practical importance, is that of the *Taylor expansion* of a function at a particular point. The idea of a Taylor expansion is to approximate a function locally by a simpler (or more convenient, or computationally cheaper) function. The accuracy of the approximation depends on the function itself, as well as the order of the Taylor expansion. Assume a function $f(x)$. The Taylor expansion of this function is then given by:

$$f(x) = f(x_0) + f^{(1)}(x_0)(x - x_0) + \frac{1}{2!} f^{(2)}(x_0)(x - x_0)^2 + \frac{1}{3!} f^{(3)}(x_0)(x - x_0)^3 + \cdots$$

Here x_0 is a point in the neighborhood of x and $f^{(n)}$ is the nth derivative of f. Often x_0 is chosen to be zero, but this is not a requirement. The Taylor approximation becomes successively better for each additional term. However, in practice, the series is often truncated after the first order term, which sometimes still gives a good local approximation.

As an example consider the second-order approximation of the exponential function:

$$\exp(x) \approx e^{x_0} + e^{x_0}(x - x_0) + 1/2\, e^{x_0}(x - x_0)^2$$

which is illustrated in Fig. A.2.

References

1. Maxima, a computer algebra system. http://maxima.sourceforge.net. Accessed 24 June 2015
2. Prism. http://www.prismmodelchecker.org/. Accessed 27 June 2015

A.2.4 Taylor Expansion

Index

© Springer-Verlag London 2015
D.J. Barnes and D. Chu, *Guide to Simulation and Modeling for Biosciences*,
Simulation Foundations, Methods and Applications,
DOI 10.1007/978-1-4471-6762-4

Printed in the United States
By Bookmasters

Printed in the United States
By Bookmasters